Lecture Notes in Computer Science

Commenced Publication in 1973
Founding and Former Series Editors:
Gerhard Goos, Juris Hartmanis, and Jan van Leeuwen

Thorsten Altenkirch   Conor McBride (Eds.)

Types for Proofs and Programs

International Workshop, TYPES 2006
Nottingham, UK, April 18-21, 2006
Revised Selected Papers

 Springer

Volume Editors

Thorsten Altenkirch
Conor McBride
University of Nottingham
School of Computer Science and Information Technology
Jubilee Campus, Wollaton Road, Nottingham NG8 1BB, UK
E-mail: {txa, ctm}@cs.nott.ac.uk

Library of Congress Control Number: 2007936170

CR Subject Classification (1998): F.3.1, F.4.1, D.3.3, I.2.3

LNCS Sublibrary: SL 1 – Theoretical Computer Science and General Issues

ISSN	0302-9743
ISBN-10	3-540-74463-0 Springer Berlin Heidelberg New York
ISBN-13	978-3-540-74463-4 Springer Berlin Heidelberg New York

Springer is a part of Springer Science+Business Media

springer.com

© Springer-Verlag Berlin Heidelberg 2007
Printed in Germany

Typesetting: Camera-ready by author, data conversion by Scientific Publishing Services, Chennai, India
Printed on acid-free paper   SPIN: 12111886   06/3180   5 4 3 2 1 0

Preface

These proceedings contain a selection of refereed papers presented at or related to the Annual Workshop of the TYPES project (EU coordination action 510996), which was held April 18–21, 2006 at the University of Nottingham, UK.

The topic of this workshop was formal reasoning and computer programming based on type theory: languages and computerized tools for reasoning, and applications in several domains such as analysis of programming languages, certified software, formalization of mathematics and mathematics education.

The workshop was attended by more than 100 researchers and included more than 60 presentations. We also had the pleasure of three invited lectures, from Bart Jacobs (University of Nijmegen), Hongwei Xi (Boston University) and Simon Peyton Jones (Microsoft Research). Simon Peyton Jones spoke in a joint session with the workshop on Trends in Functional Programming (TFP), which was co-located with the TYPES conference.

From 29 submitted papers, 17 were selected after a reviewing process. The final decisions were made by the editors.

This workshop followed a series of meetings of the TYPES working group funded by the European Union (IST project 29001, ESPRIT Working Group 21900, ESPRIT BRA 6435). The proceedings of these workshop were published in the LNCS series:

TYPES 1993 Nijmegen, The Netherlands, LNCS 806
TYPES 1994 Båstad, Sweden, LNCS 996
TYPES 1995 Turin, Italy, LNCS 1158
TYPES 1996 Aussois, France, LNCS 1512
TYPES 1998 Kloster Irsee, Germany, LNCS 1657
TYPES 1999 Lökeborg, Sweden, LNCS 1956
TYPES 2000 Durham, UK, LNCS 2277
TYPES 2002 Berg en Dal, The Netherlands, LNCS 2646
TYPES 2003 Turin, Italy, LNCS 3085
TYPES 2004 Jouy-en-Josas, France, LNCS 3839

ESPRIT BRA 6453 was a continuation of ESPRIT Action 3245, Logical Frameworks: Design, Implementation and Experiments. Proceedings for annual meetings under that action were published by Cambridge University Press in the books *Logical Frameworks* and *Logical Environments*, edited by Gérard Huet and Gordon Plotkin.

We are grateful for the support of the School of Computer Science and Information Technology at the University of Nottingham in organizing the meeting. We should like to thank James Chapman, Wouter Swierstra and Peter Morris, who helped with the administration and coordination of the meeting. We are also grateful to Peter Morris for help in the preparation of the volume.

March 2007   Thorsten Altenkirch
                                                       Conor McBride

Referees

A. Abel
P. Aczel
R. Adams
R. Atkey
S. van Bakel
C. Ballarin
S. Berardi
Y. Bertot
A. Bove
E. Brady
P. Callaghan
J. Carlstrom
J. Cheney
J. Chrząszcz
M. Coppo
J. Courant
C. Coquand
T. Coquand
R. Crole
R. Davies
J. Despeyroux
L. Dixon
G. Dowek
R. Dychhoff
M. Escardo
J-C. Filliâtre
M. Fluet
P. Fontaine
N. Gambino
H. Geuvers

N. Ghani
A. Gordon
B. Grégoire
P. Hancock
J. Harrison
M. Huisman
B. Jacobs
S. Jost
T. Kelsey
J. Lipton
Z. Luo
M.E. Maietti
J. McKinna
M. Miculan
A. Miquel
P. Morris
S. Negri
M. Oostdijk
R. Paterson
D. Pattinson
R. Pollack
T. Ridge
G. Sambin
T. Streicher
C. Urban
D. Walukiewicz-Chrząszcz
S. Weirich
A. Weiermann
B. Werner
F. Wiedijk

Table of Contents

Weyl's Predicative Classical Mathematics as a Logic-Enriched Type Theory*

Robin Adams and Zhaohui Luo

Dept of Computer Science, Royal Holloway, Univ of London
{robin,zhaohui}@cs.rhul.ac.uk

Abstract. In *Das Kontinuum*, Weyl showed how a large body of classical mathematics could be developed on a purely predicative foundation. We present a logic-enriched type theory that corresponds to Weyl's foundational system. A large part of the mathematics in Weyl's book — including Weyl's definition of the cardinality of a set and several results from real analysis — has been formalised, using the proof assistant Plastic that implements a logical framework. This case study shows how type theory can be used to represent a non-constructive foundation for mathematics.

Keywords: logic-enriched type theory, predicativism, formalisation.

1 Introduction

Type theories have proven themselves remarkably successful in the formalisation of mathematical proofs. There are several features of type theory that are of particular benefit in such formalisations, including the fact that each object carries a type which gives information about that object, and the fact that the type theory itself has an inbuilt notion of computation.

These applications of type theory have proven particularly successful for the formalisation of intuitionistic, or constructive, proofs. The correspondence between terms of a type theory and intuitionistic proofs has been well studied. The degree to which type theory can be used for the formalisation of other notions of proof has been investigated to a much lesser degree.

There have been several formalisations of classical proofs by adapting a proof checker intended for intuitionistic mathematics, say by adding the principle of excluded middle as an axiom (such as [Gon05]). But the metatheoretic properties of the type theory thus obtained, and to what degree that theory corresponds to the practice of classical mathematics, are not well known. For the more exotic schools of mathematics, such as predicativism, the situation is still worse.

We contend that the intuitions behind type theory apply outside of intuitionistic mathematics, and that the advantages of type theory would prove beneficial when applied to other forms of proof. It is equally natural in classical mathematics to divide mathematical objects into types, and it would be of as much benefit

---

* This work is partially supported by the UK EPSRC research grants GR/R84092 and GR/R72259 and EU TYPES grant 510996.

T. Altenkirch and C. McBride (Eds.): TYPES 2006, LNCS 4502, pp. 1–17, 2007.

to take advantage of the information provided by an object's type in a classical proof. The notion of computation is an important part of classical mathematics. When formally proving a property of a program, we may be perfectly satisfied with a classical proof, which could well be shorter or easier to find.

We further contend that it is worth developing and studying type theories specifically designed for non-constructive mathematical foundations. For this purpose, the systems known as *logic-enriched type theories* (LTTs), proposed by Aczel and Gambino [AG02, GA06], would seem to be particularly appropriate.

LTTs can be considered in a uniform type-theoretic framework that supports formal reasoning with different logical foundations, as proposed in [Luo06]. In particular, this may offer a uniform setting for studying and comparing different mathematical foundations, in the way that predicate logic has in traditional mathematical logic research. For example, when building a foundational system for mathematics, we must decide whether the logic shall be *classical* or *constructive* and whether *impredicative* definitions are allowed, or only *predicative*. Each of the four possible combinations of these options has been advocated as a foundation for mathematics at some point in history. The four possibilities are:

- **Impredicative classical mathematics.** This is arguably the way in which the vast majority of practising mathematicians work (although much of their work can often also be done in the other settings). Zermelo-Fraenkel Set Theory (ZF) is one such foundation.
- **Impredicative constructive mathematics.** Impredicative types theories such as CC [CH88] and UTT [Luo94], or CIC [BC04] provide its foundations.
- **Predicative classical mathematics.** This was the approach taken by Weyl in his influential monograph of 1918, *Das Kontinuum* [Wey18].
- **Predicative constructive mathematics.** Its foundations are provided, for example, by Martin-Löf's type theory. [NPS90, ML84].

Our type-theoretic framework provides a uniform setting for formalisation of these different mathematical foundations.

In this paper, we present a case study in the type-theoretic framework: to construct an LTT to represent a non-constructive approach to the foundation of mathematics; namely the predicative, classical foundational system of mathematics developed by Weyl in his monograph *Das Kontinuum* [Wey18]. We describe a formalisation in that LTT of several of the results proven in the book.

Weyl presents in his book a programme for the development of mathematics on a foundation that is *predicative*; that is, that avoids any definition which involves a 'vicious circle', where an object is defined in terms of a collection of which it is a member. The system presented in the book has attracted interest since, inspiring for example the second-order system ACA_0 [Fef00], which plays an important role in the project of Reverse Mathematics [Sim99]. It is a prominent example of a fully developed non-mainstream mathematical foundation, and so a formalisation should be of quite some interest.

We begin this paper describing in Section 2 in detail the version of Weyl's foundational system we shall be using. We then proceed in Section 3 to describe a logic-enriched type theory within a modified version of the logical framework LF[1] [Luo94]. We claim that this logic-enriched type theory faithfully corresponds to the system presented in Section 2. The formalisation itself was carried out in a modified version of the proof assistant Plastic [CL01], an implementation of LF. We describe the results proven in the formalisation in Section 4. The source code for the formalisation is available online at http://www.cs.rhul.ac.uk/~robin/weyl.

2   Weyl's Predicative Foundations for Mathematics

Hermann Weyl (1885–1955) contributed to many branches of mathematics in his lifetime. His greatest contribution to the foundations of mathematics was the book *Das Kontinuum* [Wey18] in 1918, in which he presented a predicative foundation which he showed was adequate for a large body of mathematics.

The concept of predicativity originated with Poincaré [Poi06], who advocated the *vicious circle principle*: a definition of an object is illegitimate if it is defined by reference to a totality that contains the object itself. Thus, we may not quantify over all sets when defining a set (as with Russell's famous 'set of all sets that do not contain themselves'); we may not quantify over all real numbers when defining a real number (as with the least upper bound of a set of reals); and so forth. A definition which involves such a quantification is called *impredicative*; one which does not, *predicative*. The advocacy of the exclusion of impredicative definitions has been given the name *predicativism*.

However much philosophical sympathy we may feel with predicativism, we may worry that, since impredicative definitions are so common in mathematical practice, their exclusion may require us to abandon too much of the mathematical corpus. Weyl's book provides evidence that this is not necessarily the case. In it, he shows how many results that are usually proven impredicatively can be proven predicatively; and that, even for those results that cannot, one can often prove predicatively a weaker result which in practice is just as useful. He does this by laying out a predicative foundation for mathematics, and developing a fairly large body of mathematics on this foundation.

A further discussion of the background to and content of Weyl's monograph can be found in Feferman [Fef98].

2.1   Weyl's Foundational System

We shall now present the version of Weyl's foundational system on which we based the formalisation. It differs from the semi-formal system described in *Das Kontinuum* in several details. In particular, we have extended Weyl's system with several features which are redundant in theory, but very convenient practically;

---

[1] The logical framework LF here is the typed version of Martin-Löf's logical framework [NPS90]. It is different from the Edinburgh LF [HHP93]: besides the formal differences, LF is intended to be used to specify computation rules and hence type theories such as Martin-Löf's type theory [NPS90] and UTT [Luo94]. A recent study of logical frameworks can be found in [Ada04].

these shall be described in the paragraphs headed 'Extensions to Weyl's system' below. Our notation in this section also differs considerably from Weyl's own.

Before turning to the formal details, we begin with a discussion of the intuitions behind Weyl's system, which is constructed following these principles:

1. The natural numbers are accepted as a primitive concept.
2. Sets and relations can be introduced by two methods: explicit *predicative* definitions, and definition by recursion over the natural numbers.
3. Statements about these objects are either true or false.

Regarding point 2, we are going to provide ourselves with the ability to define sets by *abstraction*: given a formula $\phi[x]$ of the system, to form the set

$$S = \{x \mid \phi[x]\} \ . \tag{1}$$

In order to ensure that every such definition is predicative, we restrict which quantifiers can occur in the formula $\phi[x]$ that can appear in (1): we may quantify over natural numbers, but we may not quantify over sets or functions. In modern terminology, we would say that $\phi[x]$ must contain only first-order quantifiers.

Weyl divides the universe of mathematical objects into collections which he calls *categories*. These categories behave very similarly to the types of a modern type theory. (This is no coincidence: Weyl was influenced by many of the ideas in Russell's theory of types when constructing his system.) For example, there shall be the category of all natural numbers, and the category of all sets of natural numbers. We give a full list of the categories present in the system below.

The categories are divided into *basic* categories, those that may be quantified over in a definition of the form (1); and the *ideal* categories, those that may not. The category of natural numbers shall be a basic category; categories of sets and categories of functions shall be ideal categories. In modern terminology, the basic categories contain first-order objects, while the ideal categories contain second-, third- and higher-order objects.

He proceeds to divide the propositions of his system into the *small*[2] propositions, those which involve quantification over basic categories only, and so may occur in a definition of the form (1); and the *large* propositions, those which involve quantification over one or more ideal category, and so may not.

In more detail, here is our version of Weyl's foundational system.

Categories. There are a number of *basic categories* and a number of *ideal categories*, each of which has *objects*.

1. There are basic categories, including the basic category $\mathbb{N}$ of natural numbers.
2. Given any categories $A_1, \ldots, A_m$ and $B_1, \ldots, B_n$, we may form the ideal category $(A_1 \times \cdots \times A_m) \to \mathrm{Set}\,(B_1 \times \cdots \times B_n)$ of functions of m arguments that take objects of categories $A_1, \ldots, A_m$, and return sets of n-tuples of

---

[2] Weyl chose the German word *finite*, which in other contexts is usually translated as 'finite'; however, we agree with Pollard and Bole [Wey87] that this would be misleading.

objects of categories $B_1, \ldots, B_n$. The number m may be zero here; the number n may not.

(These were the only categories of functions present in *Das Kontinuum*. For the purposes of our formalisation, we have added categories of functions $A \to B$ for *any* categories A and B; see 'Extensions' below.)

For example, taking $m = 0$ and $n = 1$ allows us to form the category Set (B), the category of all sets whose elements are of category B. Taking $m = 1$ and $n = 2$ allows us to form the category $A \to$ Set $(B \times C)$, the category of all functions which take an object from A and return a binary relation between the categories B and C.

Propositions

1. There are a number of *primitive relations* that hold between the objects of these categories:
   - the relation 'x is the successor of y' (Sxy) between natural numbers;
   - the relation '$x = y$' between objects of any *basic* category;
   - the relation $\langle y_1, \ldots, y_n \rangle \in F(x_1, \ldots, x_m)$ where F is of category $(A_1 \times \cdots \times A_m) \to$ Set $(B_1 \times \cdots \times B_n)$, x_i of category A_i and y_i of B_i.
2. The *small* propositions are those that can be built up from the primitive relations using the operations of substituting objects of the appropriate category for variables, the propositional connectives $\neg$, $\wedge$, $\vee$ and $\to$, and the universal and existential quantifications over the *basic* categories.
3. The *propositions* are those that can be built up from the primitive relations using substitution of objects for variables, the propositional connectives and quantification over *any* categories.

Objects

- **Explicit Definition.** Given any *small* proposition $\phi[x_1, \ldots, x_m, y_1, \ldots, y_n]$, we may introduce an object F of category $(A_1 \times \cdots \times A_m) \to$ Set $(B_1 \times \cdots \times B_n)$ by declaring

$$F(x_1, \ldots, x_m) = \{\langle y_1, \ldots, y_n \rangle \mid \phi[x_1, \ldots, x_m, y_1, \ldots, y_n]\} \quad (2)$$

Making this declaration has the effect of introducing the axiom

$$\forall \boldsymbol{x}, \boldsymbol{y}(\boldsymbol{y} \in F(\boldsymbol{x}) \leftrightarrow \phi[\boldsymbol{x}, \boldsymbol{y}]) \ . \quad (3)$$

Principle of Iteration. This principle allows us to define functions by recursion over the natural numbers; given a function F from a category $S \to S$, we can form a function G of category $S \times \mathbb{N} \to S$ by setting $G(X, n) = F^n(X)$. G is thus formed by *iterating* the function F.

More formally, let S be a category of the form Set $(B_1 \times \cdots \times B_n)$. Given an object F of category $(A_1 \times \cdots \times A_m \times S) \to S$, we may introduce an object G of category $(A_1 \times \cdots \times A_m \times S \times \mathbb{N}) \to S$ by declaring

$$\left. \begin{array}{l} G(x_1, \ldots, x_m, X, 0) = X \\ G(x_1, \ldots, x_m, X, k+1) = F(x_1, \ldots, x_m, G(x_1, \ldots, x_m, X, k)) \end{array} \right\} \quad (4)$$

where x_i is affiliated with category A_i, X with S, and k with $\mathbb{N}$.
Making these declarations has the effect of introducing the axiom

$$\forall \boldsymbol{x}, \boldsymbol{y}(\boldsymbol{y} \in G(\boldsymbol{x}, X, 0) \leftrightarrow \boldsymbol{y} \in X) \tag{5}$$

$$\forall \boldsymbol{x}, \boldsymbol{y}, a, b(Sab \rightarrow (\boldsymbol{y} \in G(\boldsymbol{x}, X, b) \leftrightarrow \boldsymbol{y} \in F(\boldsymbol{x}, G(\boldsymbol{x}, X, a))))$$

Axioms. The theorems of Weyl's system are those that can be derived via *classical* predicate logic from the following axioms:

1. The axioms for the equality relation on the basic categories.
2. Peano's axioms for the natural numbers (including proof by induction).
3. The axioms (3) and (5) associated with any definitions (2) and (4) that have been introduced.

We note that there is a one-to-one correspondence, up to the appropriate equivalence relations, between the objects of category $C = (A_1 \times \cdots \times A_m) \rightarrow$ Set $(B_1 \times \cdots \times B_n)$; and the small propositions $\phi[x_1, \ldots, x_m, y_1, \ldots, y_n]$, with distinguished free variables x_i of category A_i and y_i of category B_i. Given any F of category C, the corresponding small proposition is $\boldsymbol{y} \in F(\boldsymbol{x})$. Conversely, given any small proposition $\phi[\boldsymbol{x}, \boldsymbol{y}]$, the corresponding object F of category C is the one introduced by the declaration $F(\boldsymbol{x}) = \{\boldsymbol{y} \mid \phi[\boldsymbol{x}, \boldsymbol{y}]\}$.

Extensions to Weyl's System. For the purposes of this formalisation, we have added features which were not explicitly present in Weyl's system, but which can justifiably be seen as conservative extensions of the same. We shall allow ourselves the following.

1. We shall introduce a category $A \times B$ of *pairs* of objects, one from the category A and one from the category B. $A \times B$ shall be a basic category when A and B are both basic, and ideal otherwise. This shall allow us, for example, to talk directly about integers (which shall be pairs of natural numbers) and rationals (which shall be pairs of integers).
2. We shall introduce a category $A \rightarrow B$ of functions from A to B for all categories (not only the case where B has the form Set $(\cdots)$).
   $A \rightarrow B$ shall always be an ideal category. For the system to be predicative, quantification over functions must not be allowed in small propositions; quantifying over $A \rightarrow \mathbb{N}$, for example, would provide an effective means of quantifying over Set (A). (Recall that, classically, the power set of X and the functions from X to a two-element set are in one-to-one correspondence.)
   Weyl instead defined functions as particular sets of ordered pairs, and showed in detail how addition of natural numbers can be constructed. For the purposes of formalisation, it was much more convenient to provide ourselves with these categories of functions, and the ability to define functions by recursion, from the very beginning.

We shall permit ourselves to use a function symbol 's' for successor, rather than only the binary relation Sxy.

We have diverged from Weyl's system in two other, more minor, ways which should be noted. We choose to start the natural numbers at 0, whereas Weyl begins at 1; and, when we come to construct the real numbers, we follow the sequence of constructions $\mathbb{N} \longrightarrow \mathbb{Z} \longrightarrow \mathbb{Q} \longrightarrow \mathbb{R}$ rather than Weyl's $\mathbb{N} \longrightarrow \mathbb{Q}^+ \longrightarrow \mathbb{Q} \longrightarrow \mathbb{R}$.

3 Weyl's Foundation as a Logic-Enriched Type Theory

What immediately strikes a modern eye reading *Das Kontinuum* is how similar the system presented there is to what we now know as a type theory; almost the only change needed is to replace the word 'category' with 'type'. In particular, Weyl's system is very similar to a *logic-enriched type theory* (LTT for short).

The concept of an LTT, an extension of the notion of type theory, was proposed by Aczel and Gambino in their study of type-theoretic interpretations of constructive set theory [AG02, GA06]. A type-theoretic framework, which formulates LTTs in a logical framework, has been proposed in [Luo06] to support formal reasoning with different logical foundations. In particular, it adequately supports classical inference with a notion of predicative set, as described below.

An LTT consists of a type theory augmented with a separate, primitive mechanism for forming and proving propositions. We introduce a new syntactic class of *formulas*, and new judgement forms for a formula being a well-formed proposition, and for a proposition being provable from given hypotheses.

An LTT thus has two rigidly separated components or 'worlds': the *datatype* world of terms and types, and the *logical* world of proofs and propositions, for describing and reasoning about the datatype world[3]. In particular, we can form propositions by *quantification* over a type; and prove propositions by *induction*.

In this work, we shall also allow the datatype world to depend on the logical world in just one way: by permitting the formation of *sets*. Given a proposition $\phi[x]$, we shall allow the construction of the set $\{x \mid \phi[x]\}$ in the datatype world; thus, a set shall be a term that depends on a proposition. (Note that these sets are not themselves types.) This shall be the only way in which the datatype world may depend on the logical world; in particular, no type may depend on a proposition, and no type, term or proposition may depend on a proof.

We start by extending the logical framework LF with a kind *Prop*, standing for the world of logical propositions. Then, we introduce a type for each category: a construction in *Prop* for each method of forming propositions; a type universe U of names of the basic categories; and a propositional universe *prop* of names of the small propositions. Thus constructed, the LTT with predicative sets corresponds extremely closely to Weyl's foundational system.

---

[3] This is very much in line with the idea that there should be a clear separation between logical propositions and data types, as advocated in the development of type theories ECC and UTT [Luo94].

1. Rules of Deduction for *Type* and *El*

$$\frac{\Gamma \text{ valid}}{\Gamma \vdash Type \text{ kind}} \qquad \frac{\Gamma \vdash A : Type}{\Gamma \vdash El\,(A) \text{ kind}} \qquad \frac{\Gamma \vdash A = B : Type}{\Gamma \vdash El\,(A) = El\,(B)}$$

2. Rules of Deduction for *Prop* and *Prf*

$$\frac{\Gamma \text{ valid}}{\Gamma \vdash Prop \text{ kind}} \qquad \frac{\Gamma \vdash P : Prop}{\Gamma \vdash Prf\,(P) \text{ kind}} \qquad \frac{\Gamma \vdash P = Q : Prop}{\Gamma \vdash Prf\,(P) = Prf\,(Q)}$$

Fig. 1. Kinds *Type* and *Prop* in LF′

3.1 Logic-Enriched Type Theories in Logical Frameworks

There exist today many *logical frameworks*, designed as systems for representing many different type theories. It requires only a small change to make a logical framework capable of representing LTTs as well.

For this work, we have used the logical framework LF [Luo94], which is the basis for the proof checker Plastic [CL01]. LF provides a kind *Type* and a kind constructor *El*. To make LF capable of representing LTTs, we add a kind *Prop* and a kind constructor *Prf*. We shall refer to this extended framework as LF′.

Recall that a logical framework, such as LF or LF′, is intended as a metalanguage for constructing various type theories, the *object systems*. The frameworks consist of *kinds* and *objects*. The object systems are constructed in the framework by representing their expressions by certain objects. An LTT consists of *terms* and *types* (in the datatype world), and *propositions* and *proofs* (in the logical world). We shall build an LTT in LF′ by representing:

- the *types* by the objects of kind *Type*;
- the *terms* of type A by the objects of kind $El\,(A)$;
- the *propositions* by the objects of kind *Prop*;
- the *proofs* of the proposition ϕ by the objects of kind $Prf\,(\phi)$.

The rules of deduction for these new kinds *Prop* and $Prf\,(\cdots)$ are given in Figure 1, along with the rules those for *Type* and *El*, for comparison.

These new kinds allow us to form judgements of the following forms:

- $\Gamma \vdash \phi : Prop$, indicating that ϕ is a well-formed proposition;
- $\Gamma \vdash P : Prf\,(\phi)$, indicating that P is a proof of the proposition ϕ;
- $\Gamma, p_1 : Prf\,(\phi_1),\ldots,p_n : Prf\,(\phi_n) \vdash P : Prf\,(\psi)$, indicating that ψ is derivable from the hypotheses $\phi_1, \ldots, \phi_n$, with the object P encoding the derivation; this was denoted by $\Gamma \vdash \phi_1,\ldots,\phi_n \Rightarrow \psi$ in [AG02].

When formalizing a piece of mathematics using an LTT, the provable propositions are those ϕ for which $Prf\,(\phi)$ is inhabited. We state each theorem by forming the appropriate object ϕ of kind *Prop*, and then show that it is provable by constructing an object P of kind $Prf(\phi)$. (Due to its novelty, we shall *not* omit the constructor Prf in this paper.)

We also obtain judgements of the form $\Gamma \vdash \phi = \psi : Prop$. Judgements of this last form express that ϕ and ψ are *intensionally equal* propositions — that is, that ψ can be obtained from ϕ by a sequence of reductions and expansions of subterms. This is not to be confused with logical equivalence; intensional equality is a much stronger relation. The distinction is similar to that between judgemental equality (convertibility) and propositional equality between terms.

We recall that a type theory is specified in LF by declaring a number of *constants* with their kinds, and declaring several *computation rules* to hold between objects of some kinds of LF. An LTT can be specified in LF′ by making the above declarations for each constructor in its datatype component, and also declaring:

- for each logical constant (connective or quantifier) we wish to include, a constant of kind $(\cdots)Prop$;
- for each rule of deduction, a constant of kind $(\cdots)Prf(\phi)$
- some *computation rules for propositions*, of the form $(\cdots)(\phi = \psi : Prop)$

It was essential for this work that the logical framework we use be capable of representing computation rules. A framework such as Twelf [PS99], for example, would not be suitable for our purposes.

LTTs and Type Theories Compared. When using a type theory for formalisation, we identify each proposition with a particular type, and show that a theorem is provable by constructing a term of the corresponding type. The way we prove propositions in an LTT by constructing an object of kind $Prf(\cdots)$ is very similar. However, there are two important differences to be noted:

- We have separated the datatypes from the propositions. This allows us to add axioms without changing the datatype world. We can, for example, add the axiom Pierce (Fig. 2) without thereby causing all the function types $((A \rightarrow B) \rightarrow A) \rightarrow A$ to be inhabited.
- We do not have any computation rules on proofs. Further, a proof cannot occur inside a term, type or proposition. We are thus free to add any axioms we like to the logic: we know that, by adding the axiom Pierce (say), we shall not affect any of the properties of the reduction relation, such as decidability of convertibility or strong normalisation.

3.2   Natural Numbers, Products, Functions and Predicate Logic

We can now proceed to construct a logic-enriched type theory that corresponds to the foundational system Weyl presents in *Das Kontinuum*.

Our starting point is an LTT that contains, in its datatype component, a type $\mathbb{N}$ of natural numbers, as well as non-dependent product and function types $A \times B$ and $A \rightarrow B$; and, in its logical component, classical predicate logic. We present some of the declarations involved in its specification in Figure 2, namely those involving natural numbers (including $E_{\mathbb{N}}$, which permits the definition of

---

Natural Numbers

$\mathbb{N} : Type$
$0 : \mathbb{N}$
$s : (\mathbb{N})\mathbb{N}$
$E_{\mathbb{N}} : (C : (\mathbb{N})Type)(C0)((x : \mathbb{N})(Cx)C(sx))(n : \mathbb{N})Cn$
$Ind_{\mathbb{N}} : (P : (\mathbb{N})Prop)(Prf\,(P0))((x : \mathbb{N})(Prf\,(Px))Prf\,(P(sx)))(n : \mathbb{N})Prf\,(Pn)$

$$E_{\mathbb{N}}\,C\,a\,b\,0 = a : El\,(C0)$$
$$E_{\mathbb{N}}\,C\,a\,b\,(sn) = b\,n\,(E_{\mathbb{N}}\,C\,a\,b\,n) : El\,(C(sn))$$

Implication

$$\supset : (Prop)(Prop)Prop$$
$$\supset I : (P : Prop)(Q : Prop)((Prf\,(P))Prf\,(Q))Prf\,(\supset P\,Q)$$
$$\supset E : (P : Prop)(Q : Prop)(Prf\,(\supset P\,Q))(Prf\,(P))Prf\,(Q)$$

Pierce's Law

$$\text{Pierce} : (P : Prop)(Q : Prop)(((Prf\,(P))Prf\,(Q))Prf\,(P))Prf\,(P)$$

Fig. 2. Declaration of an LTT in LF$'$

functions by recursion, and $Ind_{\mathbb{N}}$, which permits propositions to be proven by induction) and implication. The other types and logical constants follow a similar pattern. We also include a version of Pierce's Law to ensure the logic is classical.

3.3   Type Universes and Propositional Universes

We have now introduced our collection of categories: they are the objects of kind *Type*. We still however need to divide them into the basic and ideal categories.

The device we need to do this is one with which we are familiar: that of a *type universe*. A type universe U (à la Tarski) is a type whose objects are names of types. Intuitively, the types that have a name in U are often thought of as the 'small' types, and those that do not (such as U itself) as the 'large' types. Together with U, we introduce a constant T such that, for each name $a : U$, $T(a)$ is the type named by a.

For our system, we provide ourselves with a universe whose objects are the names of the *basic* categories. We thus need a universe U that contains a name for $\mathbb{N}$, and a method for constructing a name for $A \times B$ out of a name for A and a name for B. This is done in Figure 3(1). We also introduce a relation of equality for every basic category.

Now we need to divide our propositions into the small propositions and the large propositions. To do so, we use the notion in the logical world which is analagous to a type universe: a *propositional universe*.

We wish to introduce the collection *prop* of names of the *small* propositions; that is, the propositions that only involve quantification over small types. It is not immediately obvious where this collection should live.

---

1. The Type Universe

$$U \; : \; Type$$
$$T \; : \; (U)Type$$
$$\hat{\mathbb{N}} \; : \; U$$
$$\hat{\times} \; : \; (U)(U)U$$
$$T(\hat{\mathbb{N}}) = \mathbb{N} : Type$$
$$T(\hat{\times}a\,b) = \times(Ta)\,(Tb) : Type$$

Propositional Equality

$$\simeq \; : \; (A:U)(TA)(TA)Prop$$
$$\simeq I \; : \; (A:U)(a:TA)\simeq A\,a\,a$$
$$\simeq E \; : \; (A:U)(P:(TA)Prop)(a,b:TA)$$
$$(Prf\,(\simeq A\,a\,b))(Prf\,(Pa))Prf\,(Pb)$$

2. The Propositional Universe

$$prop \; : \; Prop$$
$$V \; : \; (prop)Prop$$
$$\hat{\perp} \; : \; prop$$
$$\hat{\supset} \; : \; (prop)(prop)prop$$
$$\hat{\forall} \; : \; (a:U)((Ta)prop)prop$$
$$\hat{\simeq} \; : \; (a:U)(Ta)(Ta)prop$$
$$V(\hat{\perp}) = \perp : Prop$$
$$V(\hat{\supset}p\,q) = \supset(Vp)\,(Vq) : Prop$$
$$V(\hat{\forall}a\,p) = \forall(Ta)\,[x:Ta]V(px) : Prop$$
$$V(\hat{\simeq}a\,s\,t) = \simeq(Ta)\,s\,t$$

Fig. 3. A Type Universe and a Propositional Universe

We choose to declare *prop* : *Prop*, instead of *prop* : *Type*. Now, it must be admitted that *prop* is not conceptually a proposition; it does not assert any relation to hold between any mathematical objects. However, it seems to make little practical difference which choice is made. Choosing to place *prop* in *Prop* provides a pleasing symmetry with *U* and *Type*, and *prop* seems to belong more to the logical world than the datatype world. Until more foundational work on LTTs has been done, we accept this compromise: *prop* is a 'proposition', each of whose 'proofs' is a name of a small proposition[4].

As with the type universe, when we introduce a propositional universe *prop* we provide ourselves with a constant *V* such that, for each name $p : prop$, $V(p)$ is the proposition named by p. We also provide constants that reflect equality, the propositional connectives, and quantification over the basic types. The declarations are given in Figure 3(2). Note that the propositional universe provides us with our first examples of computation rules for propositions.

We have built *prop* as a universe à la Tarski; that is, its objects are *names* of small propositions. Plastic does not provide the necessary mechanism for defining *prop* as a universe à la Russell, where its objects would be the small propositions themselves. We suspect that the choice would make no practical difference.

3.4   The Predicative Notion of Set

We now have all the machinery necessary to be able to introduce *typed sets*. For any type *A*, we wish to introduce the type Set (A) consisting of all the sets that

---

[4] Other alternatives would be to introduce a new top-kind to hold *prop*, or to make *prop* itself a top-kind. We do not discuss these here.

$$
\begin{aligned}
\text{Set} \; &: \; (Type)Type \\
\text{set} \; &: \; (A : Type)((A)prop)\text{Set}\,(A) \\
\in \; &: \; (A : Type)(A)(\text{Set}\,(A))prop \\
\in A\,a\,(\text{set}\,A\,P) &= P\,a : prop
\end{aligned}
$$

Fig. 4. The Predicative Notion of Set

can be formed, each of whose members is an object of type A. (Thus we do not have any sets of mixed type.) We take a set to be introduced by a *small predicate* over A; that is, an object of kind $(A)prop$, a function which takes objects of A and returns (a name of) a small proposition.

We therefore make the declarations given in Figure 4:

- Given any type A, we can form the type $\text{Set}\,(A)$. The terms of $\text{Set}\,(A)$ are all the sets that can be formed whose elements are terms of type A.
- Given a *small* proposition $\phi[x]$ with variable x of type A, we can form the set $\{x : \phi[x]\}$. Formally, given a name $p[x] : prop$ of a small proposition, we can form 'set $A\,([x : A]p[x])$', which we shall write as $\{x : V(p[x])\}$.
- If $a : A$ and $X : \text{Set}\,(A)$, we can form the proposition $\in A\,a\,X$, which we shall write as $a \in X$.
- Finally, we want to ensure that the elements of the set $\{x : \phi[x]\}$ are precisely the terms a such that $\phi[a]$ is true. This is achieved by adding our second example of a computation rule on propositions, the last line on Figure 4, which we may read as: $a \in \{x : \phi[x]\}$ computes to $\phi[a]$.

As $\text{Set}\,(A)$ is always to be an ideal category, we do not provide any means for forming a name of $\text{Set}\,(A)$ in U.

These sets are not themselves types; they are terms of type $\text{Set}\,(\cdots)$. The membership condition $a \in A$ is a proposition, not a typing judgement. In particular, we distinguish between a type A and the set $\{x : A \mid \top\}$ of type $\text{Set}\,(A)$.

A similar construction could be carried out in an LTT if we wished to work in an impredicative setting, simply by replacing *prop* with *Prop* throughout Figure 4. This would allow us to form the set $\{x : \phi[x]\}$ for *any* proposition $\phi[x]$. (See [Luo06] for more details.) Thanks to the similarity of the two approaches, much if not all of the work done in the predicative system could be reused in the impredicative system. We shall return to this point in Section 4.2.

4   Formalisation in Plastic

We have formalised this work in a version of the proof assistant Plastic [CL01], modified by Paul Callaghan to be an implementation of LF'. We have produced a formalisation which includes all the definitions and proofs of several of the results from Weyl's book.

In Plastic, all lines that are to be parsed begin with the character >; any line that does not is a comment line. A constant c may be declared to have kind $(x_1 : K_1) \cdots (x_n : K_n)K$ by the input line

$$> \ [c[x_1 : K_1] \cdots [x_n : K_n] \ : \ K];$$

We can define the constant c to be the object $[x_1 : K_1] \cdots [x_n : K_n]k$ of kind $(x_1 : K_1) \cdots (x_n : K_n)K$ by writing

$$> \ [c[x_1 : K_1] \cdots [x_n : K_n] \ = \ k \ : \ K];$$

In both these lines, the kind indicator $:K$ is optional, and is usually omitted.

We can make any argument implicit by replacing it with a 'meta-variable' ?, indicating that we wish Plastic to infer its value.

These are the only features of the syntax that we shall use in this paper.

4.1   Results Proven

Cardinality of Sets. In Weyl's system, we can define the predicate 'the set X has exactly n members' in the following manner, which shows the power of the principle of iteration.

Given a basic category A, define the function $K : \mathbb{N} \to \mathrm{Set}\,(\mathrm{Set}\,(A))$ by recursion as follows. The intention is that $K(n)$ is the set of all sets $X : \mathrm{Set}\,(A)$ that have at least n members.

$$K(0) = \{X \mid \top\}$$
$$K(n+1) = \{X \mid \exists a(a \in X \wedge X \setminus \{a\} \in K(n))\}$$

In Plastic, this is done as follows:

```
> [at_least_set [tau : U] = E_Nat ([_ : Nat] Set (Set (T tau)))
>      (full (Set (T tau)))
>      [n : Nat] [Kn : Set (Set (T tau))] set (Set (T tau))
>        [X : Set (T tau)] ex tau [a : T tau]
>            and (in (T tau) a X) (in ? (setminus' tau X a) Kn)];
```

We define the proposition 'X has at least n members' to be $X \in K(n)$.

```
> [At_Least [tau : U] [X : Set (T tau)] [n : Nat]
>     = In ? X (at_least_set tau n)];
```

For n a natural number, define the *cardinal number* $\bar{n}$ to be $\{x \mid x < n\}$.

```
> [card [n : Nat] = set Nat [x : Nat] lt x n];
```

Define the *cardinality* of a set A to be $|A| = \{n \mid A \text{ has at least } sn \text{ members}\}$.

```
> [cardinality [tau : U] [A : Set (T tau)]
>     = set Nat [n : Nat] at_least tau A (succ n)];
```

We can prove the following result:

The cardinality $|X|$ of a set X is either $\{x \mid \top\}$ or $\bar{n}$ for some n.

We thus have two classes of cardinal numbers: $\bar{n}$ for finite sets, and $\{x \mid \top\}$, which we denote by ∞, for infinite sets. (There is thus only one infinite cardinality in *Das Kontinuum*.) We define 'X has exactly n members' to be $|X| \approx \overline{sn}$, where $\approx$ denotes the following equivalence relation on sets:

$$X \approx Y \equiv \forall x(x \in X \leftrightarrow x \in Y) .$$

```
> [infty = full Nat];
> [Exactly [tau : U] [A : Set (T tau)] [n : Nat]
>    = Seteq Nat (cardinality tau A) (card (succ n))];
```

With these definitions, we can prove results such as the following:

1. If A has at least n elements and $m \leq n$, then A has at least m elements.
2. If A has exactly n elements, then $m \leq n$ iff A has at least m elements.
3. If A has exactly m elements, B has exactly n elements, and A and B are disjoint, then $A \cup B$ has exactly $m + n$ elements.

We have thus provided definitions of the concepts 'having at least n members' and 'having exactly n members' in such a way that the sets

$$\{X \mid X \text{ has at least } n \text{ members}\} \text{ and } \{X \mid X \text{ has exactly } n \text{ members}\}$$

are definable predicatively. This would not be possible if we defined 'X has exactly n elements' as the existence of a bijection between X and $\overline{sn}$; we would have to quantify over the ideal category $A \to \mathbb{N}$. It also cannot be done as directly in a predicative system of second order arithmetic such as ACA_0 [Sim99].

Construction of the Reals. The set of real numbers is constructed by the following process. We first define the type of integers $\mathbb{Z}$, with a defined relation of equality $\approx_{\mathbb{Z}}$. We then define a *rational* to be a pair of integers, the second of which is non-zero. That is, for $q : \mathbb{Z} \times \mathbb{Z}$, we define '$q$ is rational' by

$$\langle x, y \rangle \text{ is rational} \equiv y \not\approx_{\mathbb{Z}} 0 .$$

We proceed to define equality of rationals $q \approx_{\mathbb{Q}} q'$, addition, multiplication and ordering on the rationals.

A *real* is a Dedekind cut of rationals; that is, an object R of the category $\text{Set} (\mathbb{Z} \times \mathbb{Z})$ that:

- is a *domain of rationals*; if $q \in R$ and $q \approx_{\mathbb{Q}} q'$, then $q' \in R$;
- is *closed downwards*; if $q \in R$, and $q' < q$, then $q' \in R$;
- has no maximal element; for every rational $q \in R$, there exists a rational $q' \in R$ such that $q < q'$;

- and is neither empty nor full; there exists a rational q such that $q \in R$, and a rational q' such that $q' \notin R$.

Equality of reals is defined to be extensional equality restricted to the rationals:

$$R \approx_{\mathbb{R}} S \equiv \forall q(q \text{ is rational} \rightarrow (q \in R \leftrightarrow q \in S))$$

We note that, in this formalisation, there was no way to define the collection of rationals as a type, say as the 'sigma-type' '$(\Sigma q : \mathbb{Z} \times \mathbb{Z})q$ is rational'. This is because our LTT offers no way to form a type from a type $\mathbb{Z} \times \mathbb{Z}$ and a proposition 'q is rational'.

Real Analysis. Weyl was keen to show that his predicative system was strong enough to be used for mathematical work by demonstrating that, while several traditional theorems cannot be proven within it, we can usually prove a version of the theorem that is only slightly weaker.

For example, it seems not to be provable predicatively the *least upper bound principle*: that every set A of real numbers bounded above has a least upper bound l. Impredicatively, we would define l to be the union of A. This cannot be done predicatively, as it involves quantification over real numbers. However, we can prove the following two statements, one of which is usually enough for any practical purpose:

1. Every set S of *rational* numbers bounded above has a unique (real) least upper bound l. Take $l = \{q \in \mathbb{Q} \mid (\exists q' \in S)q < q'\}$.
2. Every *sequence* $r_1, r_2, \dots$ of real numbers bounded above has a unique least upper bound l. Take $l = \{q \in \mathbb{Q} \mid (\exists n : \mathbb{N})q \in r_n\}$.

These involve only quantification over the rationals and the natural numbers, respectively. (We note that either of these is equivalent to the least upper bound principle in an impredicative setting.)

The first is enough to prove the classical Intermediate Value Theorem:

If $f : \text{Set}(\mathbb{Z} \times \mathbb{Z}) \rightarrow \text{Set}(\mathbb{Z} \times \mathbb{Z})$ is a continuous function from the reals to the reals, and $f(a) < v < f(b)$ for some reals a, b, v with $a < b$, then there exists a real c such that $a < c < b$ and $f(c) = v$.

Weyl proves this proposition by taking c to be the least upper bound of the set of all rationals q such that $a < q < b$ and $f(q) < v$. For the formalisation, it was more convenient to define directly: $c = \{q \in \mathbb{Q} \mid (\exists q' \in \mathbb{Q})q < q' < b \wedge f(q') < v\}$.

4.2   An Impredicative Development

As mentioned in Section 3.4, it would not be difficult to modify this formulation to get a development of the same theorems in an impredicative system. All we have to do is remove the distinction between large and small propositions.

Recall that our propositional universe was introduced by the constructors in Figure 3(2). In principle, we could simply replace these with

$$prop = Prop \qquad\qquad V = [x : Prop]Prop$$
$$\hat{\bot} = \bot \qquad\qquad\qquad \hat{\supset} = \supset$$
$$\hat{\forall} = [A : U]\forall(TA) \qquad\qquad \hat{\simeq} = \simeq$$

However, at present plastic becomes unstable when definitions are made at the top-kind level such as $prop = Prop$.

Alternatively, we can add two impredicative quantifiers to $prop$, together with their computation rules:

$$\overline{\forall} : (A : Type)((A)prop)prop \qquad V(\overline{\forall}AP) = \forall A([x : A]V(Px))$$
$$\overline{\exists} : (A : Type)((A)prop)prop \qquad V(\overline{\exists}AP) = \exists A([x : A]V(Px))$$

Now $prop$, which determines the collection of propositions over which sets can be formed, covers the whole of $Prop$. We can form the set $\{x \mid \phi[x]\}$ for any well-formed proposition $\phi[x]$. However, all our old proof files still parse[5]. Once this change has been made, we can go on to prove the statement that every set of real numbers bounded above has a least upper bound.

It would be interesting to develop impredicative analysis in our setting, and to study the reuse of proof development.

5   Conclusion

We have conducted a case study in Plastic of the use of a type-theoretic framework to construct a logic-enriched type theory as the basis for a formalisation of a non-constructive system of mathematical foundations, namely that presented in Weyl's *Das Kontinuum*. As a representation of Weyl's work, it is arguably better in some ways than such second-order systems as ACA_0 [Fef00], since we can form a definition of the cardinality of a set that is much closer to Weyl's own. The formalisation work required only a minor change to the existing logical framework implemented in Plastic, allowing us to preserve all the features of Plastic with which we were already familiar.

Future work includes comparison of the type-theoretic framework with other systems such as ACA_0. It would also be interesting to carry out the impredicative development of analysis in our setting, reusing the predicative development.

It has been brought to our attention that logic-enriched type theories may be closely related to the notion of the internal language of a topos. A precise relationship between the two is to be further explored.

Acknowledgements. Thanks go to Paul Callaghan for his efforts in extending Plastic, and Peter Aczel for his comments during his visit to Royal Holloway. Thanks also go to the anonymous referees for very detailed and helpful comments.

---

[5] The same effect could also be made by changing the construction of U, making it equal to $Type$; or by making both these changes (to $prop$ and U).

References

[Ada04]   Adams, R.: A Modular Hierarchy of Logical Frameworks. PhD thesis, University of Manchester (2004)

[AG02]   Aczel, P., Gambino, N.: Collection principles in dependent type theory. In: Callaghan, P., Luo, Z., McKinna, J., Pollack, R. (eds.) TYPES 2000. LNCS, vol. 2277, pp. 1–23. Springer, Heidelberg (2002)

[BC04]   Bertot, Y., Castéran, P.: Interactive Theorem Proving and Program Development: Coq'Art: The Calculus of Inductive Constructions. In: Texts in Theoretical Computer Science, Springer, Heidelberg (2004)

[CH88]   Coquand, Th., Huet, G.: The calculus of constructions. Information and Computation 76(2/3) (1988)

[CL01]   Callaghan, P.C., Luo, Z.: An implementation of typed LF with coercive subtyping and universes. J. of Automated Reasoning 27(1), 3–27 (2001)

[Fef98]   Feferman, S.: Weyl vindicated. In: In the Light of Logic, pp. 249–283. Oxford University Press, Oxford (1998)

[Fef00]   Feferman, S.: The significance of Hermann Weyl's Das Kontinuum. In: Hendricks, V. et al. (eds.) Proof Theory – Historical and Philosophical Significance of Synthese Library, vol. 292 (2000)

[GA06]   Gambino, N., Aczel, P.: The generalised type-theoretic interpretation of constructive set theory. J. of Symbolic Logic 71(1), 67–103 (2006)

[Gon05]   Gonthier, G.: A computer checked proof of the four colour theorem (2005)

[HHP93]   Harper, R., Honsell, F., Plotkin, G.: A framework for defining logics. Journal of the Association for Computing Machinery 40(1), 143–184 (1993)

[Luo94]   Luo, Z.: Computation and Reasoning: A Type Theory for Computer Science. Oxford University Press, Oxford (1994)

[Luo06]   Luo, Z.: A type-theoretic framework for formal reasoning with different logical foundations. In: Okada, M., Satoh, I. (eds.) Proc of the 11th Annual Asian Computing Science Conference. Tokyo (2006)

[ML84]   Martin-Löf, P.: Intuitionistic Type Theory. Bibliopolis (1984)

[NPS90]   Nordström, B., Petersson, K., Smith, J.: Programming in Martin-Löf's Type Theory: An Introduction. Oxford University Press, Oxford (1990)

[Poi06]   Poincaré, H.: Les mathématiques et la logique. Revue de Métaphysique et de Morale 13, 815–835; 14, 17–34; 14, 294–317 (1905-1906)

[PS99]   Pfenning, F., Schürmann, C.: System description: Twelf — a meta-logical framework for deductive systems. In: Ganzinger, H. (ed.) Automated Deduction - CADE-16. LNCS (LNAI), vol. 1632, pp. 202–206. Springer, Heidelberg (1999)

[Sim99]   Simpson, S.: Subsystems of Second-Order Arithmetic. Springer, Heidelberg (1999)

[Wey18]   Weyl, H.: Das Kontinuum. Translated as [Wey87] (1918)

[Wey87]   Weyl, H.: The Continuum: A Critical Examination of the Foundation of Analysis. Translated by Pollard, S., Bole, T. Thomas Jefferson University Press, Kirksville, Missouri (1987)

Crafting a Proof Assistant

Andrea Asperti, Claudio Sacerdoti Coen, Enrico Tassi, and Stefano Zacchiroli

Department of Computer Science, University of Bologna
Mura Anteo Zamboni, 7 – 40127 Bologna, Italy
{asperti,sacerdot,tassi,zacchiro}@cs.unibo.it

Abstract. Proof assistants are complex applications whose development has never been properly systematized or documented. This work is a contribution in this direction, based on our experience with the development of Matita: a new interactive theorem prover based—as Coq—on the Calculus of Inductive Constructions (CIC). In particular, we analyze its architecture focusing on the dependencies of its components, how they implement the main functionalities, and their degree of reusability.

The work is a first attempt to provide a ground for a more direct comparison between different systems and to highlight the common functionalities, not only in view of reusability but also to encourage a more systematic comparison of different softwares and architectural solutions.

1 Introduction

In contrast with automatic theorem provers, whose internal architecture is in many cases well documented (see e.g. the detailed description of Vampire in [16]), it is extremely difficult to find good system descriptions for their interactive counterpart. Traditionally, the only component of the latter systems that is suitably documented is the *kernel*, namely the part that is responsible for checking the correctness of proofs. Considering that:

1. Most systems (claim to) satisfy the so called "De Bruijn criterion", that is the principle that the correctness of the whole application should depend on the correctness of a sufficiently small (and thus reliable) kernel *and*
2. Interactive proving *looks like* a less ambitious task than fully automatic proving (eventually, this is the feeling of an external observer)

one could easily wonder where the complexity of interactive provers comes from[1]. Both points above are intentionally provocative. They are meant to emphasize that: (1) the kernel is possibly the most crucial, but surely not the most important component of interactive provers and (2) formal checking is just one of the activities of interactive provers, and probably not the most relevant one.

Of course, interactivity should be understood as a powerful integration rather than as a poor surrogate of automation: the user is supposed to interact when the system fails alone. Interaction, of course, raises a number of additional themes that are not present (or not so crucial) in automatic proving:

---

[1] e.g.: Coq is about 166,000 lines of code, to be compared with 50,000 lines of Otter.

T. Altenkirch and C. McBride (Eds.): TYPES 2006, LNCS 4502, pp. 18–32, 2007.

- library management (comprising both per-proof history and management of incomplete proofs);
- development of a strong linguistic support to enhance the human-machine communication of mathematical knowledge;
- development of user interfaces and interaction paradigms particularly suited for this kind of applications.

While the latter point has received a renewed attention in recent years, as testified by several workshops on the topic, little or no literature is available on the two former topics, hindering a real progress in the field.

In order to encourage a more systematic comparison of different software and architectural solutions we must first proceed to a more precise individuation of issues, functionalities, and software components. This work is meant to be a contribution in this direction. In particular we give in Section 2 a data-oriented high-level description of our interactive theorem prover denominated "Matita"[2]. We also try to identify the logic independent components to understand the degree of coupling between the system architecture and its logical framework. In Section 3 we provide an alternative presentation of the architecture, based on the offered functionalities. Section 4 is an estimation of the complexity of the components and of the amount of work required to implement them.

Although our architectural description comprises components that (at present) are specific to our system (such as the large use of metadata for library indexing) we believe that the overall design fits most of the existent interactive provers and could be used as a ground for a deeper software comparison of these tools.

2   Data-Driven Architectural Analysis

Formulae and proofs are the main data handled by an interactive theorem prover. Both have several possible representations according to the actions performed on them. Each representation is associated with a data type, and the components that constitute an interactive theorem prover can be classified according to the representations they act on. In this section we analyze the architecture of Matita according to this classification.

We also make the effort of identifying the components that are logic independent or that can be made such abstracting over the data types used for formulae and proofs. This study allows to quantify the efforts required in changing the underlying logic of Matita for the sake of experimenting with new logic foundations while preserving the technological achievements.

The proof and formulae representations used in Matita as well as its general architecture have been influenced by some design commitments: (1) Matita is heavily based on the Curry-Howard isomorphism. Execution of procedural and declarative scripts produce proof terms (λ-terms) that are kept for later processing. Even incomplete proofs are represented as λ-terms with typed linear

---

[2] "matita" means "pencil" in Italian: a simple, well known, and widespread authoring tool among mathematicians.

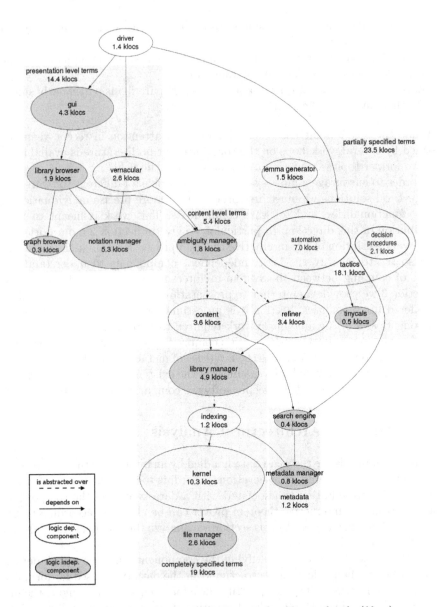

Fig. 1. Matita components with thousands of lines of code (*klocs*)

placeholders for missing subproofs. (2) The whole library, made of definitions and proof objects only, is searchable and browsable at any time. During browsing proof objects are explained in pseudo-natural language. (3) Proof authoring is performed editing either procedural or declarative scripts. Formulae are typed using ambiguous mathematical notation. Overloading is not syntactically constrained nor avoided using polymorphism.

According to the above commitments, in Matita we identified 5 term representations: presentation terms (concrete syntax), content terms (abstract syntax trees with overloaded notation), partially specified terms (λ-terms with placeholders), completely specified terms (well typed λ-terms), metadata (approximations of λ-terms).

Figure 1 shows the components of Matita organized according to the term representation they act on. For each component we show the functional dependencies on other components and the number of lines of source code. Dark gray components are either logic independent or can be made such by abstraction. Dashed arrows denote abstractions over logic dependent components. A normal arrow from a logic dependent component to a dark gray one is meant to be a dependency over the component, once it has been instantiated to the logic of the system.

We describe now each term representation together with the components of Matita acting on them.

Completely Specified Terms. Formalizing mathematics is a complex and onerous task and it is extremely important to develop large libraries of "trusted" information to rely on. At this level, the information must be completely specified in a given logical framework in order to allow formal checking. In Matita proof objects are terms of the Calculus of Inductive Constructions (CIC); terms represent both formulae and proofs. The proof-checker, implemented in the *kernel* component, is a CIC type-checker. Proof objects are saved in an XML format that is shared with the Coq Proof Assistant so that independent verification is possible.

Mathematical concepts (definitions and proof objects) are stored in a distributed library managed by the *file manager*, which acts as an abstraction layer over the concept physical locations.

Concepts stored in the library are indexed for retrieval using metadata. We conceived a logic independent metadata-set that can accommodate most logical frameworks. The logic dependent *indexing* component extracts metadata from mathematical concepts. The logic independent searching tools are described in the next section.

Finally, the *library manager* component is responsible for maintaining the coherence between related concepts (among them automatically generated *lemmas*) and between the different representations of them in the library (as completely specified terms and as metadata that approximate them).

The actual generation of lemmas is a logic dependent activity that is not directly implemented by the library manager, that is kept logical independent: the component provides hooks to register and invoke logic dependent lemma generators, whose implementation is provided in a component that we describe later and that acts on partially specified terms.

Metadata. An extensive library requires an effective and flexible search engine to retrieve concepts. Examples of flexibility are provided by queries up to instantiation or generalization of given formulae, combination of them with extra-logical constraints such as mathematical classification, and retrieval up to minor differences in the matched formula such as permutation of the hypotheses or logical equivalences. Effectiveness is required to exploit the search engine as a first step in automatic tactics. For instance, a paramodulation based procedure must first of all retrieve all the equalities in the distributed library that are likely to be exploited in the proof search. Moreover, since search is mostly logic independent, we would like to implement it on a generic representation of formulae that supports all the previous operations.

In Matita we use relational *metadata* to represent both extra-logical data and a syntactic approximation of a formula (e.g. the constant occurring in head position in the conclusion, the set of constants occurring in the rest of the conclusion and the same information for the hypotheses). The logic dependent *indexing* component, already discussed, generates the syntactic approximation from completely specified terms. The *metadata manager* component stores the metadata in a relational database for scalability and handles, for the library manager, the insertion, removal and indexing of the metadata. The *search engine* component [1] implements the approximated queries on the metadata that can be refined later on, if required, by logic dependent components.

Partially Specified Terms. In partially specified terms, subterms can be omitted replacing them with untyped linear placeholders or with typed metavariables (in the style of [8,13]). The latter are Curry-Howard isomorphic to omitted subproofs (conjectures still to be proved).

Completely specified terms are often highly redundant to keep the typechecker simple. This redundant information may be omitted during user-machine communication since it is likely to be automatically inferred by the system replacing conversion with unification [19] in the typing rules (that are relaxed to type inference rules). The *refiner* component of Matita implements unification and the type inference procedure, also inserting implicit coercions [3] to fix local type-checking errors. Coercions are particularly useful in logical systems that lack subtyping [10]. The already discussed library manager is also responsible for the management of coercions, that are constants flagged in a special way.

Subproofs are never redundant and if omitted require tactics to instantiate them with partial proofs that have simpler omitted subterms. Tactics are applied to omitted subterms until the proof object becomes completely specified and can be passed to the library manager. Higher order tactics, usually called tacticals and useful to create more complex tactics, are also implemented in the tactics component. The current implementation in Matita is based on *tinycals* [17], which supports a step-by-step execution of tacticals (usually seen as "black boxes") particularly useful for proof editing, debugging, and maintainability. Tinycals are implemented in Matita in a small but not trivial component that is completely abstracted on the representation of partial proofs.

The *lemma generator* component is responsible for the automatic generation of derived concepts (or lemmas), triggered by the insertion of new concepts in the library. The lemmas are generated automatically computing their statements and then proving them by means of tactics or by direct construction of the proof objects.

Content Level Terms. The language used to communicate proofs and especially formulae with the user must also exploit the comfortable and suggestive degree of notational abuse and overloading so typical of the mathematical language. Formalized mathematics cannot hide these ambiguities requiring terms where each symbol has a very precise and definite meaning.

Content level terms provide the (abstract) syntactic structure of the human-oriented (compact, overloaded) encoding. In the *content* component we provide translations from partially specified terms to content level terms and the other way around. The former translation, that loses information, must discriminate between terms used to represent proofs and terms used to represent formulae. Using techniques inspired by [6,7], the formers are translated to a content level representation of proof steps that can in turn easily be rendered in natural language. The representation adopted has greatly influenced the OMDoc [14] proof format that is now isomorphic to it. Terms that represent formulae are translated to MathML Content formulae [12].

The reverse translation for formulae consists in the removal of ambiguity by fixing an interpretation for each ambiguous notation and overloaded symbol used at the content level. The translation is obviously not unique and, if performed locally on each source of ambiguity, leads to a large set of partially specified terms, most of which ill-typed. To solve the problem the *ambiguity manager* component implements an algorithm [18] that drives the translation by alternating translation and refinement steps to prune out ill-typed terms as soon as possible, keeping only the refinable ones. The component is logic independent being completely abstracted over the logical system, the refinement function, and the local translation from content to partially specified terms. The local translation is implemented for occurrences of constants by means of call to the search engine.

The translation from proofs at the content level to partially specified terms is being implemented by means of special tactics following previous work [9,20] on the implementation of declarative proof styles for procedural proof assistants.

Presentation Level Terms. Presentation level captures the formatting structure (layout, styles, etc.) of proof expressions and other mathematical entities.

An important difference between the content level language and the presentation level language is that only the former is extensible. Indeed, the presentation level language has a finite vocabulary comprising standard layout schemata (fractions, sub/superscripts, matrices, ...) and the usual mathematical symbols.

The finiteness of the presentation vocabulary allows its standardization. In particular, for pretty printing of formulae we have adopted MathML Presentation [12], while editing is done using a TEX-like syntax. To visually represent proofs it is enough to embed formulae in plain text enriched with formatting boxes. Since the language of boxes is very simple, many similar specifications exist and we have adopted our own, called BoxML (but we are eager to cooperate for its standardization with other interested teams).

The *notation manager* component provides the translations from content level terms to presentation level terms and the other way around. It also provides a language [15] to associate notation to content level terms, allowing the user to extend the notation used in Matita. The notation manager is logic independent since the content level already is.

The remaining components, mostly logic independent, implement in a modular way the user interface of Matita, that is heavily based on the modern GTK+ toolkit and on standard widgets such as GTKSOURCEVIEW that implements a programming oriented text editor and GTKMATHVIEW that implements rendering of MathML Presentation formulae enabling contextual and controlled interaction with the formula.

The *graph browser* is a GTK+ widget, based on Graphviz, to render dependency graphs with the possibility of contextual interaction with them. It is mainly used in Matita to explore the dependencies between concepts, but other kind of graphs (e.g. the DAG formed by the declared coercions) are also shown.

The *library browser* is a GTK+ window that mimics a web browser, providing a centralized interface for all the searching and rendering functionalities of Matita. It is used to hierarchically browse the library, to render proofs and definitions in natural language, to query the search engine, and to inspect dependency graphs embedding the graph browser.

The *GUI* is the graphical user interface of Matita, inspired by the pioneering work on CtCoq [4] and by Proof General [2]. It differs from Proof General because the sequents are rendered in high quality MathML notation, and because it allows to open multiple library browser windows to interact with the library during proof development.

The hypertextual browsing of the library and proof-by-pointing [5] are both supported by semantic selection. Semantic selection is a technique that consists in enriching the presentation level terms with pointers to the content level terms and to the partially specified terms they correspond to. Highlight of formulae in the widget is constrained to selection of meaningful expressions, i.e. expressions that correspond to a lower level term, that is a content term or a partially or fully specified term. Once the rendering of an upper level term is selected it is possible for the application to retrieve the pointer to the lower level term. An example of applications of semantic selection is *semantic copy & paste*: the user can select an expression and paste it elsewhere preserving its semantics (i.e. the partially specified term), possibly performing some semantic transformation over it (e.g. renaming variables that would be captured or λ-lifting free variables).

Commands to the system can be given either visually (by means of buttons and menus) or textually (the preferred way to input tactics since formulae occurs as tactic arguments). The textual parser for the commands is implemented in the *vernacular* component, that is obviously system (and partially logic) dependent.

To conclude the description of the components of Matita, the *driver* component, which does not act directly on terms, is responsible for pulling together the other components, for instance to parse a command (using the vernacular component) and then triggering its execution (for instance calling the *tactics* component if the command is a tactic).

2.1 Relationship with Other Architectures

An interesting question is which components of Matita have counterparts in systems based on different architectural choices. As an example we consider how we would implement a system based on the following commitments: (1) The architecture is LCF-like. Proof objects are not recorded. (2) The system library is made of scripts. Proof concepts are indexed only after evaluation. (3) The proof language is declarative. Ambiguities in formulae are handled by the type system (e.g. type classes accounts for operator overloading).

Formulae are still represented as presentation, content, partially specified and completely specified terms. Proofs, that are distinct from formulae, exists at the presentation and content level, but do not have a counterpart as partially or completely specified terms. Since only concepts in memory can be queried, metadata are not required: formulae can be indexed using context trees or similar efficient data structures that acts on completely specified formulae.

The following components in Figure 1 have similar counterparts. The *file manager* to store environments obtained processing scripts to avoid re-execution. The *kernel*, that checks definition and theorems, is still present but now it implements the basic tactics, i.e. the tactics that implement reduction and conversion or that correspond to the introduction and elimination rules of the logics. The *indexing* component is not required since in charge of extracting metadata that are neglected. However, the *metadata manager* that used to index metadata is now provided by the context tree manager that indexes the formulae. Logic independence is lost unless the formulae are represented as sort of S-expressions, reintroducing the equivalent of the metadata data type. The *search engine* and the *library manager* are present since the corresponding functionalities (searching and management of derived notions) are still required. All the components that act on partially specified terms are present, even if basic tactics have been moved to the kernel. The *content* component is simplified since the translation (pretty printing) from completely specified terms to content level terms is pointless due to the lack of proof objects. The *ambiguity manager* that acts on content level formulae is removed or it is greatly simplified since type classes take care of overloading. Finally, all the components that act on presentation level terms are present and are likely to be reusable without major changes.

Of course the fact that many components have counterparts among the two set of architectural choices is due to the coarseness of both the description and the provided functionalities. This is wanted. In our opinion the issue of choosing the granularity level of architectures so that smaller components of different systems can be independently compared is non trivial, and was an issue we wanted to address.

3   Functional Architectural Analysis and Reusability

A different classification—other than the data-driven one given in the previous section—of the components shown in Figure 1 is along the lines of the offered functionalities. We grouped the components according to five (macro) functionalities, which are depicted in the vertical partition of Figures 2 and 3: visual interaction and browsing of a mathematical library (*GUI* column), input/output (i.e. parsing and pretty-printing) of formulae and proofs (*I/O* column), indexing and searching of concepts in a library (*search* column), management of a library of certified concepts (*library* column), and interactive development of proofs by means of tactics and decision procedures (*proof authoring* column).

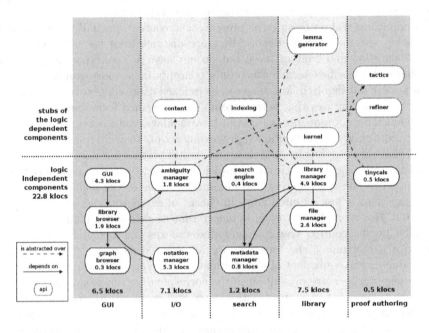

Fig. 2. Logic independent components by functionalities

In the development history of Matita this classification has been useful for the assignment of development tasks, since the knowledge required for implementing different functionalities varies substantially.

For each functionality it is interesting to assess the degree of coupling of each component with the logical framework. Having a clear separation between the logic dependent components and the logic independent ones should be one of the main guidelines for the development of interactive theorem provers, since it helps to clarify the interface of each component. Moreover, the logic independent functionalities are probably of interest to a broader community.

In Figure 2 we have isolated the logic independent components of Matita (lower part), showing the dependencies among them (solid lines). Some of them depend on "stubs" for logic dependent components, depicted in the upper part of the figure.

The effort for re-targeting Matita to a different logic amounts to provide a new implementation for the stubs. Figure 3 shows the current Matita implementation of CIC. In our case, the logic-dependent components are about 2/3 of the whole code (also due to the peculiar complexity of CIC). However, the real point is that the skills required for implementing the logic-dependent stubs are different from those needed for implementing the logic-independent components, hence potentially permitting to obtain in a reasonable time and with a limited manpower effort a first prototype for early testing. In the next section we investigate this point presenting a detailed timeline for the development of the system.

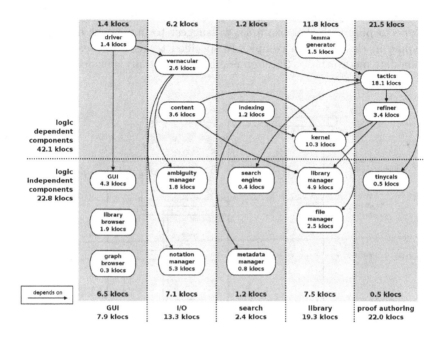

Fig. 3. Stubs implementation for CIC

A different problem is to understand if there are components that can be reused in systems based on different architectural commitments. A well known example of such a tool is the Proof General generic user interface. Having a reusable user interface is a relatively simple task since not only the user interface is logic independent, but it is also the most external component of a system. We believe that some other logic independent components of Figure 1 could be adapted to other architectures; for instance, components that deal with indexing and searching are likely to be embeddable in any system with minor efforts. This issue of reusability is one of the subject of our current research.

4   System Development

Figure 4 is an hypothetical Gantt-like diagram for the development of an interactive theorem prover with the same architectural commitments of Matita and a logic with comparable complexity. The order in which to develop the components in the figure does not reflect the development history of Matita, where we delayed a few activities, with major negative impacts on the whole schedule.

The duration of the activities in the diagram is an estimation of the time that would be required now by an independent team to re-implement Matita assuming only the knowledge derivable from the literature.

In any case, in the estimated duration of the activities we are considering the time wasted for rapid prototyping: it is not reasonable in a research community to expect the product to be developed for years without any intermediate prototype to play with. For example, we suggest to implement first reduction and typing in the kernel on completely specified terms before extending it to accommodate metavariables (later on required for partially specified terms). This

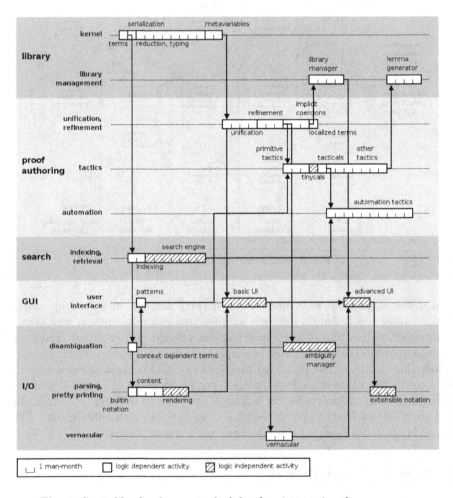

Fig. 4. Gantt-like development schedule of an interactive theorem prover

way the kernel of the type-checker can immediately be tested on a library of concepts exported from another system, and different reduction and type-checking algorithms can be compared leading to possibly interesting research results.

Activities related to logic independent components are marked as dashed in the Gantt-like diagram. If those components are reused in the implementation of the system, most functionalities but interactive proof authoring are made available very early in the development. The bad news are that the overall time required to develop the system will not change, being determined by the complexity of the logic dependent components and their dependencies that limit parallelism. Switching to a simpler logic can probably reduce in a significant way the time required to implement the kernel and the refinement component; however, it is likely to have a minor impact on the time required for tactics and decision procedures. Instead changing the initial architectural commitments (e.g. dropping proof objects and adopting an LCF-like kernel) is likely to change the Gantt in a sensible way. The overall conclusion is that the development of an interactive theorem prover is still a complex job that is unlikely to be substantially simplified in the near future.

The activities of Figure 4 refine the components already presented to improve parallel development and allow rapid prototyping. We describe now the main refinements following the timeline when possible.

We suggest to start developing the kernel omitting support for terms containing metavariables and to add it after the reduction and typing rules for completely specified terms have been debugged. The support for metavariables in the kernel should be kept minimal, only implementing typing rules and unfolding of instantiated metavariables. The core functionalities on partially specified terms, unification and refinement, are implemented in the refiner component outside the kernel. Completely omitting support for metavariables from the kernel is more compliant to the De Bruijn criterion. However, the kernel code for minimal metavariable support is really simple and small, and its omission forces an almost complete re-implementation of the kernel functionalities in the refiner that is better avoided.

Context dependent terms are a necessity for passing to tactics arguments that need to be interpreted (and disambiguated) in a context that is still unknown. In Matita context dependent terms are defined as functions from contexts to terms, but other systems adopt different representations.

Patterns are data structures to represent sequents with selected subterms. They are used as tactic arguments to localize the effect of tactics. Patterns pose a major problem to the design of textual user interfaces, that usually avoid them, but are extremely natural in graphical user interface where they correspond to visual selection (using the mouse) of subterms of the sequent.

A fixed *built-in notation* should be implemented immediately for debugging, followed by the *content* component to map completely (or even partially) specified terms to content and the other way around. Partially specified terms generated by the reverse mapping cannot be processed any further until the refiner component is implemented. Similarly, the reverse mapping of ambiguous terms is delayed until the ambiguity manager is available.

The *rendering* and *extensible notation* activities implement the notation manager component. Initially the machinery to apply extensible notation during rendering is implemented in the rendering activity. A user-friendly language to extend at run time the notation is the subject of the second activity that is better delayed until the interaction mode with the system become clear.

Handling of *implicit coercions* and *localized terms* in the refiner component can be delayed until unification and a light version of refinement are implemented. This way the implementation of tactics can start in advance. Localized terms are data structures to represent partially specified terms obtained by formulae given in input by the user. A refinement error on a localized term should be reported to the user by highlighting (possibly visually) the ill-typed subformula. Localized terms pose a serious problem since several operations such as reduction or insertion of an implicit coercion change or loose the localization information. Thus the refiner must be changed carefully to cope with the two different representations of terms.

The *basic user interface* is an interface to the library that offers browsing, searching and proof-checking, but no tactic based proof authoring. It can, however, already implement proof authoring by direct term manipulation that, once refinement is implemented, can become as advanced as Alf is [11]. The *advanced user interface* offers all the final features of the system, it can be script based and it can present the desired interaction style (procedural versus declarative).

Finally, primitive tactics, that implement the inference rules of the logic, and tacticals are requirements for the development of more advanced interactive tactics and automation tactics, that can proceed in parallel.

5   Conclusions and Future Work

We feel the need for more direct comparisons between different interactive theorem provers to highlight the common functionalities, not only in view of reusability but also to encourage a more systematic comparison of different softwares and architectural solutions. In this paper we have contributed by showing how the architecture of a system (in our case Matita) can be analyzed by classifying its software components along different axes: the representation of formulae and proofs they act on and the macro functionalities they contribute to.

Moreover, we believe that an effort should be made to clearly split the logic dependent components from those that are or can be made logic independent. In addition to be able to clarify the interfaces of the components and their dependencies, this division has immediate applications: having done this for Matita we are now able to clearly estimate the efforts, both in time and in lines of code, required to re-target the system to a different logic. This answers a frequent question posed to us by members of the Types community, since Matita presents technological innovations which are interesting for developers of systems based on logical frameworks other than CIC.

In particular, we have estimated that only one third of the code of Matita (that is still more than 22,000 lines of code) is logic independent, which can be explained by the complexity of the logical system. We have also estimated that re-targeting Matita to a logic with the same complexity as the current one would not require significantly less time than the first implementation (assuming to have enough man power to develop concurrently all parallel tasks). However, working prototypes including even advanced functionalities would be obtained quite early in the development stage, with a positive impact at least on debugging, dissemination, and system evaluation. To our knowledge, similar figures have never been presented for other systems.

A more complex issue is the independence of software components from the main architectural commitments, and consequently their degree of cross-system reusability. We believe that several of the components of Matita have counterparts in systems based on different architectures, and that at least some of them could be embedded in other systems after some modifications. This has already been proven true by Proof General for the graphical user interface. However, that is the easiest case, the graphical user interface being the most external component with no dependencies on it.

Better understanding this issue is one of our current research guidelines, but it requires an initial effort by the whole community to analyze the architectures of several systems according to common criteria in order to identify the corresponding components and to understand how they differ in the various architectures. Our next contribution will consist in a description of the API of the components presented here.

References

1. Asperti, A., Guidi, F., Coen, C.S., Tassi, E., Zacchiroli, S.: A content based mathematical search engine: Whelp. In: Filliâtre, J.-C., Paulin-Mohring, C., Werner, B. (eds.) TYPES 2004. LNCS, vol. 3839, pp. 17–32. Springer, Heidelberg (2004)
2. Aspinall, D.: Proof General: A generic tool for proof development. In: Schwartzbach, M.I., Graf, S. (eds.) ETAPS 2000 and TACAS 2000. LNCS, vol. 1785, Springer, Heidelberg (2000)
3. Barthe, G.: Implicit coercions in type systems. In: Types for Proofs and Programs: International Workshop, TYPES 1995, pp. 1–15 (1995)
4. Bertot, Y.: The CtCoq system: Design and architecture. Formal Aspects of Computing 11, 225–243 (1999)
5. Bertot, Y., Kahn, G., Théry, L.: Proof by pointing. In: Hagiya, M., Mitchell, J.C. (eds.) TACS 1994. LNCS, vol. 789, Springer, Heidelberg (1994)
6. Coscoy, Y.: Explication textuelle de preuves pour le Calcul des Constructions Inductives. PhD thesis, Université de Nice-Sophia Antipolis (2000)
7. Coscoy, Y., Kahn, G., Thery, L.: Extracting Text from Proofs. Technical Report RR-2459, Inria (Institut National de Recherche en Informatique et en Automatique), France (1995)
8. Geuvers, H., Jojgov, G.I.: Open proofs and open terms: A basis for interactive logic. In: Bradfield, J.C. (ed.) CSL 2002 and EACSL 2002. LNCS, vol. 2471, pp. 537–552. Springer, Heidelberg (2002)

9. Harrison, J.: A Mizar Mode for HOL. In: von Wright, J., Harrison, J., Grundy, J. (eds.) TPHOLs 1996. LNCS, vol. 1125, pp. 203–220. Springer, Heidelberg (1996)
10. Luo, Z.: Coercive subtyping. Journal of Logic and Computation 9(1), 105–130 (1999)
11. Magnusson, L., Nordström, B.: The ALF proof editor and its proof engine. In: Barendregt, H., Nipkow, T. (eds.) TYPES 1993. LNCS, vol. 806, pp. 213–237. Springer, Heidelberg (1994)
12. Mathematical Markup Language (MathML) Version 2.0. W3C Recommendation (February 21 2001) (2003), http://www.w3.org/TR/MathML2
13. Muñoz, C.: A Calculus of Substitutions for Incomplete-Proof Representation in Type Theory. PhD thesis, INRIA (November 1997)
14. OMDoc: An open markup format for mathematical documents (draft, version 1.2) (2005), http://www.mathweb.org/omdoc/pubs/omdoc1.2.pdf
15. Padovani, L., Zacchiroli, S.: From notation to semantics: There and back again. In: Asperti, A., Bancerek, G., Trybulec, A. (eds.) MKM 2004. LNCS (LNAI), vol. 3119, pp. 194–207. Springer, Heidelberg (2006)
16. Riazanov, A.: Implementing an Efficient Theorem Prover. PhD thesis, The University of Manchester (2003)
17. Coen, C.S., Tassi, E., Zacchiroli, S.: Tinycals: step by step tacticals. In: Proceedings of UITP 2006: User Interface for Theorem Provers. Seattle, WA, August 21, 2006. ENTCS, vol. 174(2), pp. 125–142. Elsevier Science, North-Holland (May 2007) ISSN: 1571-0661
18. Coen, C.S., Zacchiroli, S.: Efficient ambiguous parsing of mathematical formulae. In: Asperti, A., Bancerek, G., Trybulec, A. (eds.) MKM 2004. LNCS, vol. 3119, pp. 347–362. Springer, Heidelberg (2004)
19. Strecker, M.: Construction and Deduction in Type Theories. PhD thesis, Universität Ulm (1998)
20. Wiedijk, F.: Mmode, a mizar mode for the proof assistant coq. Technical Report NIII-R0333, University of Nijmegen (2003)

On Constructive Cut Admissibility in Deduction Modulo

Richard Bonichon and Olivier Hermant

Université Paris 6 - LIP6[*]
{richard.bonichon, olivier.hermant}@lip6.fr

Abstract. Deduction Modulo is a theoretical framework that allows the introduction of computational steps in deductive systems. This approach is well suited to automated theorem proving. We describe a proof-search method based upon tableaux for Gentzen's intuitionistic LJ extended with rewrite rules on propositions and terms . We prove its completeness with respect to Kripke structures. We then give a soundness proof with respect to cut-free LJ modulo. This yields a constructive proof of semantic cut elimination, which we use to characterize the relation between tableaux methods and cut elimination in the intuitionistic case.

1 Introduction

The road to automated deduction has many pitfalls. Efficient treatment of equality and equational theories is for instance a challenging area and proving $(a + b) + ((c + d) + e) = a + ((b + c) + (d + e))$ with the usual associativity and identity axioms can loop infinitely with an ineffective strategy. One would like a deterministic and terminating method where one needs only check whether the two terms are the same modulo the given axioms. We would rather use computation (i.e. blind execution) instead of deduction (non-deterministic search), thus transforming the associativity axiom into a *term-rewriting rule*. Orienting equational theories using rewriting techniques is nothing unusual, but *propositional rewrite rules* are hardly considered in the literature. It is useful to allow them. One framework to handle such rewrite rules is *deduction modulo* [4]. The rewrite rule $x * y = 0 \rightarrow x = 0 \vee y = 0$ is the oriented version of the axiom $\forall x\, \forall y\, (x * y = 0 \Leftrightarrow (x = 0 \vee y = 0))$ and can be used to prove $\exists z(a * a = z \Rightarrow a = z)$ by automated deduction methods modulo [1, 4]. This rule can not be easily turned into a term-rewriting rule.

Using rewrite rules instead of unoriented axioms is a natural speed up for automated theorem provers, as it reduces the search space. However, deduction modulo has other interesting consequences: propositional rewrite rules can be used to restart a stuck deductive process such as $P(a) \rightarrow \forall x P(x)$. Proofs modulo only contain the important deductive steps — or those a human sees as important — and no computational details, giving shorter, more readable proofs, which is important in interactive theorem proving.

---

[*] Laboratoire d'Informatique de Paris 6, 8 rue du Capitaine Scott, 75015 Paris, France.

T. Altenkirch and C. McBride (Eds.): TYPES 2006, LNCS 4502, pp. 33–47, 2007.

Classical first-order logic modulo has first been studied in [4], Intuitionistic logic is itself particularly interesting as it has the witness and the disjunction properties, and is therefore adapted to constructive mathematics and computer science, through the Curry-Howard isomorphism. Intuitionistic automated deduction procedures seem less studied, maybe because intuitionistic semantics is harder to deal with. In particular, two main different semantics exist: Kripke structures and Heyting algebras.

The focus of this paper is on obtaining a cut elimination theorem. Starting with Gentzen's result, this has turned out to be one of the most important properties in the field of logic. Cut elimination in deduction modulo is a harder problem because it is not valid in general for confluent and terminating rewrite systems. The result is generally obtained using one of the following techniques: a syntactic one, proving termination of a certain cut elimination process – its modern variant uses proof terms [5] and the reducibility method; the other proves the admissibility (or redundancy) of the cut rule by establishing the completeness of the cut-free calculus with respect to some notion of model. Known since Beth and Hintikka, the latter has been recently used by De Marco and Lipton [3]. Note that the first constructive approach for this has been developed by Dragalin [7], but it uses Heyting algebras.

This article shows the deep link between all these topics. First, we recall the intuitionistic sequent calculus modulo of [10] (Sec. 2) and its semantics.

- We give a proof-search method for LJ_{mod} (Sec. 3). Such methods are often based on tableaux when dealing with nonclassical logics. We here formulate a tableau method modulo rewrite rules on terms and propositions and conditions on their use.
- We prove that this method enjoys the usual completeness property (Sec. 4). We give a semantic proof using Kripke structures, and describe several conditions on rewrite rules for which the completeness theorem holds. Adding (propositional) rewrite rules makes this theorem harder to prove.
- Finally, we argue that our tableau method is sound with respect to cut-free intuitionistic sequents modulo (Sec. 5). Soundness is usually proved semantically, as building a tableau can be viewed as a search for a countermodel. We give a translation of tableaux proofs to single-conclusion sequent ones. Our approach is more technical but yields benefits: it entails a constructive cut elimination theorem and sheds light on the relation between intuitionistic tableaux and cut-free sequent calculus. Our final result discusses the computational content of such semantic cut-elimination methods and compares it to proof normalization.

2   Intuitionistic Sequent Calculus Modulo

Figure 1 shows a representative sample of the rules and the transformation made on the usual sequent inferences, reformulated to work on equivalent formulas modulo a rewrite system ($\mathcal{R}$) on terms and *atomic* propositions. The rewrite relation is denoted $\rightarrow$ and its reflexive symmetric transitive closure $\equiv_{\mathcal{R}}$. The full calculus can be found in [10]. In this paper, $\mathcal{R}$ is supposed to be *terminating* and *confluent*. When $\mathcal{R}$ is empty, we get back the usual LJ. A cut-free derivation of a sequent is noted $\Gamma \vdash^*_{\mathcal{R}} P$. The semantics of these inference rules is a Kripke semantics where negation is expressed in a positive way.

$$\frac{}{P \vdash_{\mathcal{R}} Q} \text{ axiom if } P \equiv_{\mathcal{R}} Q \qquad \frac{\Gamma \vdash_{\mathcal{R}} P \quad \Gamma, Q \vdash_{\mathcal{R}} S}{\Gamma, R \vdash_{\mathcal{R}} S} \Rightarrow\text{-l if } R \equiv_{\mathcal{R}} (P \Rightarrow Q)$$

$$\frac{\Gamma, P \vdash_{\mathcal{R}} Q}{\Gamma \vdash_{\mathcal{R}} R} \Rightarrow\text{-r if } R \equiv_{\mathcal{R}} (P \Rightarrow Q) \qquad \frac{\Gamma, P \vdash_{\mathcal{R}} S \quad \Gamma \vdash_{\mathcal{R}} Q}{\Gamma \vdash_{\mathcal{R}} S} \text{ cut if } P \equiv_{\mathcal{R}} Q$$

Fig. 1. Some rules of LJ modulo

Definition 1 (modified Kripke structures). *A modified Kripke structure (in short Kripke structure) $\mathcal{K}$ is a quadruple $\langle K, \leq, D, \Vdash \rangle$ such that:*

- *K is a non-empty set (the worlds), partially ordered by $\leq$.*
- *D (the domain) is a monotonous function over K: if $\alpha \leq \beta$ then $D_\alpha \subseteq D_\beta$.*
- *Predicate symbols are interpreted syntactically, which is enough for the scope of this paper.*
- *To each function symbol f we associate a function: $\hat{f} : K \longrightarrow D_\alpha^n \longrightarrow D_\alpha$, such that when $\alpha \leq \beta$, $\hat{f}(\alpha)(a_1, ..., a_n) = \hat{f}(\beta)(a_1, ..., a_n)$ on D_α. (f in β extends f in α).*
- *The interpretation of a term t under a substitution σ, $|t|^\alpha_\sigma$ is defined by induction on t as usual.*
- *$\Vdash$ is a relation between worlds and interpreted predicates: $\alpha \Vdash P(a_1, ..., a_n)$ for $a_i \in D_\alpha$. We extend it as a relation between worlds and propositions, under a substitution σ mapping variables to elements of D_α. It must satisfy:*
  1. *atomic relation: let $P(x_1, ..., x_n)$ be a predicate. We let $\alpha \Vdash_\sigma P(t_1, ..., t_n)$ iff $\alpha \Vdash P(|t_1|^\alpha_\sigma, ..., |t_n|^\alpha_\sigma)$.*
  2. *Monotonicity on atoms: for any predicate $P(x_1, ..., x_n)$, any worlds $\beta \geq \alpha$ and terms $t_1, ..., t_n$: $\alpha \Vdash_\sigma P(t_1, ..., t_n)$ implies $\beta \Vdash_\sigma P(t_1, ..., t_n)$.*
  3. *$\alpha \Vdash_\sigma A \vee B$ iff $\alpha \Vdash_\sigma A$ or $\alpha \Vdash_\sigma B$.*
  4. *$\alpha \Vdash_\sigma A \wedge B$ iff $\alpha \Vdash_\sigma A$ and $\alpha \Vdash_\sigma B$.*
  5. *$\alpha \Vdash_\sigma A \Rightarrow B$ iff for any $\beta \geq \alpha$, $\beta \Vdash_\sigma A$ implies $\beta \Vdash_\sigma B$.*
  6. *$\alpha \Vdash_\sigma \neg A$ iff for any $\beta \geq \alpha$, $\beta \Vdash_\sigma A$ implies $\beta \Vdash_\sigma \perp$ (denoted $\beta \nVdash_\sigma A$).*
  7. *$\alpha \Vdash_\sigma \exists x A$ iff there exists an element $a \in D(\alpha)$ such that $\alpha \Vdash_{\sigma + \langle a/x \rangle} A$.*
  8. *$\alpha \Vdash_\sigma \forall x A$ iff for any $\beta \geq \alpha$, for any element $a \in D(\beta)$, $\beta \Vdash_{\sigma + \langle a/x \rangle} A$.*
  9. *The explosion property: if $\beta \Vdash_\sigma \perp$ then for any $\alpha \in K$, any proposition P, any substitution θ, $\alpha \Vdash P$. Such a Kripke structure is called improper.*
  10. *Validity of $\mathcal{R}$: for any world $p \in K$, formulas $P \equiv_{\mathcal{R}} Q$, $p \Vdash P$ iff $p \Vdash Q$.*

The positive treatment of the negation is essential for constructivity: we do not have to consider proper Kripke structures, so we avoid the use of König's lemma [16] to identify an infinite branch. Our definition is inspired by Veldman's [17] and Friedman's [16], but may be closer to Krivine's classical one [11], as we add *only the improper* Kripke structure to the usual definition.

Condition 10 is equivalent [9] to considering all instances of rewrite rules $P \rightarrow A$ where P is atomic. For any world $p \in K$, we must have $p \Vdash A$ iff $p \Vdash P$.

3 Intuitionistic Predicate Tableaux Modulo

Proof-search methods for deduction modulo have been developed using both resolution [4] and tableaux [1] for first-order classical logic. We present here a simple ground tableau method for intuitionistic deduction modulo. A tableau is an attempt to define a model of the formulas at its root. Working with Kripke structures forces us to build worlds, therefore a truth statement in a tableau will be a truth statement in a particular world.

Any branch represents a possible model. A closed branch represents a contradiction in the attempt to define a particular model (the one laying on this branch), thus it is improper. If all the branches can be closed, the root formulas should be provable (Sec. 5). The systematic tableau generation of Sec. 4 is a systematic model search.

$A\downarrow$ stands for the normalization of the formula A with respect to the rewrite rules of $\mathcal{R}$. We keep the basic expansion rules of [12] in Fig. 2, working on statements representing signed forced formulas at some world p (identical Kripke world, hence partially ordered by $\geq$). A statement $Tp \Vdash P$ (resp. $Fp \Vdash P$) should be read as the commitment to set P forced (resp. unforced) at world p. P is unforced at world p means that if P is forced, then $\bot$ should also be forced (the Kripke structure we try to define is improper). Notice that we use the same forcing symbol $\Vdash$ when we deal with Kripke structures and that nothing is said about the partial order $\geq$ on worlds. When the sign of a forcing statement is irrelevant, B is a shortcut for "T or F".

We extend intuitionistic tableaux to use term and propositional rewriting. Expansion rules should be read as follows: when the premise formula is anywhere in the considered path (i.e. not necessarily the leaf), the tableau tree can be expanded with the consequence formula(s). *Branch closure* is the usual binary closure: we close a branch if both $Tp \Vdash P$ and $Fq \Vdash P$ occurs on it for $p \leq q$. A tableau is *closed* and yields a refutation of the input formula when every branch is closed. We choose to pre-normalize every formula with respect to the rewrite system before entering the tableau construction. Therefore we need to be careful when using any rule handling certain quantifiers. Each time a formula is produce by the positive $\forall$ or the negative $\exists$ quantifier rule, we re-normalize it. In the other two quantifier rules, we need not to normalize as the constant is fresh also with respect to rewrite rules.

As we focus on the generation of a model, we define a systematic complete search within our tableau modulo framework using the rules of Fig. 2.

Definition 2 (Systematic tableau generation). *We introduce here the notion of complete systematic tableau. We construct a tree representing an intuitionistic variation of a Hintikka set (see for example [14] for a definition),* *satisfying Lem. 1. The construction is similar to those in [3, 12] therefore we only sketch it here. Our set of worlds K is made of finite sequences of natural numbers, ordered by the prefix ordering. For any world, we define by induction a base set $\mathcal{D}(p)$:*

- *$\mathcal{D}(\emptyset)$ is the language constructed over $\mathcal{L}_0 \cup C_\emptyset$, $\mathcal{L}_0$ is the ground terms of the first-order language.*
- *If q is the concatenation of the sequence p and the natural number k (i.e. $p * k$), we define $\mathcal{D}(q)$ as the language constructed over $\mathcal{D}(p) \cup C_q$, where we have countably many disjoint sets C_p of fresh constants for each finite sequence p.*

In each of those sets, we define an enumeration of the terms as well as an enumeration of the pairs (p, t) where $p \in K$ and $t \in \mathcal{D}(p)$. The tableau is then constructed step by step by expanding each time the least unused node on it (any unused node on the leftmost smallest branch); this node thus becomes used. Let π the path from the root to this node. We detail some cases among the most significant:

- *If on the path π we have two contradictory statements, then close all the branches containing this path.*
- *$Tp \Vdash \forall x P(x)$: Let (q, t) be the least pair such that $q \geq p$, $t \in \mathcal{D}(q)$, q occurs on π, and the statement $Tq \Vdash P(t)\downarrow$ does not appear on π. Attach at each leaf of the branches having π as an initial segment the two statements $\{Tq \Vdash P(t)\downarrow, Tp \Vdash \forall x P(x)\}$. We keep an unused copy of $Tp \Vdash \forall x P(x)$ in order to go on (later) in the enumeration of the terms and worlds.*
- *$Fp \Vdash \forall x P(x)$. Let k be the least number such that $p * k$ does not occur on any branch having π as initial segment. It is incomparable with any world, except p. Let $c \in C_q$ be a fresh constant. Attach at each leaf of the branches extending π the statement $Fq \Vdash P(c)$.*

$Tp \Vdash A \vee B$	$Fp \Vdash A \wedge B$	$Tp \Vdash A \wedge B$	$Fp \Vdash A \vee B$
$Tp \Vdash A \mid Tp \Vdash B$	$Fp \Vdash A \mid Fp \Vdash B$	$Tp \Vdash A, Tp \Vdash B$	$Fp \Vdash A, Fp \Vdash B$
$Tp \Vdash A \Rightarrow B$	$Fp \Vdash A \Rightarrow B$	$Tp \Vdash \neg A$	$Fp \Vdash \neg A$
$Fp' \Vdash A \mid Tp' \Vdash B$	$Tp' \Vdash A, Fp' \Vdash B$	$Fp' \Vdash A$	$Tp' \Vdash A$
for any $p' \geq p$	for some new $p' \geq p$	for any $p' \geq p$	for some new $p' \geq p$
$Tp \Vdash \exists x P(x)$	$Fp \Vdash \exists x P(x)$	$Tp \Vdash \forall x P(x)$	$Fp \Vdash \forall x P(x)$
$Tp \Vdash P(c)$	$Fp \Vdash P(t)\downarrow$	$Tp' \Vdash P(t)\downarrow$	$Fp' \Vdash P(c)$
for some new c	for any t	for any $p' \geq p$ and any t	for some new $p' \geq p$ and some new c

Fig. 2. Rules for intuitionistic predicate tableaux modulo

From Def. 1, we adopt the convention that on a closed branch, every statement appears (since the Kripke structure will be improper). The tableau construction of Def. 2 satisfies the following lemma:

Lemma 1. *Let π be a branch of a fully expanded tableau, generated by Def. 2. Then, on π:*

- *if $Bp \Vdash P$ appears, then $Bp \Vdash P{\downarrow}$ appears.*
- *if $Tp \Vdash A \wedge B$ (resp. $Fp \Vdash A \vee B$) appears, then $Tp \Vdash A$ and $Tp \Vdash B$ (resp. $Fp \Vdash A$ and $Fp \Vdash B$) appears.*
- *if $Fp \Vdash A \wedge B$ (resp. $Tp \Vdash A \vee B$) appears, then either $Fp \Vdash A$ or $Fp \Vdash B$ (resp. $Tp \Vdash A$ or $Tp \Vdash B$) appears.*
- *if $Tp \Vdash \neg P$ (resp. $Tp \Vdash A \Rightarrow B$) and a world $p' \geq p$ appear then $Fp' \Vdash P$ (resp. $Fp' \Vdash A$ or $Tp' \Vdash B$) appears*
- *if $Fp \Vdash \neg P$ (resp. $Fp \Vdash A \Rightarrow B$) appears then for some world $p' \geq p$, $Tp' \Vdash P$ (resp. $Tp' \Vdash A$ and $Fp' \Vdash B$) appears.*
- *if $Fp \Vdash \exists x P(x)$ appears then for every $t \in \mathcal{D}(p)$, $Fp \Vdash P(t){\downarrow}$ appears.*
- *if $Tp \Vdash \exists x P(x)$ appears then for some fresh constant $c \in \mathcal{D}(p)$, $P(c)$ appears.*
- *if $Fp \Vdash \forall x P(x)$ appears then for some world $p' \geq p$ and some fresh constant $c \in D(p')$, $Tp' \Vdash P(c)$ appears.*
- *if $Tp \Vdash \forall x P(x)$ and a world $p' \geq p$ appear on the branch, then for every term $t \in D(p')$, $Tp' \Vdash P([x := t]){\downarrow}$ appears.*

Proof. We call $\tau = \bigcup \tau_n$ the tableau generated by Def. 2. All nodes of τ are used, and for the $\forall$ positive statements, any world $q \geq p$ and term of $\mathcal{D}(q)$ has been enumerated. The convention we adopted ensures this also on a closed branch. $\square$

Remark: Lem. 1 defines the intuitionistic version of a Hintikka set (equivalently, a Schütte *semi-valuation*). Noe that it may be inconsistent.   •

4   Completeness

The extended liberality of LJ_{mod} rules entails a harder completeness proof, as the constructed Kripke structure must also be a model of the rewrite rules.

Theorem 1 (Completeness of the tableau method). *Let $\mathcal{R}$ be a rewrite system verifying one of the conditions below, Γ be a set of propositions, P be a proposition. If for any node α of any Kripke structure $\mathcal{K}$, $\alpha \Vdash \Gamma$ implies $\alpha \Vdash P$, then any branch of the complete systematic tableau for $T\emptyset \Vdash \Gamma, F\emptyset \Vdash P$ is closed.*

From an (open) branch of the tableau, one usually defines a model according to the statements on atoms, extending it "naturally" to compound formulas. But we here must additionally ensure that the built Kripke structure validates the rewrite rules. We detail such constructions in the following sections for large classes of rewrite systems, and the completeness proofs associated.

Moreover, our proof differs from usual tableau completeness proofs in that we do not consider open branches. Our approach is the following: given any branch π of a completely developed tableau, we define a Kripke structure. We prove that it agrees with the statements on π and that it validates $\mathcal{R}$. By hypothesis $\emptyset \Vdash \Gamma$, $\emptyset \Vdash P$, whereas by construction $(\emptyset \Vdash P) \Rightarrow (\emptyset \Vdash \bot)$. The Kripke structure is then improper, and on π we shall meet $Tp \Vdash \bot$ for some p. So π is closed.

4.1   An Order Condition

It is shown in [10] how to build a model for an order condition. We again give the construction. Although the Kripke structure built is rather different, the proofs are almost the same, so we do not give them.

Definition 3 (Order condition on rewrite systems). *We consider rewrite systems compatible with a well-founded order $\prec$ on formulas such that if $P \rightarrow Q$ then $Q \prec P$ and if A is a subformula of B then $A \prec B$.*

Given a branch π, we define the Kripke structure $\mathcal{K} = \langle K, \leq, D, \Vdash \rangle$:

- $K = \{p \ : \ p \text{ is a sequence of integers}\}$. $\leq$ is the prefix order (as in τ).
- $\mathcal{D}(p)$ is the set of closed terms appearing in all the forcing statements involving some world $q \leq p$.
- The forcing relation $\Vdash$ is defined by induction on $\prec$. For *normal* atomic formulas we let $q \Vdash A$ iff $Tp \Vdash A$ appears on the branch for some $p \leq q$. We extend $\Vdash$ to non atomic formulas according to the definition of a Kripke structure. There are three non standard cases: we let $p \Vdash \neg P$ if for any $q \geq p$ we do not have $q \Vdash P$. If A is a non-normal atom, we set $p \Vdash A$ iff $p \Vdash A\downarrow$. At last, if $Tp \Vdash \bot$ is on the branch, then we add $q \Vdash P$ for every q and P.

This definition is well-founded as $\prec$ is well-founded. It obviously defines a Kripke structure. We now prove a result that would not be needed with the usual definition of Kripke structures:

Lemma 2. *If the Kripke structure $\mathcal{K}$ is improper, then the branch π is closed.*

Proof. $p \Vdash \bot$ can hold for two reasons. First if the statement $Tp \Vdash \bot$ appears: the branch is closed. Second if we have both $p \Vdash P$ and $(p \Vdash P) \Rightarrow (p \Vdash \bot)$, for some world p and formula P. This last statement can only be derived from $q \Vdash \neg P$ for some $q \leq p$, but this statement never appears unless we already know that $p \Vdash \bot$ - since $p \Vdash P$.

The Kripke structure $\mathcal{K}$ agrees with the branch π: if $Tp \Vdash A$ (resp. $Fp \Vdash A$) appears on π, then $p \Vdash A$ (resp. $p \nVdash A$). This is obvious for normal atoms, and extended by induction on $\prec$. The case analysis is slightly different, since we now interpret $p \nVdash A$ as "if $p \Vdash A$ then $p \Vdash \bot$". We detail some cases below.

- if $Fp \Vdash P$ appears, with P a normal atom. $p \nVdash P$ (*i.e.* $p \Vdash P \Rightarrow p \Vdash \bot$) holds, since $p \Vdash A$ is defined in $\mathcal{K}$ only when π is closed.
- if $Fp \Vdash \neg A$ appears, then from the tableau generation, the statement $Tp*k \Vdash A$ appears on π, for some new $p * k$ – recall that if $T q \Vdash \bot$ appears, then every statement appears. By induction hypothesis, $p * k \Vdash A$. If $p \Vdash \neg A$ then by monotonicity $p * k \Vdash \neg A$, $\mathcal{K}$ is improper, hence $p \nVdash \neg A$.

Moreover, the Kripke structure is a model of the rewrite rules (by induction on $\prec$). We then have: $\emptyset \Vdash \Gamma$, $\emptyset \nVdash P$, and $\mathcal{K}$ is a Kripke model of $\mathcal{R}$.

4.2 A Positivity Condition

We now focus on another (new) condition on propositional rewrite rules.

Definition 4 (Positivity condition on rewrite systems). *A rewrite system is positive if every rewrite rule $P \to Q$, the negation normal form of Q does not contain any negation (see also [9] for a longer equivalent inductive definition).*

If the positivity condition holds, we first need to saturate the branch, in order to decide the truth value of as many formulas as possible. We define a saturation process by following the definition of a Kripke structure. We enumerate the pairs (p, P) where $p \in K$ and P is a formula over the language $\mathcal{D}(p)$ and add the following statements to the branch:

- $Bp \Vdash P$ if $Bp \Vdash P{\downarrow}$ appears.
- $Tp \Vdash P$ (resp. $Fp \Vdash P$) if $Tq \Vdash P$ (resp. $Fq \Vdash P$) appears for $q \leq p$ (resp. $q \geq p$). Truth propagates upwards and falsity downwards, from Def. 1. .
- $Tp \Vdash P \wedge Q$ (resp. $Fp \Vdash P \wedge Q$) if $Tp \Vdash P$ and $Tp \Vdash Q$ (resp. either $Fp \Vdash P$ or $Fp \Vdash Q$) appear.
- $Tp \Vdash \neg P$ (resp. $Fp \Vdash \neg P$) if for any $q \geq p$, $Fq \Vdash P$ (resp. for some $q \geq p$, $Tq \Vdash P$) appears.
- $Tp \Vdash \forall x P(x)$ (resp. $Fp \Vdash \forall x P(x)$) if for any $q \geq p$, any term $t \in \mathcal{D}(q)$, $Tq \Vdash P(t){\downarrow}$ (resp. for some $q \geq p$ and some $t \in \mathcal{D}(q)$, $Fq \Vdash P(t){\downarrow}$) appears.

During this completion process, two opposite forcing statements $Tp \Vdash P$ and $Fq \Vdash P$ for some $q \geq p$ appear only if the branch was already closed, and Lem. 1 remains valid.

Since the number of formulas having an interpretation increases at each iteration of this process, this operation has a least fixpoint that we take as the branch in the rest of this section.

Lemma 3
- If $Tp \Vdash P$ appears, then for any $q \geq p$, $Tq \Vdash P$ appears.
- If $Fp \Vdash P$ appears, then for any $q \leq p$, $Fq \Vdash P$ appears.
- If $P \equiv_{\mathcal{R}} Q$ and $Bp \Vdash P$ appears, then $Bp \Vdash Q$ appears.
- The new branch is closed iff the original one is closed.
- The new branch verifies lemma 1

Proof. The completion process entails the two first claims and we have just proved the two last ones. The third one stands because $Bp \Vdash P$ appears only if $Bp \Vdash P{\downarrow}$ does (which is proved by induction on the size of P). □

This process is necessary to define a Kripke structure. We need to know as much as possible about every formula. It is really absolutely necessary for Lem. 5. This is the intuitionistic counterpart of Schütte's *partial valuation*, since it satisfies more than Lem. 1. Indeed, we satisfy equivalence between left and right hand sides. For instance: $Fp \Vdash \forall x P(x)$ appears *iff* for some world $p' \geq p$ and some fresh constant $c \in D(p')$, $Tp' \Vdash P(c){\downarrow}$ appears. The only difference with usual partial valuations is that we could be in a degenerate case. However, the valuation is not yet total (some formulas can be left uninterpreted), and we still have no model. So we build the Kripke structure $\mathcal{K} = \langle K, \leq, D, \Vdash \rangle$ as in Sec. 4.1 except that the forcing relation $\Vdash$ is defined by induction on the size of formulas. For *every* atomic predicate (over the language $\mathcal{D}(q)$) we let $q \Vdash A$ if $Tq \Vdash A$ appears on the branch. If $Fp \Vdash A$ does *not* appear we also let $p \Vdash A$. We extend this forcing relation to non atomic formulas as before. This model is trivially a Kripke structure. We now prove that $\mathcal{K}$ agrees with the branch:

Lemma 4. *If a statement $Tp \Vdash P$ (resp. $Fp \Vdash P$) appears on the branch, then $p \Vdash P$ (resp. $p \nVdash P$) in the Kripke structure $\mathcal{K}$.*

Proof. By induction on the structure of P. The base case (atomic) is trivial from the definition. Other cases are immediate as the branch satisfies Lem. 1.   □

As the Kripke structure agrees with the branch, $\emptyset \Vdash \Gamma$ and $\emptyset \nVdash P$. We now need to show that the Kripke structure is a model of $\mathcal{R}$. We know (Lem. 4) that if $A \to P$ and $P\!\downarrow = A\!\downarrow$ appear in the branch as $Bp \Vdash A\!\downarrow$, then all three formulas $(A, P, P\!\downarrow)$ share the same forcing relation with p. But what if $P\!\downarrow$ does not appear? Recall then that the rewrite system is positive. Hence P is positive. Let us prove the following lemma:

Lemma 5. *Let P^+ be a positive formula and Q^- be a negative formula (i.e. $\neg Q$ is positive) defined over $\mathcal{D}(p)$. If $Bp \Vdash P^+$ (resp. $Bp \Vdash Q^-$) does not appear (whether $B = T$ or $B = F$) in the branch, then $p \Vdash P^+$ (resp. $p \nVdash Q^-$).*

Proof. We suppose that no statement $Tp \Vdash \bot$ appears in the branch, otherwise $Bp \Vdash R$ appears for any p and R. Therefore, in the (proper) Kripke structure defined, $p \nVdash P$ means in particular that we do not have $p \Vdash P$. We proceed by induction on the structure of P and Q and detail only some key cases. If P is an atom, even non normal, then it is positive, and in the constructed Kripke structure, $p \Vdash P$.

If $P^+ = A^+ \vee B^+$, then since $Tp \Vdash P^+$ does not appear, neither $Tp \Vdash A$ nor $Tp \Vdash B$ appears. Otherwise $Tp \Vdash P$ would have been set by the saturation process. Similarly, either $Fp \Vdash A$ or $Fp \Vdash B$ does not appear. Suppose the first statement does not appear, then we apply the induction hypothesis to A and get that $p \Vdash A$, therefore $p \Vdash P$. Now if $P^- = A^- \vee B^-$, we have the same results. We have to prove $p \nVdash A$ and $p \nVdash B$. There are two cases: if $Fp \Vdash A$ appears, conclude by Lemma 4 otherwise use the induction hypothesis.

If $P = \forall x R^+(x)$, let $q \geq p$ be a world and $t \in \mathcal{D}(q)$. $Fp \Vdash P^+$ does not appear, hence no statement $Fq \Vdash R(t)$ appear (otherwise $Fp \Vdash P$ would have been set by the saturation process). If $Tq \Vdash R(t)$ appears, $q \Vdash R(t)$ by Lem. 4. Otherwise $q \Vdash R(t)$ by the induction hypothesis. Therefore, by the Kripke structure definition, $p \Vdash \forall x R^+(x)$. If $Q = \forall x R^-(x)$ then similarly there is at least one world $q \geq p$ and one term $t \in \mathcal{D}(q)$ for which $Tq \Vdash R(t)$ does not appear. If $Fq \Vdash R(t)$ appears, we apply Lem. 4, otherwise we use the induction hypothesis. In both cases, $q \nVdash R(t)$. Thus, by the Kripke structure definition, $p \nVdash \forall x R^-(x)$. *The other connectors are treated in exactly the same way.*   □

Now let A be an atom, p a world, and $A \to P$. If A appears in a statement $Bp \Vdash A$, then $Bp \Vdash P$ (by Lem. 3) and by Lem. 4 A and P have the same interpretation. Otherwise, since P is positive by hypothesis, $p \Vdash P$, and $p \Vdash A$ by definition. Either way, the rewrite rules are valid in $\mathcal{K}$ which is thus a model.

4.3   Mixing the Two Conditions

Consider two rewrite systems $\mathcal{R}_>$ and $\mathcal{R}_+$. Under the confluence and termination of $\mathcal{R} = \mathcal{R}_> \cup \mathcal{R}_+$ and the condition that $\mathcal{R}_+$ is right-normal for $\mathcal{R}_>$, we are able to prove completeness of the tableau method:

Definition 5 (Right normality). *Take two rewrite systems $\mathcal{R}'$ and $\mathcal{R}$. $\mathcal{R}'$ is right normal for $\mathcal{R}$ if, for any propositional rule $l \to r \in \mathcal{R}'$, all the instances of atoms of r by $\mathcal{R}$-normal substitutions σ are in normal form for $\mathcal{R}$.*

This condition has never been studied before. The model is built as follows: given a branch, saturate it as in Sec. 4.2 and define the model *by induction on the well-founded order*. We interpret non $\mathcal{R}_+$-normal atoms exactly as in Sec. 4.2. The Kripke structure $\mathcal{K}$ agrees as before with the branch and is a model of $\mathcal{R}_>$ Both claims are proved by induction over the well-founded order $>$. Furthermore, $\mathcal{K}$ is also a model of $\mathcal{R}_+$.

Lemma 6. *Let P^+ be a positive formula and Q^- be a negative formula defined over $\mathcal{D}(p)$. Suppose that all instances (by R-normal substitutions) of atoms from P, Q are normal for $\mathcal{R}_>$.*

If $Bp \Vdash P^+$ (resp. $Bp \Vdash Q^-$) does not appear (whether $B = T$ or $B = F$) in the branch, then $p \Vdash P^+$ (resp. $p \nVdash Q^-$).

Proof. By induction on the formula structure, as in Lem. 5. Note that we cannot apply the rewrite rules of $\mathcal{R}_>$.   □

We can then conclude that every $P \to Q \in \mathcal{R}_+$ is valid in the Kripke structure.

4.4   On Computational Content

We exhibit a result that will be important in the discussion of the relations between constructive semantic cut elimination and proof normalization. This rewrite rule is already discussed in [10], in a nonconstructive setting. Consider this rewrite system, where A is any atomic formula, and $y \simeq z$ stands for $\forall x(y \in x \Rightarrow z \in x)$:

$$R \in R \to \forall y(y \simeq R \Rightarrow (y \in R \Rightarrow (A \Rightarrow A))) \tag{1}$$

Theorem 2. *The tableau modulo method for this rewrite system is complete.*

Proof. Given a branch, define the Kripke structure $\mathcal{K}$ as in Sec. 4.2: it agrees with this branch (proved as in Sec. 4.2). If the Kripke structure is improper, it means that the branch is closed. Moreover the rewrite rule 1 is valid. Indeed, the formula $R \in R \Leftrightarrow \forall y(y \simeq R \Rightarrow (y \in R \Rightarrow (A \Rightarrow A)))$ is always forced at any node of any Kripke structure (it is an intuitionistic tautology). This completeness proof leads to a cut elimination theorem for this rewrite system.   □

5   Soundness

We will now prove the soundness of the tableau method w.r.t. cut-free (single-conclusion) LJ_{mod}. In classical logic, it is common knowledge that a ground tableau proof corresponds to a cut-free proof of the sequent calculus. In the intuitionistic case, it is not obvious since a tableau proof roughly corresponds to a multi-succedent sequent proof [8, 20, 19], while a single-conclusion sequent calculus has at most one right member. The soundness of intuitionistic tableaux (and of multi-succedent calculi) is *always* proved with respect to semantics [7, 8, 12, 19, 20]. [3] attempts a syntactic soundness proof but some details seem rather erroneous ($\vee$ case). For that, we first state some definitions, building upon those of [3].

Definition 6. *Let p be a world. We define the sets $T_p(\pi) = \{P \mid Tq \Vdash P \in \pi$ for some $q \leq p\}$ and $F_p(\pi) = \{P \mid Fp \Vdash P \in \pi\}$. Let $\bigvee S$ stand for the disjunction of some elements of S. A path π is consistent if for any world p, any finite disjunction $\bigvee F_p(\pi)$, $T_p(\pi) \nvdash^*_{\mathcal{R}} \bigvee F_p(\pi)$.*

$Bp \Vdash A \in \pi$ means that this very statement appears on the path π. $T_p(\pi)$ contains all true formulas at any world below p, while $F_p(\pi)$ contains the false formulas *only at world p*. This is due to the Kripke structure definition: unforced formulas at p can be forced in future worlds, whereas truth is a definitive commitment. The major difference between Def. 6 and the one of [3] is the definition of consistency of a path.

Forbidding the cut rule forces us to prove again associativity of $\vee$:

Lemma 7. *Let A, B, C be formulas and Γ be a set of formulas. If we have a proof θ of $\Gamma \vdash^*_{\mathcal{R}} A \vee (B \vee C)$ then we can construct a proof θ' of $\Gamma \vdash^*_{\mathcal{R}} (A \vee B) \vee C$.*

Proof. The proof proceeds by induction on θ. The induction hypothesis needs a strengthening: θ is a proof of $\Gamma \vdash^*_{\mathcal{R}} Q$ where Q is one of $A \vee (B \vee C)$ or $B \vee C$.

If the last rule is axiom, then replace it by a proof of $P \vdash^*_{\mathcal{R}} (A \vee B) \vee C$. If it is $\vee$-r, we get a proof θ' of $\Gamma \vdash^*_{\mathcal{R}} A$, $\Gamma \vdash^*_{\mathcal{R}} B$, $\Gamma \vdash^*_{\mathcal{R}} C$ or $\Gamma \vdash^*_{\mathcal{R}} B \vee C$. In the first three cases, plug *two* $\vee$-r rules for a proof of $\Gamma \vdash^*_{\mathcal{R}} (A \vee B) \vee C$. In the last one, apply the induction hypothesis to θ'. $\mathcal{R}$ is confluent, so no other rule can apply to Q [9]. Thus, otherwise, apply the induction hypothesis to the premise(s) (unless Q becomes erased) and use the same rule to get $\Gamma \vdash^*_{\mathcal{R}} (A \vee B) \vee C$.   □

The commutativity of $\vee$ is immediate (switching premises). Hence, we now note $\bigvee S$ the disjunction of some subset of S, disregarding order and parentheses. We can also weaken the conclusion, adding $\vee$-r rules to get the disjunction of the whole S. The following lemma "strengthens" some rules of the sequent calculus, allowing multiple right propositions. Such properties are trivial with the cut rule. As we want a cut elimination theorem, we need more elaborate proofs.

Lemma 8. *Let A, B, C be formulas and Γ_1, Γ_2 be sets of formulas. From proofs of $\Gamma_1 \vdash^*_{\mathcal{R}} A \vee C$ and $\Gamma_2 \vdash^*_{\mathcal{R}} B \vee C$ (resp. $\Gamma_1 \vdash^*_{\mathcal{R}} P(t) \vee C$, resp. $\Gamma_1, B \vdash^*_{\mathcal{R}} C$ and $\Gamma_2 \vdash^*_{\mathcal{R}} A \vee C$) we can construct a proof of $\Gamma_1, \Gamma_2 \vdash^*_{\mathcal{R}} (A \wedge B) \vee C$ (resp. $\Gamma_1 \vdash^*_{\mathcal{R}} (\exists x P(x)) \vee C$, resp. $\Gamma_1, \Gamma_2, A \Rightarrow B \vdash^*_{\mathcal{R}} C$).*

Proof. We focus on the first part of the lemma. The other parts are proved using the same pattern. We construct bottom-up from the two proofs π_1 and π_2 of $\Gamma_1 \vdash_{\mathcal{R}}^* A \vee C$ and $\Gamma_2 \vdash_{\mathcal{R}}^* B \vee C$ a proof of the sequent $\Gamma_1, \Gamma_2 \vdash_{\mathcal{R}}^* (A \wedge B) \vee C$. We also consider axiom rules applying only on atoms. This harmless restriction of sequent calculus is standard and this also holds in deduction modulo [9].

The idea is simple: first include a copy of π_1 using Γ_1, then, at the leaves of π_1, when necessary, take a copy of π_2 using Γ_2 (unchanged by the first induction on π_1). Let us detail a bit. We construct a proof of $\Gamma_1, \Gamma_2 \vdash_{\mathcal{R}}^* (A \wedge B) \vee C$ by induction on π_1. If the first rule is:

- a rule with Γ_1 as an active formula, apply the induction hypothesis to the premise(s) and then apply the same rule. For instance, for the $\Rightarrow$-l rule:

$$\frac{\dfrac{\pi_1'}{\Gamma_1, Q \vdash_{\mathcal{R}}^* A \vee C} \qquad \dfrac{\pi_1''}{\Gamma_1 \vdash_{\mathcal{R}}^* P}}{\Gamma_1, P \Rightarrow Q \vdash_{\mathcal{R}}^* A \vee C}$$

we apply the induction hypothesis to π_1', get a proof π' of $\Gamma_1, Q, \Gamma_2 \vdash_{\mathcal{R}}^*$ $(A \wedge B) \vee C$, and we then apply the $\Rightarrow$-l rule:

$$\frac{\dfrac{\pi'}{\Gamma_1, Q, \Gamma_2 \vdash_{\mathcal{R}}^* (A \wedge B) \vee C} \qquad \dfrac{\pi_1''}{\Gamma_1, \Gamma_2 \vdash_{\mathcal{R}}^* P} \text{ weakenings}}{\Gamma_1, P \Rightarrow Q, \Gamma_2 \vdash_{\mathcal{R}}^* (A \wedge B) \vee C}$$

- a right weakening (on $A \vee C$). We instead weaken on $(A \wedge B) \vee C$ and add left weakenings to introduce Γ_2. We get a proof of $\Gamma_1, \Gamma_2 \vdash_{\mathcal{R}}^* (A \wedge B) \vee C$.
- a $\vee$-r rule ($\mathcal{V}_1$). By assumption, we can not have axiom rule, since the considered proposition is not an atom. This case is the most interesting. We stop the induction on π_1 and initiate an induction on π_2. As usual, we rename the fresh constants of π_2 in order for them to be fresh for Γ_1. If the first rule is:
  - a left rule r. Apply r the the proof(s) obtained by induction hypothesis.
  - a right weakening. Similar as in the induction on π_1.
  - a $\vee$-r rule ($\mathcal{V}_2$). There are two subcases. If the premise is a proof π_1' of $\Gamma_1 \vdash_{\mathcal{R}}^* C'$ with $C' \equiv_{\mathcal{R}} C$, construct the following proof (ignoring π_2):

$$\frac{\dfrac{\pi_1'}{\Gamma_1, \Gamma_2 \vdash_{\mathcal{R}}^* C'} \text{ weakenings}}{\Gamma_1, \Gamma_2 \vdash_{\mathcal{R}}^* (A \wedge B) \vee C} \vee\text{-right}$$

Otherwise the premise is a proof π_1' of $\Gamma_1 \vdash_{\mathcal{R}}^* A'$ with $A' \equiv_{\mathcal{R}} A$. If on π_2 the premise is a proof $\Gamma_2 \vdash_{\mathcal{R}}^* C'$, we construct the above proof, switching indexes 1 and 2. Otherwise, it is a proof of $\Gamma_2 \vdash_{\mathcal{R}}^* B'$ with $B' \equiv_{\mathcal{R}} B$, and we construct the proof:

$$\frac{\dfrac{\dfrac{\pi_2'}{\Gamma_1, \Gamma_2 \vdash_{\mathcal{R}}^* B'} \text{ weakenings} \qquad \dfrac{\pi_1'}{\Gamma_1, \Gamma_2 \vdash_{\mathcal{R}}^* A'} \text{ weakenings}}{\Gamma_1, B' \vdash_{\mathcal{R}}^* (A \wedge B)} \wedge\text{-r}}{\Gamma_1, \Gamma_2 \vdash_{\mathcal{R}}^* (A \wedge B) \vee C} \vee\text{-r}$$

where $B' \equiv_{\mathcal{R}} B$, $C' \equiv_{\mathcal{R}} C$ and $B' \vee C'$ is the formula used in $\mathcal{A}_2$. It is a disjunction because $\mathcal{R}$ is confluent and left members are atomic, so main connectors of two equivalent compound formulas are the same ([10, 9]).

The treatment of $\exists$ needs only one induction, as does $\Rightarrow$: the sequent $\Gamma_1, B \vdash^*_{\mathcal{R}} C$ contains no disjunction. $\qquad\qquad\qquad\qquad\qquad\qquad\qquad\qquad\qquad\qquad\qquad\qquad\qquad\qquad$ □

Note. This result has to be compared with the LB sequent calculus [20], where the very same rules are allowed. However, soundness of LB is proved semantically. Our result is a syntactic proof.

We will make use of Lem. 8 with $\Gamma_1 = \Gamma_2 = \Gamma$. Contracting then Γ_1, Γ_2 gives us a proof of $\Gamma \vdash^*_{\mathcal{R}} (A \wedge B) \vee C$. We are now ready to prove soundness of the intuitionistic tableau construction with respect to cut-free sequent calculus.

Theorem 3 (Tableaux syntactic cut-free soundness). *Let Γ be a set of formulas and P be a formula. If $\Gamma \nvdash^*_{\mathcal{R}} P$ then there is a consistent path π in the complete systematic tableau developement of $T\emptyset \Vdash \Gamma\!\downarrow, F\emptyset \Vdash P\!\downarrow$.*

Remark: The contrapositive of this theorem has exactly the same proof, complicated by some uninteresting additional cases. $\qquad\qquad\qquad\qquad\qquad\qquad\qquad\qquad$ •

Proof. We show that if π is a consistent branch in a partially developed tableau, the method of Sec. 3 extends it (at some step) in at least one consistent path.

The root of the tableau is consistent: having $\Gamma \vdash^*_{\mathcal{R}} P$ is the same as having $\Gamma\!\downarrow\vdash^*_{\mathcal{R}} P\!\downarrow$. This is a classical result of deduction modulo (see for instance [4, 9, 10]). Now let $Bp \Vdash P$ the least unused statement in the tableau developement appearing on π (and P is normal by construction). If $Bp \Vdash P$ is:

- $Tp \Vdash Q \wedge R$, π is extended following the rules of figure 2 with $Tp \Vdash Q$ and $Tp \Vdash R$. If the new path is inconsistent, the added statement must be involved, and we have a proof of $T_p(\pi') \vdash^*_{\mathcal{R}} F_p(\pi')$ But $T_p(\pi') = T_p(\pi) \cup \{Q, R\}$ and $F_p(\pi') = F_p(\pi)$. We apply $\wedge$-l and obtain a proof of $T_p(\pi), P \vdash^*_{\mathcal{R}} F_p(\pi)$ contradicting $P \in T_p(\pi)$.
- $Fp \Vdash Q \wedge R$, π is extended with two paths π_0 and π_1. If both new paths are inconsistent, we get the two proofs $T_p(\pi) \vdash^*_{\mathcal{R}} Q \vee \bigvee F_p(\pi)$ and $T_p(\pi) \vdash^*_{\mathcal{R}} R \vee \bigvee F_p(\pi)$ with $T_p(\pi) = T_p(\pi_0) = T_p(\pi_1)$, $F_p(\pi_0) = F_p(\pi) \cup \{Q\}$ and $F_p(\pi_1) = F_p(\pi) \cup \{R\}$. Potentially weakening (Lem. 7), we consider both occurrences of $\bigvee F_p(\pi)$ to be equal and we apply Lem. 8 to get a proof of $T_p(\pi) \vdash^*_{\mathcal{R}} (Q \wedge R) \vee \bigvee F_p(\pi)$ i.e. a contradiction, since $Q \wedge R \in F_p(\pi)$
- $Tp \Vdash Q \vee R$. If both new paths are inconsistent, combine with $\vee$-l the proofs $T_p(\pi), Q \vdash^*_{\mathcal{R}} \bigvee F_p(\pi)$ and $T_p(\pi), R \vdash^*_{\mathcal{R}} \bigvee F_p(\pi)$ to get a contradiction.
- $Fp \Vdash Q \vee R$. If the new path is inconsistent, we have a proof of $T_p(\pi) \vdash^*_{\mathcal{R}} (Q \vee R) \vee \bigvee F_p(\pi)$ (using Lem. 7). But $Q \vee R \in F_p(\pi)$.
- $Tp \Vdash Q \Rightarrow R$, then if both new paths are inconsistent we have proofs of $T_{p'}(\pi_0) \vdash^*_{\mathcal{R}} \bigvee F_{p'}(\pi_0)$ and $T_{p'}(\pi_1) \vdash^*_{\mathcal{R}} \bigvee F_{p'}(\pi_1)$ since things changing from π change at world p'. By definitions of $T_{p'}$ and $F_{p'}$, we have proofs of $T_{p'}(\pi) \vdash^*_{\mathcal{R}} Q \vee \bigvee F_{p'}(\pi)$ and $T_{p'}(\pi), R \vdash^*_{\mathcal{R}} \bigvee F_{p'}(\pi)$ By Lem. 8 we get a proof of $T_{p'}(\pi), Q \Rightarrow R \vdash^*_{\mathcal{R}} \bigvee F_{p'}(\pi)$, which contradicts $Q \Rightarrow R \in T_{p'}(\pi)$.

- $Fp \Vdash Q \Rightarrow R$. If the new path is inconsistent, we have a proof θ of $T_{p'}(\pi') \vdash_{\mathcal{R}}^*$ $\bigvee F_{p'}(\pi')$. Since p' is a new world, comparable only with the $q \leq p$ on π, $T_{p'}(\pi') = T_p(\pi) \cup \{Q\}$ and $F_{p'}(\pi') = \{R\}$. Hence, we can apply the $\Rightarrow$-r rule to θ, and we obtain a proof of $T_p(\pi) \vdash_{\mathcal{R}}^* Q \Rightarrow R$, yielding the inconsistency of π since $Q \Rightarrow R \in F_p(\pi)$.

  It is extremely important to have no choice for $F_{p'}(\pi')$ but R. It is here that the logic gets intuitionistic. Other tableaux methods (like [20]) have also a special treatement of the $\Rightarrow$ and $\forall$ connectors: we need a sequent with only one member on the right side.
- $\neg P$ behaves as $P \Rightarrow \bot$. So both cases are consequences of the previous.
- $Tp \Vdash \exists x Q(x)$. If the new path is inconsistent, we have a proof of $T_p(\pi), Q(c) \vdash_{\mathcal{R}}^*$ $\bigvee F_p(\pi)$. We apply the $\exists$-l rule as c is fresh, yielding the inconsistency of π.
- $Fp \Vdash \exists x Q(x)$. If the new path is inconsistent, we have a proof of $T_p(\pi) \vdash_{\mathcal{R}}^*$ $Q(t) \downarrow \vee \bigvee F_p(\pi)$. We transform this proof into a proof of $T_p(\pi) \vdash_{\mathcal{R}}^* Q(t) \vee$ $\bigvee F_p(\pi)$ since $Q(t) \equiv_{\mathcal{R}} Q(t) \downarrow$. Then using lemma 8 we get a proof of: $T_p(\pi) \vdash_{\mathcal{R}}^* \exists x Q(x) \vee \bigvee F_p(\pi)$, thereby contradicting the consistency of π.
- $Tp \Vdash \forall x Q(x)$. If the new path is inconsistent, we have a proof of $T_{p'}(\pi), Q(t) \downarrow$ $\vdash_{\mathcal{R}}^* F_{p'}(\pi)$, then converted into a proof of $T_{p'}(\pi), Q(t) \vdash_{\mathcal{R}}^* F_{p'}(\pi)$. Apply the $\forall$-l rule to get $T_{p'}(\pi), \forall x Q(x) \vdash_{\mathcal{R}}^* F_{p'}(\pi)$, and the inconsistency of π.
- $Fp \Vdash \forall x Q(x)$. If the new path is inconsistent, we must have a proof of $T_{p'}(\pi') \vdash_{\mathcal{R}}^* F_{p'}(\pi')$. As for $Fp \Vdash Q \Rightarrow R$, p' is a new world, comparable only with p. So, we have in fact a proof of $T_p(\pi) \vdash_{\mathcal{R}}^* Q(c)$. We apply the $\forall$-r rule, since c is fresh. This yields the inconsistency of π.   □

We have established that a closed tableau *is* a cut-free proof of LJ_{mod}. This result is new, even in LJ. The combination of the soundness theorem of sequent calculus w.r.t. modified Kripke structures, Th. 1 and Th. 3 yields *a constructive semantic cut elimination theorem*, holding for the conditions on rewrite rules seen in Sec. 4:

Theorem 4 (Cut elimination for LJ_{mod}). *If* $\Gamma \vdash_{\mathcal{R}} P$ *then* $\Gamma \vdash_{\mathcal{R}}^* P$.

6   Conclusion and Further Work

We have formulated a simple tableau procedure for intuitionistic logic modulo and proved its completeness and syntactic soundness to show that the computational content of the semantic cut elimination theorem actually *is* a tableau method.

The method itself could be much improved with a better handling of the rewrite steps (normalizing possibly includes unnecessary steps). We could also treat quantifiers differently : free-variable tableaux based upon [19, 20] or the introduction of Skolem symbols (more tricky in intuitionistic logic, see [13, 18]) would indeed improve efficiency.

The rewrite system 1 of Sec. 4.4 does not possess the proof normalization property (see [10]): any attempt to normalize the proof (with a cut on $R \in R$) of $\vdash_{\mathcal{R}} A \Rightarrow A$ can only fail. We can semantically eliminate this cut, because we have the *semantic* information that $A \Rightarrow A$ is a tautology. The proof normalization method however does not. In this case, the semantic analysis is sharper, and it shows the gap between the two methods. Finally, the link between semantic cut elimination and normalization by evaluation methods as in [2], where a Kripke-style framework is described, seems a promising field of investigation.

References

[1] Bonichon, R.: TaMeD: A tableau method for deduction modulo. In: Basin, D., Rusinowitch, M. (eds.) IJCAR 2004. LNCS (LNAI), vol. 3097, pp. 445–459. Springer, Heidelberg (2004)
[2] Coquand, C.: From semantic to rules: a machine assisted analysis. In: Meinke, K., Börger, E., Gurevich, Y. (eds.) CSL 1993. LNCS, vol. 832, pp. 91–105. Springer, Heidelberg (1994)
[3] De Marco, M., Lipton, J.: Completeness and cut elimination in Church's intuitionistic theory of types. Journal of Logic and Computation 15, 821–854 (2005)
[4] Dowek, G., Hardin, T., Kirchner, C.: Theorem proving modulo. Journal of Automated Reasoning 31, 33–72 (2003)
[5] Dowek, G., Werner, B.: Proof normalization modulo. The Journal of Symbolic Logic 68(4), 1289–1316 (2003)
[6] Dragalin, A.G.: A completeness theorem for higher-order intuitionistic logic: an intuitionistic proof. In: Skordev, D.G. (ed.) Mathematical Logic and Its Applications, pp. 107–124. Plenum, New York (1987)
[7] Dragalin, A.G.: Mathematical Intuitionism: Introduction to Proof Theory. Translation of Mathematical Monographs. American Mathematical Society vol. 67 (1988)
[8] Dummett, M.: Elements of Intuitionism. Oxford University Press, Oxford (2000)
[9] Hermant, O.: Méthodes Sémantiques en Déduction Modulo. PhD thesis, Université Paris 7 - Denis Diderot (2005)
[10] Hermant, O.: Semantic cut elimination in the intuitionistic sequent calculus. In: Urzyczyn, P. (ed.) TLCA 2005. LNCS, vol. 3461, pp. 221–233. Springer, Heidelberg (2005)
[11] Krivine, J.-L.: Une preuve formelle et intuitionniste du théorème de complétude de la logique classique. The Bulletin of Symbolic Logic 2, 405–421 (1996)
[12] Nerode, A., Shore, R.A.: Logic for Applications. Springer, Heidelberg (1993)
[13] Shankar, N.: Proof search in the intuitionistic sequent calculus. In: Kapur, D. (ed.) Automated Deduction - CADE-11. LNCS, vol. 607, pp. 522–536. Springer, Heidelberg (1992)
[14] Smullyan, R.: First Order Logic. Springer, Heidelberg (1968)
[15] Takeuti, G.: Proof Theory. Studies in Logic and The Foundations of Mathematics, 2nd edn., vol. 81, North-Holland, Amsterdam (1987)
[16] Troelstra, A.S., Van Dalen, D.: Constructivism in Mathematics. Studies in Logic and The Foundations of Mathematics, vol. 2. North Holland, Amsterdam (1988)
[17] Veldman, W.: An intuitionistic completeness theorem for intuitionistic predicate logic. Journal of Symbolic Logic 41, 159–166 (1976)
[18] Voronkov, A.: Proof-search in intuitionistic logic based on constraint satisfaction. In: Miglioli, P., Moscato, U., Ornaghi, M., Mundici, D. (eds.) TABLEAUX 1996. LNCS, vol. 1071, pp. 312–329. Springer, Heidelberg (1996)
[19] Waaler, A.: Connection in Nonclassical Logics. In: Handbook of Automated Reasoning, vol. II. North Holland, Amsterdam (2001)
[20] Waaler, A., Wallen, L.: Tableaux for Intuitionistic Logics. In: Handbook of Tableau Methods, pp. 255–296. Kluwer Academic Publishers, Boston (1999)

Fast Reflexive Arithmetic Tactics
the Linear Case and Beyond

Frédéric Besson[*]

Irisa/Inria, Campus de Beaulieu, 35042 Rennes Cedex, France

Abstract. When goals fall in decidable logic fragments, users of proof-assistants expect automation. However, despite the availability of decision procedures, automation does not come for free. The reason is that decision procedures do not generate proof terms. In this paper, we show how to design efficient and lightweight reflexive tactics for a hierarchy of quantifier-free fragments of integer arithmetics. The tactics can cope with a wide class of linear and non-linear goals. For each logic fragment, off-the-shelf algorithms generate *certificates* of infeasibility that are then validated by straightforward *reflexive checkers* proved correct inside the proof-assistant. This approach has been prototyped using the Coq proof-assistant. Preliminary experiments are promising as the tactics run fast and produce small proof terms.

1 Introduction

In an ideal world, proof assistants would be theorem provers. They would be fed with theorems and would either generate a proof of them (if one exists) or reject them (if none exists). Unfortunately, in real life, theorems can be undecidable Yet, theorems often fall in decidable fragments. For those, users of proof assistants expect the proof process to be discharged to dedicated efficient decision procedures. However, using off-the-shelf provers is complicated by the fact that they cannot be trusted. A decision procedure, that when given as input a supposedly-so theorem, laconically answers back *yes* is useless. Proof assistants only accept proofs which they can check by their own means.

Approaches to obtain proofs from decision procedures usually require a substantial engineering efforts and a deep understanding of the internals of the decision procedure. A common approach consists in instrumenting the procedure so that it generates proof traces that are *replayed* in the proof-assistant. The Coq omega tactic by Pierre Crégut ([9] chapter 17) is representative of this trend. The tactics is a decision procedure for quantifier-free linear integer arithmetics. It generates Coq proof terms from traces obtained from an instrumented version of the Omega test [22]. Another approach, implemented by the Coq ring tactics [13], is to prove correct the decision procedure inside the proof-assistant and

---

[*] This work was partly funded by the IST-FET programme of the European Commission, under the IST-2005-015905 MOBIUS project.

T. Altenkirch and C. McBride (Eds.): TYPES 2006, LNCS 4502, pp. 48–62, 2007.
© Springer-Verlag Berlin Heidelberg 2007

use computational reflection. In this case, both the computational complexity of the decision procedure and the complexity of proving it correct are limiting factors.

In this paper, we adhere to the so-called *sceptical approach* advocated by Harrison and Théry [16]. The key insight is to separate proof-search from proof-checking. Proof search is delegated to fine-tuned external tools which produce certificates to be checked by the proof-assistant. In this paper, we present the design, in the Coq proof-assistant, of a tactics for a hierarchy of quantifier-free fragments of integer arithmetics. The originality of the approach is that proof witnesses, *i.e.*, certificates, are computed by black-box off-the-shelf provers. The soundness of a *reflexive* certificate checker is then proved correct inside the proof-assistant. For the logic fragments we consider, checkers are considerably simpler than provers. Hence, using a pair (untrusted prover, proved checker) is a very lightweight and efficient implementation technique to make decision procedures available to proof-assistants.

The contributions of this paper are both theoretical and practical. On the theoretical side, we put the shed on mathematical theorems that provide infeasibility certificates for linear and non-linear fragments of integer arithmetics. On the practical side, we show how to use these theorems to design powerful and space-and-time efficient checkers for these certificates. The implementation has been carried out for the Coq proof-assistant. Experiments show that our new reflexive tactic for linear arithmetics outperforms state-of-the-art Coq tactics.

The rest of this paper is organised as follows. Section 2 recalls the principles of reflection proofs. Section 3 presents the mathematical results on which our certificate checkers are based on. Section 4 describe our implementation of these checkers in the Coq proof-assistant. Section 5 compares to related work and concludes.

2   Principle of Reflection Proofs

Reflection proofs are a feature of proof-assistants embedding a programming language. (See Chapter 16 of the Coq'Art book [2] for a presentation of reflection proofs in Coq.) In essence, this technique is reducing a proof to a computation. Degenerated examples of this proof pattern are equality proofs of *ground*, *i.e.*, variable free, arithmetic expressions. Suppose that we are given the proof goal

$$4 + 8 + 15 + 16 + 23 + 42 = 108$$

Its proof is quite simple: evaluate $4 + 8 + 15 + 16 + 23 + 42$; check that the result is indeed 108. Reflection proofs become more challenging when goals involves variables. Consider, for instance, the following goal where x is universally quantified:

$$4 \times x + 8 \times x + 15 \times x + 16 \times x + 23 \times x + 42 \times x = 108 \times x$$

Because the expression contain variables, evaluation alone is unable to prove the equality. The above goal requires a more elaborate reflection scheme.

2.1 Prover-Based Reflection

Reflection proofs are set up by the following steps:

1. encode logical propositions into symbolic expressions F;
2. provide an evaluation function $[\![.]\!] : Env \to F \to Prop$ that given an environment binding variables and a symbolic expression returns a logical proposition;
3. implement a semantically sound and computable function $prover : F \to bool$ verifying $\forall ef, prover(ef) = true \Rightarrow \forall env, [\![ef]\!]_{env}$.

A reflexive proof proceeds in the following way. First, we construct an environment env binding variables. (Typically, variables in symbolic expressions are indexes and environments map indexes to variables.) Then, the goal formula f is replaced by $[\![ef]\!]_{env}$ such that, by computation, $[\![ef]\!]_{env}$ evaluates to f. As the prover is sound, if $prover(ef)$ returns $true$, we conclude that f holds.

Example 1. *Following the methodology described above, we show how goals of the form $c_1 \times x + \ldots c_n \times x = c \times x$ (where the c_is are integer constants and x is the only universally quantified variable) can be solved by reflection.*

- *Such formulae can be coded by pairs $([c_1; \ldots; c_n], c) \in F = \mathbb{Z}^* \times \mathbb{Z}$.*
- *The semantics function $[\![.]\!]$ is defined by $[\![l, c]\!]_x \stackrel{\triangle}{=} listExpr(x, l) = c \times x$ where $listExpr : \mathbb{Z} \times \mathbb{Z}^* \to \mathbb{Z}$ is defined by*

$$
\begin{aligned}
listExpr(x, [\,]) &\stackrel{\triangle}{=} 0 \\
listExpr(x, [c]) &\stackrel{\triangle}{=} c \times x \\
listExpr(x, c :: l) &\stackrel{\triangle}{=} c \times x + listExpr(x, l)
\end{aligned}
$$

- *Given a pair (l, c), the prover computes the sum of the elements in l and checks equality with the constant c.*

$$
prover(l, c) \stackrel{\triangle}{=} (fold\ +\ l\ 0) = c
$$

To make decision procedure available to proof-assistants, the reflexive prover-based approach is very appealing. It is conceptually simple and can be applied to any textbook decision procedure. Moreover, besides soundness, it allows to reason about the completeness of the prover. As a result, the end-user of the proof-assistant gets maximum confidence. Upon success, the theorem holds; upon failure, the theorem is wrong.

2.2 Checker-Based Reflection

On the one hand, provers efficiency is due to efficient data-structures, clever algorithms and fine-tuned heuristics. On the other hand, manageable soundness proofs usually hinge upon simple algorithms. In any case, reflection proofs require runtime efficiency. Obviously, these facts are difficult to reconcile. To obtain a good trade-off between computational efficiency and proof simplicity, we advocate implementing and proving correct *certificate checkers* instead of genuine provers. Compared to provers, checkers take a certificate as an additional input and verify the following property:

$$\forall ef, (\exists cert, checker(cert, ef) = true) \Rightarrow \forall env, [\![ef]\!]_{env}$$

The benefits are twofold : checkers are simpler and faster. Complexity theory ascertains that checkers run faster than provers. In particular, contrary to all known provers, checkers for NP-complete decision problems have polynomial complexity. A reflexive proof of $[\![ef]\!]_{env}$ now amounts to providing a certificate *cert* such that $checker(cert, t)$ evaluates to *true*. As they are checked inside a proof-assistant, certificates can be generated by any untrusted optimised procedure.

Using certificates and reflexive checkers to design automated tactics is not a new idea. For instance, proof traces generated by instrumented decision procedures can be understood as certificates. In this case, reflexive checkers are trace validators which verify the logical soundness of the proof steps recorded in the trace. The Coq `romega` tactic [8] is representative of this trace-based approach: traces generated by an instrumented version of the Omega test [22] act as certificates that are validated by a reflexive checker. The drawback of this method is that instrumentation is an intrusive task and require to dig into the internals of the decision procedure. In the following, we present conjunctive fragments of integer arithmetics for which certificate generators are genuine off-the-shelf provers. The advantage is immediate: provers are now black-boxes. Moreover, the checkers are quite simple to implement and prove correct.

3   Certificates for Integer Arithmetics

In this part, we study a hierarchy of three quantifier-free fragments of integer arithmetics. We describe certificates, off-the-shelf provers and certificate checkers associated to them. We consider formulae that are conjunctions of inequalities and we are interested in proving the unsatisfiability of these inequalities. Formally, formulae of interest have the form:

$$\neg \left(\bigwedge_{i=1}^{k} e_i(x_1, \ldots, x_n) \geq 0 \right)$$

where the e_is are fragment-specific integer expressions and the x_is are universally quantified variables.

For each fragment, we shall prove a theorem of the following general form:

$$(\exists cert, Cond(cert, e_1, \ldots, e_k)) \Rightarrow \forall(x_1, \ldots, x_n), \neg \left(\bigwedge_{i=1}^{k} e_i(x_1, \ldots, x_n) \right)$$

In essence, such a theorem establish that *cert* is a certificate of the infeasibility of the e_is. We then show that certificates can be generated by off-the-shelf algorithms and that *Cond* is decidable and can be efficiently implemented by a *checker* algorithm.

3.1   Potential Constraints

To begin with, consider *potential constraints*. These are constraints of the form $x - y + c \geq 0$. Deciding the infeasibility of conjunctions of such constraints amounts to finding a cycle of negative weight in a graph such that a edge $x \xrightarrow{c} y$ corresponds to a constraint $x - y + c \geq 0$ [1,21][1].

Theorem 1

$$\exists \pi \in Path, \bigwedge \begin{pmatrix} isCycle(\pi) \\ weight(\pi) < 0 \\ \pi \subseteq (\bigcup_{i=1}^{k} \{x_{i_1} \xrightarrow{c_i} x_{i_2}\}) \end{pmatrix} \Rightarrow \forall x_1, \ldots, x_n, \neg(\bigwedge_{i=1}^{k} x_{i_1} - x_{i_2} + c_i \geq 0)$$

Proof. Ad absurdum, we suppose that we have $\bigwedge_{i=1}^{k} x_{i_1} - x_{i_2} + c_i \geq 0$ for some $x_1, \ldots, x_n$. If we sum the constraints over a cycle π, variables cancel and the result is the total weight of the path $\sum_{x \xrightarrow{c} y \in \pi} x - y + c = \sum_{x \xrightarrow{c} y \in \pi} c$. Moreover, by hypothesis, we also have that $\left(\sum_{x \xrightarrow{c} y \in \pi} x - y + c\right) \geq 0$ (each element of the sum being positive). We conclude that the total weight of cycles is necessarily positive. It follows that the existence of a cycle of negative weight c yields a contradiction.   □

As a result, a negative cycle is a certificate of infeasibility of a conjunction of potential constraints.

Bellmann-Ford shortest path algorithm is a certificate generator which runs in complexity $O(n \times k)$ where n is the number of nodes (or variables) and k is the number of edges (or constraints). However, this algorithm does not find the best certificate *i.e.*, the negative cycle of shortest length. Certificates, *i.e.*, graph cycles, can be coded by a list of binary indexes – each of them identifying one of the k constraints. The worst-case certificate is then a Hamiltonian circuit which is a permutation of the k constraint indexes. Its asymptotic size is therefore $k \times log(k)$:

$$size(i_1, \ldots, i_k) = \sum_{j=1}^{k} log(i_j) = \sum_{j=1}^{k} log(j) = log(\Pi_{j=1}^{k} j) = log(k!) \sim k \times log(k)$$

Verifying a certificate consists in checking that:

1. indexes are bound to genuine expressions;
2. verify that expressions form a cycle;
3. compute the total weight of the cycle and check its negativity

This can be implemented in time linear in the size of the certificate.

---

[1] As shown by Shostak [24], this graph-based approach generalises to constraints of the form $a \times x - b \times y + c \geq 0$.

3.2   Linear Constraints

The linear fragment of arithmetics might be the most widely used. It consists of formulae built over the following expressions:

$$Expr ::= c_1 \times x_1 + \ldots + c_n \times x_n + c_{n+1}$$

A well-known result of linear programming is Farkas's Lemma which states a strong duality result.

Lemma 1 (Farkas's Lemma (Variant)). *Let $A : \mathbb{Q}^{m \times n}$ be a rational-valued matrix and $b : \mathbb{Q}^m$ be a rational-valued vector. Exactly one of the following statement holds:*

- *$\exists(y \in \mathbb{Q}^n), y \geq \bar{0}, b^t \cdot y < 0, A^t \cdot y = \bar{0}$*
- *$\exists(x \in \mathbb{Q}^m), A \cdot x \geq b$*

Over $\mathbb{Z}$, Farkas's Lemma is sufficient to provide infeasibility certificates for systems of inequalities.

Lemma 2 (Weakened Farkas's Lemma (over $\mathbb{Z}$)). *Let $A : \mathbb{Z}^{m \times n}$ be a integer-valued matrix and $b : \mathbb{Z}^m$ be a integer-valued vector.*

$$\exists(y \in \mathbb{Z}^n), y \geq \bar{0}, b^t \cdot y < 0, A^t \cdot y = \bar{0} \Rightarrow \forall(x \in \mathbb{Z}^n), \neg A \cdot x \geq b$$

Proof. *Ad absurdum,* we suppose that we have $A \cdot x \geq b$ for some vector x. Since y is a positive vector, we have that $y^t \cdot (A \cdot x) \geq y^t \cdot b$. However, $y^t \cdot (A \cdot x) = (y^t \cdot A) \cdot x = (A^t \cdot y)^t \cdot x$. Because $A^t \cdot y = 0$, we conclude that $0 \geq y^t \cdot b$ which contradicts the hypothesis stating that $y^t \cdot b$ is strictly negative.   □

Over $\mathbb{Z}$, Farkas's lemma is not complete. Incompleteness is a consequence of the discreetness of $\mathbb{Z}$: there are systems that have solutions over $\mathbb{Q}$ but not over $\mathbb{Z}$. A canonical example is the equation $2.x = 1$. The unique solution is the rational $1/2$ which obviously is not an integer. Yet, the loss of completeness is balanced by a gain in efficiency. Whereas deciding infeasibility of system of integer constraints is NP-complete; the same problem can be solved over the rationals in polynomial time.

Indeed, an infeasibility certificate is produced as the solution of the *linear program*

$$min\{y^t \cdot \bar{1} \mid y \geq \bar{0}, b^t \cdot y < 0, A^t \cdot y = \bar{0}\}$$

Note that linear programming also *optimises* the certificate. To get *small* certificates, we propose to minimise the sum of the elements of the solution vector.

Linear programs can be solved in polynomial time using interior point methods [17]. The Simplex method – despite its worst-case exponential complexity – is nonetheless a practical competitive choice.

Linear programs are efficiently solved over the rationals. Nonetheless, an integer certificate can be obtained from any rational certificate.

Proposition 1 (Integer certificate). *For any rational certificate of the form* $cert_{\mathbb{Q}} = [p_1/q_1; \ldots; p_k/q_k]$, *an integer certificate is*

$$cert_{\mathbb{Z}} = [p'_1; \ldots; p'_k]$$

where $p'_i = p_i \times lcm/q_i$ *and* lcm *is the least common multiple of the* $q_i s$.

Worst-case estimates of the size of the certificates are inherited from the theory of integer and linear programming (see for instance [23]).

Theorem 2 (from [23] Corollary 10.2a). *The bit size of the rational solution of a linear program is at most* $4d^2(d+1)(\sigma+1)$ *where*

- d *is the dimension of the problem;*
- σ *is the number of bits of the biggest coefficient of the linear program.*

Using Lemma 1 and Theorem 2, the next Corollary gives a coarse upper-bound of the bit size of integer certificates.

Corollary 1 (Bit size of integer certificates). *The bit size of integer certificates is bounded by* $4k^3(k+1)(\sigma+1)$

Proof. Let $cert_{\mathbb{Z}} = [p'_1; \ldots; p'_k]$ be the certificate obtained from a rational certificate $cert_{\mathbb{Q}} = [p_1/q_1; \ldots; p_k/q_k]$.

$$
\begin{aligned}
\mid cert_{\mathbb{Z}} \mid &= \sum_{i=1}^{k} log(p'_i) \\
&= \sum_{i=1}^{k}(log(p_i) - log(q_i)) + \sum_{i=1}^{k} log(lcm) \\
&= \sum_{i=1}^{k}(log(p_i) - log(q_i)) + k \times log(lcm)
\end{aligned}
$$

At worse, the $q_i s$ are relatively prime and $lcm = \Pi_{i=1}^{n} q_i$.

$$
\mid cert_{\mathbb{Z}} \mid \leq k \times \sum_{i=1}^{k} log(q_i) + \sum_{i=1}^{k}(log(p_i) - log(q_i))
$$

As $\mid cert_{\mathbb{Q}} \mid = \sum_{i=1}^{k} log(p_i) + log(q_i)$, we have that $\mid cert_{\mathbb{Z}} \mid \leq k \times \mid cert_{\mathbb{Q}} \mid$. By Theorem 2, we conclude the proof and obtain the $4k^3(k+1)(\sigma+1)$ bound. $\square$

Optimising certificates over the rationals is reasonable. Rational certificates are produced in polynomial time. Moreover, the worst-case size of the integer certificates is kept reasonable.

Checking a certificate *cert* amounts to

1. checking the positiveness of the integers in *cert*;
2. computing the matrix-vector product $A^t \cdot cert$ and verifying that the result is the null vector;
3. computing the scalar product $b^t \cdot cert$ and verifying its strict negativity

Overall, this leads to a quadratic-time $O(n \times k)$ checker in the number of arithmetic operations.

3.3   Polynomial Constraints

For our last fragment, we consider unrestricted expressions built over variables, integer constants, addition and multiplication.

$$e \in Expr ::= x \mid c \mid e_1 + e_2 \mid e_1 \times e_2$$

As it reduces to solving diophantine equations, the logical fragment we consider is not decidable over the integers. However, it is a result by Tarski [26] that the first order logic $\langle \mathbb{R}, +, *, 0 \rangle$ is decidable. In the previous section, by lifting our problem over the rationals, we traded incompleteness for efficiency. Here, we trade incompleteness for decidability.

In 1974, Stengle generalises Hilbert's *nullstellenstaz* to systems of polynomial inequalities [25]. As a matter of fact, this provides a *positivstellensatz*, *i.e.*, a theorem of positivity, which states a necessary and sufficient condition for the existence of a solution to systems of polynomial inequalities. Over the integers, unlike Farkas's lemma, Stengle's *positivstellensatz* yields sound infeasibility certificates for conjunctions of polynomial inequalities.

Definition 1 (Cone). *Let $P \subseteq \mathbb{Z}[\bar{x}]$ be a finite set of polynomials. The cone of P (Cone(P)) is the smallest set such that*

1. $\forall p \in P, p \in Cone(P)$
2. $\forall p_1, p_2 \in Cone(P), p_1 + p_2 \in Cone(P)$
3. $\forall p_1, p_2 \in Cone(P), p_1 \times p_2 \in Cone(P)$
4. $\forall p \in \mathbb{Z}[\bar{x}], p^2 \in Cone(P)$

Theorem 3 states sufficient conditions for infeasibility certificates:

Theorem 3 (Weakened Positivstellensatz). *Let $P \subseteq \mathbb{Z}[x_1, \ldots, x_n]$ be a finite set of polynomials.*

$$\exists cert \in Cone(P), cert \equiv -1 \Rightarrow \forall x_1, \ldots, x_n, \neg \bigwedge_{p \in P} p(x_1, \ldots, x_n) \geq 0$$

Proof. By adbsurdum, we suppose that we have $\bigwedge_{p \in P} p(x_1, \ldots, x_n) \geq 0$ for some $x_1, \ldots, x_n$. By routine induction over the definition of a *Cone*, we prove that any polynomial $p \in Cone(P)$ is such that $p(x_1, \ldots, x_n)$ is positive. This contradicts the existence of the polynomial *cert* which uniformly evaluates to -1.   □

Certificate generators explore the cone to pick a certificate. Stengle's result [25] shows that only a restricted (though infinite) part of the cone needs to be considered. A certificate *cert* can be decomposed into a finite sum of products of the following form:

$$cert \in \sum_{s \in 2^P} \left(q_s \times \prod_{p \in s} p \right)$$

where $q_s = p_1^2 + \ldots + p_i^2$ is a *sum of squares* polynomial.

As pointed out by Parrilo [20], a layered certificate search can be carried out by increasing the formal degree of the certificate. For a given degree, finding a certificate amounts to finding polynomials (of known degree) that are sums of squares. This is a problem that can be solved efficiently (in polynomial time) by recasting it as a *semidefinite program* [27]. The key insight is that a polynomial q is a *sum of square* if and only if it can be written as

$$ q = \begin{pmatrix} m_1 \\ \dots \\ m_n \end{pmatrix}^t \cdot Q \cdot \begin{pmatrix} m_1 \\ \dots \\ m_n \end{pmatrix} $$

for some positive semidefinite matrix Q and some vector $(m_1, \dots, m_n)$ of linearly independent monomials.

An infeasibility certificates is a polynomial which belongs to the cone and is equivalent to -1. Using a suitable encoding, cone membership can be tested in linear time. Equivalence with -1 can be checked by putting the polynomial in Horner's normal form.

4   Implementation in the Coq Proof-Assistant

In this part, we present the design of the Coq reflexive tactics micromega[2]. This tactics solves linear and non-linear goals using the certificates and certificate generators described in Section 3. For the linear case, experiments show that micromega outperforms the existing Coq (r)omega tactics both in term of proof-term size and checking time.

4.1   Encoding of Formulae

As already mentioned in section 2.1, to set up a reflection proof, logical sentences are encoded into syntactic terms. Arithmetic expressions are represented by the following inductive type:

```
Inductive Expr : Set :=
  | V       (v:Var)
  | C       (c:Z)
  | Mult    (e1:Expr) (e2:Expr)
  | Add     (e1:Expr) (e2:Expr)
  | UMinus  (e:Expr).
```

The eval_expr function maps syntactic expressions to arithmetic expressions. It is defined by structural induction over the structure of Expr.

```
Fixpoint eval_expr (env:Env) (p:Expr) {struct p}: Z :=
  match p with
    | V v        ⇒ get_env env v
```

---

[2] Micromega is available at http://www.irisa.fr/lande/fbesson.html

```
  | C c          ⇒ c
  | Mult p q  ⇒ (eval_expr env p) * (eval_expr env q)
  | Add p q   ⇒ (eval_expr env p) + (eval_expr env q)
  | UMinus p  ⇒ - (eval_expr env p)
  end.
```

The environment binds variable identifiers to their integer value. For efficiency, variables identifiers are binary indexes and environments are binary trees. As a result, the function `eval_expr` runs in time linear in the size of the input expression.

Formulae are lists of expressions `Formulae := list Expr` and are equipped with an evaluation function `eval : Env→Formulae→Prop`

```
Fixpoint eval (env:Env)(f:Formulae){struct f}: Prop :=
  match f with
  | nil   ⇒ False
  | e::rf ⇒ ((eval_expr env e) ≥ 0) → (eval env rf)
  end.
```

The `eval` function generates formulae of the form $e_1(x_1, \ldots, x_n) \geq 0 \rightarrow \ldots \rightarrow e_k(x_1, \ldots, x_n) \geq 0 \rightarrow False$. By simple propositional reasoning, such a formula is equivalent to $\neg \left(\bigwedge_{i=1}^{k} e_i(x_1, \ldots, x_n) \geq 0 \right)$ which is exactly the logical fragment studied in Section 3.

4.2   Proving the Infeasibility Criterion

At the core of our tactics are theorems which are reducing infeasibility of formulae to the existence of certificates. In the following, we present our formalisation of Stengle's Positivstellensatz in Coq. The cone of a set of polynomials is defined by an inductive predicate:

```
Inductive Cone (P: list Expr) : Expr → Prop :=
  | IsGen    : ∀ p, In p P→Cone P p
  | IsSquare: ∀ p, Cone P (Power p 2)
  | IsMult  : ∀ p q, Cone P p→Cone P q→Cone P (Mult p q)
  | IsAdd   : ∀ p q, Cone P p→Cone P q→Cone P (Add p q)
  | IsPos   : ∀ c, c ≥ 0→Cone P (C c).
```

The fifth rule `IsPos` is redundant and absent from the formal definition of a cone (Definition 1). Indeed, any positive integer can be decomposed into a sum of square. It is added for convenience and to allow a simpler and faster decoding of certificates.

We are then able to state (and prove) our weakened *positivstellensatz*.

```
Theorem positivstellensatz :  ∀ (f:Formulae),
  (∃ (e:Expr),
        Cone f e ∧
        (∀ env', eval_expr env' e = -1))  →
  ∀ env, eval env f.
```

4.3   Checking Certificates

Given a certificate *cert*, we need an algorithm to check that

1. the certificate belongs to the cone Cone(f,cert);
2. the certificate always evaluate to -1;

If certificates were terms of type Expr, proving cone membership would be a tricky task. This would be complicated and inefficient to reconstruct the cone decomposition of the expression. To avoid this pitfall, by construction, our certificates always belong to the cone. To do that, the data-structure of the certificates mimics the definition of the cone predicate:

```
Inductive Certificate : Set :=
  | Cert_IsGen     (n:nat)
  | Cert_IsSquare  (e:Expr)
  | Cert_IsMult    (e1:Expr) (e2:Expr)
  | Cert_IsAdd     (e1:Expr) (e2:Expr)
  | Cert_IsZpos    (p:positive)
  | Cert_IsZ0.
```

Given the generators of a cone, *i.e.*, a list of expressions, a certificate is decoded into an expression:

```
Fixpoint decode
  (P: list Expr) (c: Certificate) {struct c} : Expr :=
  match c with
  |Cert_IsGen n    ⇒ nth n P (C 0)
  |Cert_IsSquare p ⇒ Mult p p
  |Cert_IsMult p q ⇒ Mult (decode P p)(decode P q)
  |Cert_Add p q    ⇒ Add (decode P p)(decode P q)
  |Cert_IsZpos p   ⇒ C (Zpos p)
  |Cert_IsZ0       ⇒ C Z0
  end.
```

This construction ensures that certificates are always mapped to expressions that belong to the cone as stated by the following lemma.

```
Lemma cert_in_cone : ∀ P cert, Cone P (decode 1 cert).
```

Because our certificate encoding ensures cone membership, it remains to test that a polynomial always evaluates to a negative constant. To do that, we reuse the algorithm developed by the Coq ring tactics which normalises polynomial expressions. In the end, our checker is implemented by the following algorithm:

```
Let checker (c:Certificate) (P: list Expr) : bool :=
    (polynomial_simplify (decode P c)) == -1
```

4.4   Certificate Generation

In Section 3, we shed the light on three different arithmetic fragments, namely, potential constraints (Section 3.1), linear constraints (Section 3.2) and polynomial constraints (Section 3.3). Obviously, these logic fragments form a strict hierarchy: polynomial constraints subsume linear and potential constraints. It appears that this hierarchy is also apparent at the level of certificates: both negative-weighted cycles and Farkas's Lemma certificates can be interpreted as *Positivstellensatz* certificates.

For linear goals, our certificates are produced by a handcrafted linear solver. For non-linear goals, we are using the full-fledged semidefinite programming solver Csdp [5] through its HOL Light interface [15]. Anyhow, whatever their origin, certificates are translated into *Positivstellensatz* certificates.

4.5   Experiments

We have assessed the efficiency of `micromega` with respect to the existing Coq tactic `romega`[3]. As mentioned earlier, the `romega` is a reflexive tactics which solves linear goals by checking traces obtained from an instrumented version of the Omega test. Our benchmarks are the smallest SMALLINT problems of the Pseudo Boolean Evaluation 2005/2006 contest[4]. The number of variables is ranging from 220 to 2800 while the number of constraints is ranging from 42 to 160. The benchmarks are run on a 2.4 Ghz Intel Xeon desktop with 4GB of memory. The graph of Figure 1 presents the running time of the Coq typechecking of the certificates generated by `romega` ■ and `micromega` ▲. For this

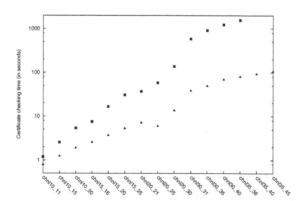

Fig. 1. Evolution of the type-checking time

experiment, the certificates produced by `micromega` are always faster to check. Moreover, `micromega` scales far better than `romega`. It is also worth noting than

---

[3] `Romega` already outperforms the `omega` tactics.
[4] http://www.cril.univ-artois.fr/PB06

romega fails to complete the last two benchmarks. For the last benchmark, the origin of the failure is not fully elucidated. For the penultimate one, a stack-overflow exception is thrown while type-checking the certificate.

Figure 2 plots the size of compiled proof-terms (.vo files) generated by romega ■ and micromega ▲ together with the textual size of the problems •. For small instances, the two tactics generate proof-terms of similar size. For big

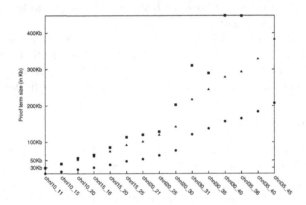

Fig. 2. Evolution of the size of proof-terms

instances, proof-terms of generated by micromega are smaller than those produced by romega. Moreover, their size is more closely correlated to the problem size.

On the negative side, both tactics are using a huge amount of memory. (For the biggest problems, the memory skyrockets up to 2.5 GB.) Further investigations is needed to fully understand this behaviour.

5   Related Work and Conclusion

In this paper, we have identified logical fragments for which certificate generators are off-the-shelf decision procedures (or algorithms) and reflexive certificate checkers are proved correct inside the proof-assistant. Using the same approach, Grégoire *et al.,* [14] check Pocklington certificates to get efficient reflexive Coq proofs that a number is prime. In both cases, the checkers benefit from the performance of the novel Coq virtual machine [12].

For Isabelle/HOL, recent works attest the efficiency of reflexive approaches. Chaieb and Nipkow have proved correct Cooper's decision procedure for Presburger arithmetics [7]. To obtain fast reflexive proofs, the HOL program is compiled into ML code and run inside the HOL kernel. Most related to ours is the work by Obua [19] which is using a reflexive checker to verify certificates generated by the Simplex. Our work extends this approach by considering more general certificates, namely *positivstellensatz* certificates.

To prove non-linear goals, Harrison mentions (chapter 9.2 of the HOL Light tutorial [15]) his use of semidefinite programming. The difference with our approach is that the HOL Light checker needs not to be proved correct but is a Caml program of type `cert → term → thm`. Dynamically, the HOL Light kernel ensures that theorems can only be constructed using sound logical inferences.

As decision procedures get more and more sophisticated and fine-tuned, the need for trustworthy checkers has surged. For instance, the state-of-the-art zChaff SAT solver is now generating proof traces [29]. When proof traces exist, experiments show that they can be efficiently rerun inside a proof-assistant. Using Isabelle/HOL, Weber [28] reruns zChaff traces to solve problems that Isabelle/HOL decision procedure could not cope with. Fontaine *et al.*, [11] are using a similar approach to solve quantifier-free formulae with uninterpreted symbols by rerunning proof traces generated by the Harvey SMT prover [10].

In Proof Carrying Code [18], a piece of code is downloaded packed with a checkable certificate – a proof accessing that it is not malicious. Certificate generation is done ahead-of-time while certificate checking is done at download time. Previous work has shown how to bootstrap a PCC infrastructure using a general-purpose proof-assistant like Coq [6,3,4]. In this context, the triples (certificate,checkers,prover) defined here could be used to efficiently check arithmetic verification conditions arising from the analysis of programs.

Acknowledgements. Thanks are due to the anonymous referees for putting this work into perspective and pointing out relevant related work.

References

1. Bellman, R.: On a routing problem. In Quarterly of Applied Mathematics 16, 87–90 (1958)
2. Bertot, Y., Casteran, P.: Interactive Theorem Proving and Program Development. Coq'Art: the Calculus of Inductive Constructions. Springer, Heidelberg (2004)
3. Besson, F., Jensen, T., Pichardie, D.: A PCC Architecture based on Certified Abstract Interpretation. In: Proc. of 1st Int. Workshop on Emerging Applications of Abstract Interpretation, ENTCS, Springer, Heidelberg (2006)
4. Besson, F., Jensen, T., Pichardie, D.: Proof-Carrying Code from Certified Abstract Interpretation and Fixpoint Compression. Theoretical Computer Science 364, 273–291 (2006)
5. Borchers, B.: Csdp, 2.3 user's guide. Optimization Methods and Software 11(2), 597–611 (1999)
6. Cachera, D., Jensen, T., Pichardie, D., Rusu, V.: Extracting a data flow analyser in constructive logic. Theor. Comput. Sci. 342(1), 56–78 (2005)
7. Chaieb, A., Nipkow, T.: Verifying and reflecting quantifier elimination for presburger arithmetic. In: Sutcliffe, G., Voronkov, A. (eds.) LPAR 2005. LNCS (LNAI), vol. 3835, pp. 367–380. Springer, Heidelberg (2005)
8. Crégut, P.: Une procédure de décision réflexive pour un fragment de l'arithmétique de presburger. In Journées Francophones des Langages Applicatifs (2004)
9. The Coq development team. The coq proof assistant - reference manual v 8.1

10. Déharbe, D., Ranise, S.: Light-weight theorem proving for debugging and verifying units of code. In: 1st IEEE Int. Conf. on Software Engineering and Formal Methods, IEEE Computer Society, Los Alamitos (2003)
11. Fontaine, P., Marion, J-Y., Merz, S., Nieto, L., Tiu, A.: Expressiveness + automation + soundness: Towards combining SMT solvers and interactive proof assistants. In: Hermanns, H., Palsberg, J. (eds.) TACAS 2006 and ETAPS 2006. LNCS, vol. 3920, pp. 167–181. Springer, Heidelberg (2006)
12. Grégoire, B., Leroy, X.: A compiled implementation of strong reduction. In: Proc. of the 7th Int. Conf. on Functional Programming, pp. 235–246. ACM Press, New York (2002)
13. Grégoire, B., Mahboubi, A.: Proving equalities in a commutative ring done right in coq. In: Hurd, J., Melham, T. (eds.) TPHOLs 2005. LNCS, vol. 3603, pp. 98–113. Springer, Heidelberg (2005)
14. Grégoire, B., Théry, L., Werner, B.: A computational approach to pocklington certificates in type theory. In: Hagiya, M., Wadler, P. (eds.) FLOPS 2006. LNCS, vol. 3945, pp. 97–113. Springer, Heidelberg (2006)
15. Harrison, J.: HOL light tutorial (for version 2.20)
16. Harrison, J., Théry, L.: A skeptic's approach to combining HOL and Maple. Journal of Automated Reasoning 21, 279–294 (1998)
17. Karmarkar, N.: A new polynomial-time algorithm for linear programming. In: Proc. of the 16th ACM Symp. on Theory of Computing, pp. 302–311. ACM Press, New York (1984)
18. Necula, G.: Proof-carrying code. In: Proc. of the 24th ACM Symp. on Principles of Programming Languages, pp. 106–119. ACM Press, New York (1997)
19. Obua, S.: Proving bounds for real linear programs in isabelle/hol. In: Hurd, J., Melham, T. (eds.) TPHOLs 2005. LNCS, vol. 3603, pp. 227–244. Springer, Heidelberg (2005)
20. Parrilo, P.A.: Semidefinite programming relaxations for semialgebraic problems. Math. Program. 96(2), 293–320 (2003)
21. Pratt, V.: Two easy theories whose combination is hard. Technical report, Massachusetts Institute of Technology (1977)
22. Pugh, W.: The omega test: a fast and practical integer programming algorithm for dependence analysis. In: Proc. of the 1991 ACM/IEEE conference on Supercomputing, pp. 4–13. ACM Press, New York (1991)
23. Schrijver, A.: Theory of Linear and Integer Programming. Wiley, Chichester (1998)
24. Shostak, R.: Deciding linear inequalities by computing loop residues. J. ACM 28(4), 769–779 (1981)
25. Stengle, G.: A nullstellensatz and a positivstellensatz in semialgebraic geometry. Mathematische Annalen 207(2), 87–97 (1973)
26. Tarski, A.: A Decision Method for Elementary Algebra and Geometry, 2nd edn. University of California Press (1951)
27. Vandenberghe, L., Boyd, S.: Semidefinite programming. SIAM Rev. 38(1), 49–95 (1996)
28. Weber, T.: Using a SAT solver as a fast decision procedure for propositional logic in an LCF-style theorem prover. In: Proc. of 18th Int. Conf. on the Theorem Proving in Higher Order Logics, pp. 180–189 (August 2005)
29. Zhang, L., Malik, S.: Validating sat solvers using an independent resolution-based checker: Practical implementations and other applications. In: Design, Automation and Test in Europe, pp. 10880–10885. IEEE Computer Society, Los Alamitos (2003)

Combining de Bruijn Indices and Higher-Order Abstract Syntax in Coq

Venanzio Capretta[*] and Amy P. Felty

School of Information Technology and Engineering
and Department of Mathematics and Statistics
University of Ottawa, Canada
venanzio@cs.ru.nl, afelty@site.uottawa.ca

Abstract. The use of *higher-order abstract syntax* is an important approach for the representation of binding constructs in encodings of languages and logics in a logical framework. Formal meta-reasoning about such object languages is a particular challenge. We present a mechanism for such reasoning, formalized in Coq, inspired by the Hybrid tool in Isabelle. At the base level, we define a de Bruijn representation of terms with basic operations and a reasoning framework. At a higher level, we can represent languages and reason about them using higher-order syntax. We take advantage of Coq's constructive logic by formulating many definitions as Coq programs. We illustrate the method on two examples: the untyped lambda calculus and quantified propositional logic. For each language, we can define recursion and induction principles that work directly on the higher-order syntax.

1 Introduction

There are well-known challenges in reasoning within a logical framework about languages encoded using higher-order syntax to represent binding constructs. To illustrate, consider a simple example of an object language – the untyped λ-calculus – encoded in a typed meta-language. We encode λ-terms, in higher-order syntax, by a type term with constructors: abs of type (<u>term</u> → term) → term and app of type term → term → term. We represent binding by negative occurrences of the defined type. (Here, the single negative occurrence is underlined.) The Coq system [3,4] implements the Calculus of Inductive Constructions (CIC) [5,27]: Like many other systems, it does not allow negative occurrences in constructors of inductive types.

Our approach realizes higher-order syntax encodings of terms with an underlying de Bruijn representation [6]. De Bruijn syntax has two advantages: α-convertibility is just equality and there is no variable capture in substitution. A main advantage of higher-order syntax is that it allows substitution by function application at the meta-level. We define higher-order syntax encodings on top of the base level so that they expand to de Bruijn terms.

---

[*] Now at the Computer Science Institute (iCIS), Radboud University Nijmegen, The Netherlands.

T. Altenkirch and C. McBride (Eds.): TYPES 2006, LNCS 4502, pp. 63–77, 2007.

We provide libraries of operations and lemmas to reason on the higher-order syntax, hiding the details of the de Bruijn representation. This approach is inspired by the Hybrid system [1], implemented in Isabelle [18]. The general structure is the same, but our basic definitions and operators to build a higher level on top of de Bruijn terms are quite different.

Coq's constructive logic allows us to define operators as functions, rather than relations as in Hybrid. This simplifies some of the reasoning and provides more flexibility in specifying object languages. We obtain new induction principles to reason directly on the higher-order syntax, as well as non-dependent recursion principles to define programs on the higher-order syntax.

Our framework includes two parts. The first part is a general library of definitions and lemmas used by any object language. It includes the definition of the de Bruijn representation and of several recursive functions on de Bruijn terms, e.g., substitution. The second part is a methodology to instantiate the library to a particular object language. It includes definitions and lemmas that follow a general pattern and can easily be adapted from one object language to the next. An important result is the validation of induction and recursion principles on the higher-order syntax of the language.

We illustrate our framework on two examples, the untyped λ-calculus (LC) and quantified propositional logic (QPL); the same languages were implemented in Hybrid [1]. For example, we give a Coq function computing the negation normal form of formulas in QPL. This definition uses the recursion principle for the higher-order syntax of QPL. The proof that it does indeed produce normal forms is quite simple, making direct use of our induction principle for QPL. In Hybrid, negation normal forms and many other definitions are given as relations instead. In Coq, they are functions and our higher-order recursion principles allow us to provide simpler, more direct proofs.

Section 2 presents the general part of our framework starting with the de Bruijn syntax and Sect. 3 illustrates how we instantiate this general framework to LC. Section 4 uses this instantiation to prove properties of this object language. In Sect. 5, we apply our framework to QPL. In Sect. 6, we discuss related work, and in Sect. 7, we conclude and discuss future work.

We formalized all the results in this paper in the proof assistant Coq. The files of the formalization are available at: `http://www.site.uottawa.ca/~afelty/coq/types06_coq.html`

2   De Bruijn Syntax

We describe the lowest level formalization of the syntax. We define a generic type of expressions built on a parameter type of constants con. Expressions are built from variables and constants through application and abstraction. There are two kinds of variables, free and bound. Two types, var and bnd, both instantiated as natural numbers, are used for indexing the two kinds.

Bound variables are treated as de Bruijn indices: this notation eliminates the need to specify the name of the abstracted variable. Thus, the abstraction operator is a simple unary constructor. Expressions are an inductive type:

> **Inductive** expr : Set :=
> CON : con → expr
> VAR : var → expr
> BND : bnd → expr
> APP : expr → expr → expr
> ABS : expr → expr

This is the same definition used in Hybrid [1]. The idea is that the variable (BND i) is bound by the i-th occurrence of ABS above it in the syntax tree. In the following example, we underline all the variable occurrences bound by the first ABS:

$$\underline{\text{ABS}}\,(\text{APP}\,(\text{ABS}\,(\text{APP}\,(\underline{\text{BND}\,1})\,(\text{BND}\,0)))\,(\underline{\text{BND}\,0})).$$

Written in the usual λ-calculus notation, this expression would be $\lambda x.(\lambda y.x\,y)\,x$.

There may occur BND variables with indices higher than the total number of ABSs above them. These are called *dangling* variables. They should not occur in correct terms, but we need to handle them in higher-order binding.

Substitution can be defined for both kinds of variables, but we need it only for the BND variables (substitution of VAR variables can be obtained by first swapping the variable with a fresh BND variable, as shown below, and then substituting the latter).

If j is a de Bruijn index, we define the term $e[j/x]$ (written as `(bsubst e j x)` in the Coq code), obtained by substituting the dangling BND j variable with x. This operation is slightly complicated by the fact that the identity of a BND variable depends on how many ABSs are above it: Every time we go under an ABS, we need to increment the index of the substituted variable and of all the dangling variables in x. In ordinary λ-calculus notation we have $((\lambda y.x\,y)\,x)[x/y] = (\lambda y'.y\,y')\,y$.[1] In de Bruijn syntax, we have, using BND 0 for x and BND 1 for y:

$$(\text{APP}\,(\text{ABS}\,(\text{APP}\,\underline{(\text{BND}\,1)}\,(\text{BND}\,0)))\,\underline{(\text{BND}\,0)})[0/\text{BND}\,1]$$
$$= (\text{APP}\,(\text{ABS}\,(\text{APP}\,\underline{(\text{BND}\,2)}\,(\text{BND}\,0)))\,\underline{(\text{BND}\,1)})$$

The two underlined occurrences of BND 1 and BND 0 on the left-hand side represent the same dangling variable, the one with index 0 at the top level. Similarly, the two occurrences of BND 2 and BND 1 on the right-hand side represent the same dangling variable, the one with index 1 at the top level.

---

[1] Here, y has been renamed y' to avoid capture. Using de Bruijn notation, there is never any variable capture, since renaming is implicit in the syntax.

Call $\hat{x}$ (written as (bshift x) in the Coq code) the result of incrementing by one all the dangling variables in x. We define:

$$(\mathsf{CON}\,c)[j/x] = \mathsf{CON}\,c$$
$$(\mathsf{VAR}\,v)[j/x] = \mathsf{VAR}\,v$$
$$(\mathsf{BND}\,i)[j/x] = \textbf{if}\ j = i\ \textbf{then}\ x\ \textbf{else}\ \mathsf{BND}\,i$$
$$(\mathsf{APP}\,e_1\,e_2)[j/x] = \mathsf{APP}\,(e_1[j/x])\,(e_2[j/x])$$
$$(\mathsf{ABS}\,e)[j/x] = \mathsf{ABS}\,(e[j+1/\hat{x}])$$

To define binding operators, we need to turn a free variable into a bound one. If e is an expression, j a bound variable, and v a free variable, then we denote by $e[j \leftrightarrow v]$ (written as (ebind v j e) in the Coq code) the result of swapping $\mathsf{BND}\,j$ with $\mathsf{VAR}\,v$ in e, taking into account the change in indexing caused by the occurrences of ABS:

$$(\mathsf{CON}\,c)[j \leftrightarrow v] = \mathsf{CON}\,c$$
$$(\mathsf{VAR}\,w)[j \leftrightarrow v] = \textbf{if}\ w = v\ \textbf{then}\ \mathsf{BND}\,j\ \textbf{else}\ \mathsf{VAR}\,w$$
$$(\mathsf{BND}\,i)[j \leftrightarrow v] = \textbf{if}\ i = j\ \textbf{then}\ \mathsf{VAR}\,v\ \textbf{else}\ \mathsf{BND}\,i$$
$$(\mathsf{APP}\,e_1\,e_2)[j \leftrightarrow v] = \mathsf{APP}\,(e_1[j \leftrightarrow v])\,(e_2[j \leftrightarrow v])$$
$$(\mathsf{ABS}\,e)[j \leftrightarrow v] = \mathsf{ABS}\,(e[j+1 \leftrightarrow v])$$

We can easily prove that the operation is its own inverse: $e[j \leftrightarrow v][j \leftrightarrow v] = e$.

Finally, the operation newvar e gives the index of the first free variable not occurring in e. Because it is a new variable, if we replace any dangling BND variable with it and then swap the two, we obtain the original term:

Lemma 1. *For e : expr, j : bnd, and $n =$ newvar e; $e[j/(\mathsf{VAR}\,n)][j \leftrightarrow n] = e$.*

3   Higher-Order Syntax: The Untyped λ-Calculus

The first object language that we try to encode is the untyped λ-calculus. Its higher-order definition would be:

$$\textbf{Inductive term : Set :=}$$
$$\text{abs} : (\text{term} \rightarrow \text{term}) \rightarrow \text{term}$$
$$\text{app} : \text{term} \rightarrow \text{term} \rightarrow \text{term}$$

This definition is not accepted by Coq, because of the negative occurrence of term in the type of abs. However, we can simulate it on top of the de Bruijn framework. Define lexpr to be the type of expressions obtained by instantiating the type of constants con by the two element type LCcon = {app, abs}. Not all expressions in lexpr represent valid λ-terms: **app** can be applied only to two expressions and **abs** can be applied only to an abstraction. Therefore there are just two correct forms for a λ-term, besides variables:

$$\mathsf{APP}\,(\mathsf{APP}\,(\mathsf{CON}\,\textbf{app})\,e_1)\,e_2 \quad \text{and} \quad \mathsf{APP}\,(\mathsf{CON}\,\textbf{abs})\,(\mathsf{ABS}\,e).[2]$$

---

[2] Do not confuse APP and ABS with **app** and **abs**: the first are constructors for expressions, the second are constants in the syntax of the λ-calculus.

This is easily captured by a boolean function on expressions:

$\text{term}_{\text{check}} : \text{lexpr} \to \mathbb{B}$
  $\text{term}_{\text{check}} (\text{VAR} \, v) = \text{true}$
  $\text{term}_{\text{check}} (\text{BND} \, i) = \text{true}$
  $\text{term}_{\text{check}} (\text{APP} \, (\text{APP} \, (\text{CON app}) \, e_1) \, e_2) = (\text{term}_{\text{check}} \, e_1) \text{ and } (\text{term}_{\text{check}} \, e_2)$
  $\text{term}_{\text{check}} (\text{APP} \, (\text{CON abs}) \, (\text{ABS} \, e)) = \text{term}_{\text{check}} \, e$
  $\text{term}_{\text{check}} \_ = \text{false}$

In constructive systems like Coq, a boolean function is not the same as a predicate. However, it can be turned into a predicate Tcheck by $(\text{Tcheck} \, e) = \text{Is\_true} (\text{term}_{\text{check}} \, e)$, where $\text{Is\_true true} = \text{True}$ and $\text{Is\_true false} = \text{False}$[3].

Well-formed λ-terms are those expressions satisfying Tcheck. They can be defined in type theory by a record type[4]:

$$\textbf{Record} \text{ term} := \text{mk\_term}$$
$$\{\, \text{t\_expr} : \text{lexpr};$$
$$\text{t\_check} : \text{Tcheck} \, \text{t\_expr}\}$$

The advantage of our definition of Tcheck is that its value is always True for well-formed expressions. This implies that the t_check component of a term must always be I, the only proof of True. This ensures that terms are completely defined by their t_expr component:

Lemma 2 (term unicity). $\forall t_1, t_2 : \text{term}, (\text{t\_expr} \, t_1) = (\text{t\_expr} \, t_2) \to t_1 = t_2$.

Now our aim is to define higher-order syntax for term. It is easy to define notation for variables and application:

$\text{Var} \, v = \text{mk\_term} \, (\text{VAR} \, v) \, \text{I}$
$\text{Bind} \, i = \text{mk\_term} \, (\text{BND} \, i) \, \text{I}$
$t_1 \, @ \, t_2 = \text{mk\_term} \, (\text{APP} \, (\text{APP} \, (\text{CON app}) \, (\text{t\_expr} \, t_1)) \, (\text{t\_expr} \, t_2)) \, \bigstar$

(The $\bigstar$ symbol stands for a proof of Tcheck that can easily be constructed from t_check t_1 and t_check t_2. In the rest of this paper, we often omit the details of Tcheck proofs and use this symbol instead.)

The crucial problem is the definition of a higher-order notation for abstraction. Let $f : \text{term} \to \text{term}$; we want to define a term $(\text{Fun} \, x, f \, x)$ representing the λ-term $(\lambda x. f \, x)$. The underlying expression of this term must be an abstraction, i.e., it must be in the form $(\text{APP} \, (\text{CON abs}) \, (\text{ABS} \, e))$. The idea is that e should be the result of replacing the metavariable x in $f \, x$ by the bound variable Bind 0. However, the simple solution of applying f to Bind 0 is incorrect: Different occurrences of the metavariable should be replaced by different de Bruijn indices, according to the number of abstractions above them. The solution is: First apply f to a new free variable $\text{Var} \, n$, and then replace $\text{VAR} \, n$ with

---

[3] The values true and false are in $\mathbb{B}$ which is a member of Set, while True and False are propositions, i.e., members of Prop.

[4] Here, mk_term is the constructor of term; t_expr and t_check are field names.

BND 0 in the underlying expression. Formally: tbind $f = $ t_expr $(f$ (Var $n))[0 \leftrightarrow n]$. The proof tbind_check f of (Tcheck (tbind f)) can be constructed from the proof t_check $(f$ (Var n)). We can then define a term that we call the *body* of the function f: tbody $f = $ mk_term (tbind f) (tbind_check f). Finally, we can define the higher-order notation for λ-abstraction:

$$\text{Fun } x, f\, x = \text{mk\_term}(\text{APP (CON abs) (ABS (tbind } (\lambda x, f\, x)))) \bigstar$$

We must clarify what it means for n to be a new variable for a function f : term $\rightarrow$ term. In Coq, the function space term $\rightarrow$ term includes meta-terms that do not encode terms of the object language LC, often called *exotic terms* (see [7]). Functions that do encode terms are those that work uniformly on all arguments. Since we do not require uniformity, $(f\, x)$ may have a different set of free variables for each argument x. It is in general not possible to find a variable that is new for all the results. We could have, e.g., $f\, x = $ Var (size x) where (size x) is the total number of variables and constants occurring in x. Only if f is uniform, e.g., if $f\, x = $ (Var 1) @ (x @ (Var 0)), we can determine objectively an authentic free variable, in this case $n = 2$. For the general case, we simply define n to be a free variable for $(f$ (Bind 0)). This definition gives an authentic free variable when f is uniform.

To prove some results, we must require that functions are uniform, so we must define this notion formally. Intuitively, uniformity means that all values $(f\, x)$ are defined *by the same expression*. We say that f *is an abstraction* if this happens. We already defined the body of f, (tbody f). We now state that f is an abstraction if, for every term x, $(f\, x)$ is obtained by *applying* the body of f to x. We define the result of applying a function body to an argument by the use of the operation of substitution of bound variables.

$$\text{If } t = \text{mk\_term } e_t\, h_t \text{ and } x = \text{mk\_term } e_x\, h_x,$$
$$\text{then tapp } t\, x = \text{mk\_term } (e_t[0/e_x]) \bigstar$$

We can now link the higher-order abstraction operator Fun to the application of the constant **abs** at the de Bruijn level in an exact sense.

Lemma 3. *For all* e : t_expr *and* h : Tcheck e, *we have*

$$\text{Fun } x, \text{tapp } (\text{mk\_term } e\, h)\, x = \text{mk\_term (APP (CON abs) (ABS } e)) \bigstar$$

Proof. Unfold the definitions of Fun, tbind and tapp and then use Lemma 1.

The body of the application of a term is always the term itself. We define a function to be an abstraction if it is equal to the application of its body.

Lemma 4. $\forall t$: term, $t = $ tbody (tapp t)

Definition 1. *Given a function* f : term $\rightarrow$ term, *we define its* canonical form *as* funt $f = $ tapp (tbody f) : term $\rightarrow$ term. *We say that* f *is an* abstraction, is_abst f, *if* $\forall x, f\, x = $ funt $f\, x$.

Some definitions and results will hold only in the case that the function is an abstraction. For example, if we want to formalize β-reduction, we should add such a hypothesis: $\mathsf{is\_abst}\, f \;\to\; ((\mathsf{Fun}\, x, f\, x)\; @\; t)\; \leadsto_\beta\; f\, t$. This assumption does not appear in informal reasoning, so we would like it to be automatically provable. It would be relatively easy to define a tactic to dispose of such hypotheses mechanically. A different solution is to use always the canonical form in place of the bare function. In the case of β-reduction we would write: $((\mathsf{Fun}\, x, f\, x)\,@\,t)\; \leadsto_\beta\; (\mathsf{funt}\, f)\, t$. Note that if f happens to be an abstraction, then $(\mathsf{funt}\, f)\, t$ and $f\, t$ are convertible at the meta-level, so the two formulations are equivalent and in the second one we are exempted from proving the uniformity of f. In Sect. 4 we follow Hybrid in adopting the first definition, but using the second one would be equally easy.

Coq provides an automatic induction principle on expressions. It would be more convenient to have an induction principle tailored to the higher-order syntax of terms. The first step in this direction is an induction principle on expressions satisfying Tcheck:

Theorem 1 (Induction on well-formed expressions). *Let $P : \mathsf{lexpr} \to \mathsf{Prop}$ be a predicate on expressions such that the following hypotheses hold:*

$$\forall v : \mathsf{var}, P\,(\mathsf{VAR}\, v)$$
$$\forall i : \mathsf{bnd}, P\,(\mathsf{BND}\, i)$$
$$\forall e_1, e_2 : \mathsf{lexpr}, P\, e_1 \to P\, e_2 \to P\,(\mathsf{APP}\,(\mathsf{APP}\,(\mathsf{CON\, app})\, e_1)\, e_2)$$
$$\forall e : \mathsf{lexpr}, P\, e \to P\,(\mathsf{APP}\,(\mathsf{CON\, abs})\,(\mathsf{ABS}\, e))$$

Then $(P\, e)$ is true for every $e : \mathsf{lexpr}$ such that $(\mathsf{Tcheck}\, e)$ holds.

Proof. By induction on the structure of e. The assumptions provide us with derivations of $(P\, e)$ from the inductive hypotheses that P holds for all subterms of e satisfying Tcheck, but only if e is in one of the four allowed forms. If e is in a different form, the result is obtained by *reductio ad absurdum* from the assumption $(\mathsf{Tcheck}\, e) = \mathsf{False}$. To apply the induction hypotheses to subterms, we need a proof that Tcheck holds for them. This is easily derivable from $(\mathsf{Tcheck}\, e)$.

This induction principle was used to prove several results about terms. A fully higher-order induction principle can be derived from it.

Theorem 2 (Induction on terms). *Let $P : \mathsf{term} \to \mathsf{Prop}$ be a predicate on terms such that the following hypotheses hold:*

$$\forall v : \mathsf{var}, P\,(\mathsf{Var}\, v)$$
$$\forall i : \mathsf{bnd}, P\,(\mathsf{Bind}\, i)$$
$$\forall t_1, t_2 : \mathsf{term}, P\, t_1 \to P\, t_2 \to P\,(t_1\, @\, t_2)$$
$$\forall f : \mathsf{term} \to \mathsf{term}, P\,(\mathsf{tbody}\,(\lambda x, f\, x)) \to P\,(\mathsf{Fun}\, x, f\, x)$$

Then $(P\, t)$ is true for every $t : \mathsf{term}$[5].

---

[5] The induction hypothesis is formulated for $(\mathsf{tbody}\,(\lambda x, f\, x))$ rather than for $(\mathsf{tbody}\, f)$ because, in Coq, extensionally equal functions are not always provably equal, so we may need to use their η-expansion.

Proof. Since t must be in the form $t = (\mathsf{mk\_term}\, e\, h)$ where e : lexpr and h : $(\mathsf{Tcheck}\, e)$, we apply Theorem 1. This requires solving two problems.

First, P is a predicate on terms, while Theorem 1 requires a predicate on expressions. We define it by: $\bar{P}\, e = \forall h : (\mathsf{Tcheck}\, e), P\,(\mathsf{mk\_term}\, e\, h)$.

Second, we have to prove the four assumptions about $\bar{P}$ of Theorem 1. The first three are easily derived from the corresponding assumptions in the statement of this theorem. The fourth requires a little more work. We need to prove that $\forall e : \mathsf{lexpr}, \bar{P}\, e \to \bar{P}\,(\mathsf{APP}\,(\mathsf{CON}\,\mathbf{abs})\,(\mathsf{ABS}\, e))$. Let then e be an expression and assume $(\bar{P}\, e) = \forall h : (\mathsf{Tcheck}\, e), P\,(\mathsf{mk\_term}\, e\, h)$ holds. The conclusion of the statement is unfolded to:

$$\forall h : (\mathsf{Tcheck}\, e), P\,(\mathsf{mk\_term}\,(\mathsf{APP}\,(\mathsf{CON}\,\mathbf{abs})\,(\mathsf{ABS}\, e))\, h).$$

By Lemma 3, the above expression has a higher-order equivalent:

$$(\mathsf{mk\_term}\,(\mathsf{APP}\,(\mathsf{CON}\,\mathbf{abs})\,(\mathsf{ABS}\, e))\, h) = \mathsf{Fun}\, x, \mathsf{tapp}\,(\mathsf{mk\_term}\, e\, \bigstar)\, x.$$

By the fourth assumption in the statement, P holds for this term if it holds for $(\mathsf{tbody}\,(\lambda x, \mathsf{tapp}\,(\mathsf{mk\_term}\, e\, \bigstar)\, x))$, which is $(\mathsf{tbody}\,(\mathsf{tapp}\,(\mathsf{mk\_term}\, e\, \bigstar)))$ by extensionality of tbody. This follows from $(P\,(\mathsf{mk\_term}\, e\, \bigstar))$ by Lemma 4, and this last proposition holds by assumption, so we are done.

In Sect. 5 we give a similar induction principle for Quantified Propositional Logic and explain its use on an example.

As stated earlier, although inspired by Hybrid [1], our basic definitions and operators to build the higher-order level on top of de Bruijn terms are quite different. For example, our higher-order notation for λ-abstraction $(\mathsf{Fun}\, x, f\, x)$ is defined only on well-formed terms. In Hybrid, the corresponding lambda is defined on all expressions. A predicate is defined identifying all non-well-formed terms, which are mapped to a default expression. Hybrid also defines an induction principle similar to our Theorem 1. Here we go a step further and show that a fully higher-order induction principle can be derived from it (Theorem 2). In Sect. 5, we also provide a non-dependent recursion principle (Theorem 4).

4   Lazy Evaluation of Untyped λ-Terms

Using the higher-order syntax from the previous section, we give a direct definition of lazy evaluation of closed λ-terms as an inductive relation:

> **Inductive** $\Downarrow$: term $\to$ term $\to$ Prop :=
> $\forall f$: term $\to$ term, $(\mathsf{Fun}\, x, f\, x) \Downarrow (\mathsf{Fun}\, x, f\, x)$
> $\forall e_1, e_2, v$: term, $\forall f$: term $\to$ term,
> $\mathsf{is\_abst}\, f \to e_1 \Downarrow (\mathsf{Fun}\, x, f\, x) \to (f\, e_2) \Downarrow v \to (e_1\, @\, e_2) \Downarrow v$

Notice that in the second rule, expressing β-reduction, substitution is obtained simply by higher-order application: $(f\, e_2)$.

By a straightforward induction on this definition, we can prove that evaluation is a functional relation. The uniformity hypothesis (is_abst f) in the application case is important (see Definition 1); without it, the property does not hold. In particular, the proof uses the following lemma, which holds only for functions which satisfy the is_abst predicate:

Lemma 5 (Fun injectivity). *Let f_1, f_2 : term $\rightarrow$ term such that* is_abst f_1 *and* is_abst f_2; *if* (Fun $x, f_1\,x$) = (Fun $x, f_2\,x$), *then* $\forall x$: term, $(f_1\,x) = (f_2\,x)$.

Our main theorem follows by direct structural induction on the proof of the evaluation relation.

Theorem 3 (Unique values). *Let e, v_1, v_2 : term; if both $e \Downarrow v_1$ and $e \Downarrow v_2$ hold, then $v_1 = v_2$.*

5 Quantified Propositional Logic

Our second example of an object language, Quantified Propositional Logic (QPL), is also inspired by the Hybrid system [1]. The informal higher-order syntax representation of QPL would be the following:

> **Inductive** formula : Set :=
> Not : formula $\rightarrow$ formula
> Imp : formula $\rightarrow$ formula $\rightarrow$ formula
> And : formula $\rightarrow$ formula $\rightarrow$ formula
> Or : formula $\rightarrow$ formula $\rightarrow$ formula
> All : (formula $\rightarrow$ formula) $\rightarrow$ formula
> Ex : (formula $\rightarrow$ formula) $\rightarrow$ formula

As for the case of LC, this definition is not acceptable in type theory because of the negative occurrences of formula in the types of the two quantifiers.

As in Sect. 3, we instantiate the type of de Bruijn expressions with the type of constants QPLcon = {**not, imp, and, or, all, ex**}. We call oo the type of expressions built on these constants.

Similarly to the definition of term$_{\text{check}}$ and Tcheck, we define a boolean function formula$_{\text{check}}$ and a predicate Fcheck to restrict the well-formed expressions to those in one of the following forms:

VAR v	APP (APP (CON **and**) e_1) e_2
BND i	APP (APP (CON **or**) e_1) e_2
APP (CON **not**) e	APP (CON **all**) (ABS e)
APP (APP (CON **imp**) e_1) e_2	APP (CON **ex**) (ABS e)

Then the type of formulas can be defined by:

> **Record** formula := mk_formula
> { f_expr : oo;
>   f_check : Fcheck f_expr}

The higher-order syntax of QPL is defined similarly to that of LC:

$$\mathsf{Var}\, v = \mathsf{mk\_formula}\,(\mathsf{VAR}\, v)\, |$$
$$\mathsf{Bind}\, i = \mathsf{mk\_formula}\,(\mathsf{BND}\, i)\, |$$
$$\mathsf{Not}\, a = \mathsf{mk\_formula}\,(\mathsf{APP}\,\mathbf{not}\,(\mathsf{f\_expr}\, a))\, \bigstar$$
$$a_1\,\mathsf{Imp}\, a_2 = \mathsf{mk\_formula}\,(\mathsf{APP}\,(\mathsf{APP}\,\mathbf{imp}\,(\mathsf{f\_expr}\, a_1))\,(\mathsf{f\_expr}\, a_2))\, \bigstar$$
$$a_1\,\mathsf{And}\, a_2 = \mathsf{mk\_formula}\,(\mathsf{APP}\,(\mathsf{APP}\,\mathbf{and}\,(\mathsf{f\_expr}\, a_1))\,(\mathsf{f\_expr}\, a_2))\, \bigstar$$
$$a_1\,\mathsf{Or}\, a_2 = \mathsf{mk\_formula}\,(\mathsf{APP}\,(\mathsf{APP}\,\mathbf{or}\,(\mathsf{f\_expr}\, a_1))\,(\mathsf{f\_expr}\, a_2))\, \bigstar$$
$$\mathsf{All}\, x, f\, x = \mathsf{mk\_formula}\,(\mathsf{APP}\,(\mathsf{CON}\,\mathbf{all})\,(\mathsf{ABS}\,(\mathsf{fbind}\,(\lambda x, f\, x))))\, \bigstar$$
$$\mathsf{Ex}\, x, f\, x = \mathsf{mk\_formula}\,(\mathsf{APP}\,(\mathsf{CON}\,\mathbf{ex})\,(\mathsf{ABS}\,(\mathsf{fbind}\,(\lambda x, f\, x))))\, \bigstar$$

where fbind is defined exactly as tbind in Sect. 3. As in that case, fbind f satisfies Fcheck, giving us the formula fbody f, the body of the function f. As before, the *canonical form* funf f is the application of the body of f (following Definition 1).

For this example, we carry the encoding of higher-order syntax further by defining a non-dependent recursion principle.

Theorem 4. *For any type B, we can define a function of type* formula $\to B$ *by recursively specifying its results on the higher-order form of formulas:*

$$\mathsf{Hvar} : \mathsf{var} \to B$$
$$\mathsf{Hbind} : \mathsf{bnd} \to B$$
$$\mathsf{Hnot} : \mathsf{formula} \to B \to B$$
$$\mathsf{Himp} : \mathsf{formula} \to \mathsf{formula} \to B \to B \to B$$
$$\mathsf{Hand} : \mathsf{formula} \to \mathsf{formula} \to B \to B \to B$$
$$\mathsf{Hor} : \mathsf{formula} \to \mathsf{formula} \to B \to B \to B$$
$$\mathsf{Hall} : (\mathsf{formula} \to \mathsf{formula}) \to B \to B$$
$$\mathsf{Hex} : (\mathsf{formula} \to \mathsf{formula}) \to B \to B$$

If $\mathbf{f} = \mathsf{form\_rec}_{\mathsf{Hvar},\mathsf{Hbind},\mathsf{Hnot},\mathsf{Himp},\mathsf{Hand},\mathsf{Hor},\mathsf{Hall},\mathsf{Hex}} : \mathsf{formula} \to B$ *is the function so defined, then the following reduction equations hold:*

$$\mathbf{f}\,(\mathsf{Var}\, v) = \mathsf{Hvar}\, v \qquad \mathbf{f}\,(a_1\,\mathsf{Imp}\, a_2) = \mathsf{Himp}\, a_1\, a_2\,(\mathbf{f}\, a_1)\,(\mathbf{f}\, a_2)$$
$$\mathbf{f}\,(\mathsf{Bind}\, i) = \mathsf{Hbind}\, i \qquad \mathbf{f}\,(a_1\,\mathsf{And}\, a_2) = \mathsf{Hand}\, a_1\, a_2\,(\mathbf{f}\, a_1)\,(\mathbf{f}\, a_2)$$
$$\mathbf{f}\,(\mathsf{Not}\, a) = \mathsf{Hnot}\, a\,(\mathbf{f}\, a) \qquad \mathbf{f}\,(a_1\,\mathsf{Or}\, a_2) = \mathsf{Hor}\, a_1\, a_2\,(\mathbf{f}\, a_1)\,(\mathbf{f}\, a_2)$$
$$\mathbf{f}\,(\mathsf{All}\, x, f\, x) = \mathsf{Hall}\,(\mathsf{funf}\,(\lambda x. f\, x))\,(\mathbf{f}\,(\mathsf{fbody}\,(\lambda x. f\, x)))$$
$$\mathbf{f}\,(\mathsf{Ex}\, x, f\, x) = \mathsf{Hex}\,(\mathsf{funf}\,(\lambda x. f\, x))\,(\mathbf{f}\,(\mathsf{fbody}\,(\lambda x. f\, x)))$$

Proof. Let a : formula; it must be of the form (mk_formula $e\, h$). The definition is by recursion on the structure of e. The assumptions give the inductive steps when e is in one of the allowed forms. If e is in a different form, we can give an arbitrary output, since h : (Fcheck e) = False is absurd.

For the allowed expressions, we transform e into its equivalent higher-order form in the same manner as was done for λ-terms in the proof of Theorem 2. The reduction equations follow from unfolding definitions.

As illustration, we use this recursion principle to define the negation normal form of a formula. Intuitively, (nnf a) recursively moves negations inside connectives and quantifiers, eliminating double negations. For example, here are four of the several possible cases:

$$\mathsf{nnf}\,(\mathsf{Not}\,(\mathsf{Not}\,a)) = a$$
$$\mathsf{nnf}\,(\mathsf{Not}\,(a_1\,\mathsf{Imp}\,a_2)) = (\mathsf{nnf}\,a_1)\,\mathsf{And}\,(\mathsf{nnf}\,(\mathsf{Not}\,a_2))$$
$$\mathsf{nnf}\,(\mathsf{All}\,x, f\,x) = \mathsf{All}\,x, \mathsf{fapp}\,(\mathsf{nnf}\,(\mathsf{fbody}\,(\lambda x, f\,x)))\,x$$
$$\mathsf{nnf}\,(\mathsf{Not}\,(\mathsf{All}\,x, f\,x)) = \mathsf{Ex}\,x, \mathsf{fapp}\,(\mathsf{nnf}\,(\mathsf{Not}\,(\mathsf{fbody}\,(\lambda x, f\,x))))\,x$$

Notice, in particular, the way nnf is defined on quantifiers. It would be incorrect to try to define it as $\mathsf{nnf}\,(\mathsf{All}\,x, f\,x) = \mathsf{All}\,x, (\mathsf{nnf}\,(f\,x))$, because this would imply that in order to define nnf on $(\mathsf{All}\,x, x)$, it must already be recursively defined on every x, in particular on $x = (\mathsf{All}\,x, f\,x)$ itself. This definition is circular and therefore incorrect. Instead, we recursively apply nnf only to the body of f and then quantify over the application of the result.

To define nnf formally, we need to give its result simultaneously on a formula a and its negation $(\mathsf{Not}\,a)$. Therefore, we define an auxiliary function $\mathsf{nnf\_aux}$: formula $\to$ formula $\times$ formula in such a way that $\mathsf{nnf\_aux}\,a = \langle \mathsf{nnf}\,a, \mathsf{nnf}\,(\mathsf{Not}\,a)\rangle$. We apply form_rec with $B :=$ formula $\times$ formula:

$$\mathsf{Hvar}\,v = \langle \mathsf{Var}\,v, \mathsf{Not}\,(\mathsf{Var}\,v)\rangle$$
$$\mathsf{Hbind}\,i = \langle \mathsf{Bind}\,i, \mathsf{Not}\,(\mathsf{Bind}\,i)\rangle$$
$$\mathsf{Hnot}\,a\,\langle u, v\rangle = \langle v, u\rangle$$
$$\mathsf{Himp}\,a_1\,a_2\,\langle u_1, v_1\rangle\,\langle u_2, v_2\rangle = \langle v_1\,\mathsf{Or}\,u_2, u_1\,\mathsf{And}\,v_2\rangle$$
$$\mathsf{Hand}\,a_1\,a_2\,\langle u_1, v_1\rangle\,\langle u_2, v_2\rangle = \langle u_1\,\mathsf{And}\,u_2, v_1\,\mathsf{Or}\,v_2\rangle$$
$$\mathsf{Hor}\,a_1\,a_2\,\langle u_1, v_1\rangle\,\langle u_2, v_2\rangle = \langle u_1\,\mathsf{Or}\,u_2, v_1\,\mathsf{And}\,v_2\rangle$$
$$\mathsf{Hall}\,f\,\langle u, v\rangle = \langle (\mathsf{All}\,x, \mathsf{fapp}\,u\,x), (\mathsf{Ex}\,x, \mathsf{fapp}\,v\,x)\rangle$$
$$\mathsf{Hex}\,f\,\langle u, v\rangle = \langle (\mathsf{Ex}\,x, \mathsf{fapp}\,u\,x), (\mathsf{All}\,x, \mathsf{fapp}\,v\,x)\rangle$$

and then define $\mathsf{nnf}\,a = \pi_1\,(\mathsf{nnf\_aux}\,a)$. The arguments in the form $\langle u, v\rangle$ represent the result of the recursive calls on the formula and its negation. For example, in the definition of Hnot, u represents $(\mathsf{nnf}\,a)$ and v represents $(\mathsf{nnf}\,(\mathsf{Not}\,a))$. In the definition of Hall, u represents $(\mathsf{nnf}\,(\mathsf{fbody}\,(\lambda x, f\,x)))$ and v represents $\mathsf{nnf}\,(\mathsf{Not}\,(\mathsf{fbody}\,(\lambda x, f\,x)))$.

We also prove an induction principle on formulas, similar to Theorem 2.

Theorem 5 (Induction on formulas). *Let P : formula $\to$ Prop be a predicate on formulas such that the following hypotheses hold:*

$$\forall v : \mathsf{var}, P\,(\mathsf{Var}\,v)$$
$$\forall i : \mathsf{bnd}, P\,(\mathsf{Bind}\,i)$$
$$\forall a : \mathsf{formula}, P\,a \to P\,(\mathsf{Not}\,a)$$
$$\forall a_1, a_2 : \mathsf{formula}, P\,a_1 \to P\,a_2 \to P\,(a_1\,\mathsf{Imp}\,a_2)$$
$$\forall a_1, a_2 : \mathsf{formula}, P\,a_1 \to P\,a_2 \to P\,(a_1\,\mathsf{And}\,a_2)$$
$$\forall a_1, a_2 : \mathsf{formula}, P\,a_1 \to P\,a_2 \to P\,(a_1\,\mathsf{Or}\,a_2)$$
$$\forall f : \mathsf{formula} \to \mathsf{formula}, P\,(\mathsf{fbody}\,(\lambda x, f\,x)) \to P\,(\mathsf{All}\,x, f\,x)$$
$$\forall f : \mathsf{formula} \to \mathsf{formula}, P\,(\mathsf{fbody}\,(\lambda x, f\,x)) \to P\,(\mathsf{Ex}\,x, f\,x)$$

Then $(P\,a)$ is true for every a : formula.

As an application of this principle, we defined an inductive predicate is_Nnf stating that a formula is in negation normal form, i.e., negation can occur only on variables, and proved that the result of nnf is always in normal form.

6   Related Work

There is extensive literature on approaches to representing object languages with higher-order syntax and reasoning about them within the same framework. Pollack's notes on the problem of reasoning about binding [22] give a high-level summary of many of them. Some of them were used to solve the POPLmark challenge problem set [2]. We mention a few here.

Several approaches have used Coq. These include the use of *weak higher-order abstract syntax* [7,14]. In weak higher-order syntax, the problem of negative occurrences in syntax encodings is handled by replacing them by a new type. For example, the abs constructor for the untyped λ-terms introduced in Sect. 1 has type (var → term) → term, where var is a type of variables. Some additional operations are needed to encode and reason about this new type, which at times is inconvenient. Miculan's approach [14,15] introduces a "theory of contexts" to handle this representation of variables, with extensive use of axioms whose soundness must be justified independently.

McDowell and Miller [12] introduce a new logic specifically designed for reasoning with higher-order syntax. Their logic is intuitionistic and higher-order with support for natural number induction and definitions. In general, higher-order syntax mainly addresses encodings of term-level abstraction. More recent work by Miller and Tiu [16] includes a new quantifier for this style of logic, which provides an elegant way to handle abstractions at the level of proofs. Another approach uses multi-level encodings [8,17]. This approach also aims to capture more than term-level abstraction, and is inspired by the work of McDowell and Miller but uses Coq and Isabelle, respectively.

Gabbay and Pitts [9] define a variant of classical set theory that includes primitives for variable renaming and variable freshness, and a new "freshness quantifier." Using this set theory, it is possible to prove properties by structural induction and also to define functions by recursion over syntax.

The Twelf system [21], which implements the Logical Framework (LF) has also been used as a framework for reasoning using higher-order syntax. In particular Schürmann [23] has developed a logic which extends LF with support for meta-reasoning about object logics expressed in LF. The design of the component for reasoning by induction does not include induction principles for higher-order encodings. Instead, it is based on a realizability interpretation of proof terms. The Twelf implementation of this approach includes powerful automated support for inductive proofs.

Schürmann, Despeyroux, and Pfenning [24] develop a modal metatheory that allows the formalization of higher-order abstract syntax with a primitive recursion principle. They introduce a modal operator □. Intuitively, for every type A there is a type $□A$ of *closed* objects of type A. In addition to the regular function type

$A \to B$, there is a more restricted type $A \Rightarrow B \equiv \Box A \to B$ of uniform functions. Functions used as arguments for higher-order constructors are of this kind. This allows them to define a recursion principle that avoids the usual circularity problems. The system has not yet been extended to a framework with dependent types.

Schürmann et. al. have also worked on designing a new calculus for defining recursive functions directly on higher-order syntax [25]. Built-in primitives are provided for the reduction equations for the higher-order case, in contrast to our approach where we define the recursion principle on top of the base level de Bruijn encoding, and prove the reduction equations as lemmas.

Nogin et. al. [19] build a theory in MetaPRL that includes both a higher-order syntax and a de Bruijn representation of terms, with a translation between the two. Induction principles are defined at the de Bruijn level. Their basic library is more extensive than ours; it provides syntactic infrastructure for reflective reasoning and variable-length bindings. Instantiating the basic theory and proving properties about specific object logics is left as future work.

Solutions to the POPLmark challenge also include first-order approaches which adopt de Bruijn representations, such as the one by Stump [26] that uses named bound variables and indices for free variables, and solves part 1a of POPLmark. Another earlier first-order approach by Melham avoids de Bruijn syntax altogether and encodes abstractions using names paired with expressions [13]. Working at this level requires dealing with low-level details about α-conversion, free and bound variables, substitution, etc. Gordon and Melham [11] generalize this name-carrying syntax approach and develop a general theory of untyped λ-terms up to α-conversion, including induction and recursion principles. They illustrate that their theory can be used as a meta-language for representing object languages in such a way that the user is free from concerns of α-conversion. Norrish [20] improves the recursion principles, allowing greater flexibility in defining recursive functions on this syntax. Gordon [10] was able to take a step further in improving the name-carrying syntax approach by defining this kind of syntax in terms of an underlying de Bruijn notation. Gordon's work was the starting point for Hybrid [1], which kept the underlying de Bruijn notation, but used a higher-order representation at the higher-level.

7   Conclusion

We have presented a new method for reasoning in Coq about object languages represented using higher-order syntax. We have shown how to structure proof development so that reasoning about object languages takes place at the level of higher-order syntax, even though the underlying syntax uses de Bruijn notation. An important advantage of our approach is the ability to define induction and recursion principles directly on the higher-order syntax representation of terms. Our examples illustrate the use of this framework for encoding object languages and their properties in a manner that allows direct and simple reasoning.

Future work includes considering a wider variety of object languages and completing more extensive proofs. It also includes adapting the preliminary ideas from Hybrid [1] to show that our representation of the λ-calculus is an adequate one. It is unlikely that these results will carry over directly to our constructive setting, so further work will be required. We also plan to generalize the methodology to define object languages with binding and prove their higher-order recursion and induction principles: the user should be able to define a new language and reason about it using only higher-order syntax, without having to look at all at the lower de Bruijn level. In a forthcoming article, we develop a higher-order universal algebra in which the user can define a language by giving typing rules for the constants. Higher-order terms and associated induction and recursion principles are then automatically derived.

Acknowledgments

The authors would like to thank Alberto Momigliano for useful comments on an earlier draft of this paper. Thanks also go to the Natural Sciences and Engineering Research Council of Canada for providing support for this work.

References

1. Ambler, S.J., Crole, R.L., Momigliano, A.: Combining higher order abstract syntax with tactical theorem proving and (co)induction. In: Carreño, V.A., Muñoz, C.A., Tahar, S. (eds.) TPHOLs 2002. LNCS, vol. 2410, pp. 13–30. Springer, Heidelberg (2002)
2. Aydemir, B.E., Bohannon, A., Fairbairn, M., Foster, J.N., Pierce, B.C., Sewell, P., Vytiniotis, D., Washburn, G., Weirich, S., Zdancewic, S.: Mechanized metatheory for the masses: The POPLmark challenge. In: RSCTC 2000. LNCS, pp. 50–65. Springer, Heidelberg (2005), http://fling-l.seas.upenn.edu/~plclub/cgi-bin/poplmark/index.php?title=The_POPLmark_Challenge
3. Bertot, Y., Castéran, P.: Interactive Theorem Proving and Program Development. Coq'Art: The Calculus of Inductive Constructions. Springer, Heidelberg (2004)
4. Coq Development Team, LogiCal Project. The Coq Proof Assistant reference manual: Version 8.0. Technical report, INRIA (2006)
5. Coquand, T., Huet, G.: The Calculus of Constructions. Information and Computation 76, 95–120 (1988)
6. de Bruijn, N.G.: Lambda-calculus notation with nameless dummies: a tool for automatic formula manipulation with application to the Church-Rosser theorem. Indag. Math. 34, 381–392 (1972)
7. Despeyroux, J., Felty, A., Hirschowitz, A.: Higher-order abstract syntax in Coq. In: Dezani-Ciancaglini, M., Plotkin, G. (eds.) TLCA 1995. LNCS, vol. 902, pp. 124–138. Springer, Heidelberg (1995)
8. Felty, A.P.: Two-level meta-reasoning in Coq. In: Carreño, V.A., Muñoz, C.A., Tahar, S. (eds.) TPHOLs 2002. LNCS, vol. 2410, pp. 198–213. Springer, Heidelberg (2002)
9. Gabbay, M.J., Pitts, A.M.: A new approach to abstract syntax with variable binding. Formal Aspects of Computing 13, 341–363 (2001)
10. Gordon, A.D.: A mechanisation of name-carrying syntax up to alpha-conversion. In: Sixth International Workshop on Higher-Order Logic Theorem Proving and its Applications. LNCS, pp. 413–425. Springer, Heidelberg (1993)

11. Gordon, A.D., Melham, T.: Five axioms of alpha-conversion. In: Lanzi, P.L., Stolz-mann, W., Wilson, S.W. (eds.) IWLCS 2000. LNCS (LNAI), vol. 1996, pp. 173–190. Springer, Heidelberg (2001)

12. McDowell, R., Miller, D.: Reasoning with higher-order abstract syntax in a logical framework. ACM Transactions on Computational Logic 3(1), 80–136 (2002)

13. Melham, T.F.: A mechanized theory of the Π-calculus in HOL. Nordic Journal of Computing 1(1), 50–76 (1994)

14. Miculan, M.: Developing (meta)theory of λ-calculus in the theory of contexts. Electronic Notes in Theoretical Computer Science 58(1), 37–58 (2001) (MERLIN 2001: Mechanized Reasoning about Languages with Variable Binding)

15. Miculan, M.: On the formalization of the modal μ-calculus in the calculus of inductive constructions. Information and Computation 164(1), 199–231 (2001)

16. Miller, D., Tiu, A.: A proof theory for generic judgments. ACM Transactions on Computational Logic 6(4), 749–783 (2005)

17. Momigliano, A., Ambler, S.J.: Multi-level meta-reasoning with higher-order abstract syntax. In: Sixth International Conference on Foundations of Software Science and Computational Structures, April 2003. LNCS, pp. 375–391. Springer, Heidelberg (2003)

18. Nipkow, T., Paulson, L.C., Wenzel, M.: Isabelle/HOL — A Proof Assistant for Higher-Order Logic. LNCS. Springer, Heidelberg (2002)

19. Nogin, A., Kopylov, A., Yu, X., Hickey, J.: A computational approach to reflective meta-reasoning about languages with bindings. In: MERLIN '05: Proceedings of the 3rd ACM SIGPLAN Workshop on MEchanized Reasoning about Languages with varIable biNding, pp. 2–12. ACM Press, New York (2005)

20. Norrish, M.: Recursive function definition for types with binders. In: Slind, K., Bunker, A., Gopalakrishnan, G.C. (eds.) TPHOLs 2004. LNCS, vol. 3223, pp. 241–256. Springer, Heidelberg (2004)

21. Pfenning, F., Schürmann, C.: System description: Twelf — a meta-logical framework for deductive systems. In: Ganzinger, H. (ed.) Automated Deduction - CADE-16. LNCS (LNAI), vol. 1632, pp. 202–206. Springer, Heidelberg (1999)

22. Pollack, R.: Reasoning about languages with binding. Presentation (2006), available at http://homepages.inf.ed.ac.uk/rap/export/bindingChallenge_slides.pdf

23. Schürmann, C.: Automating the Meta Theory of Deductive Systems. PhD thesis, Carnegie Mellon University (2000)

24. Schürmann, C., Despeyroux, J., Pfenning, F.: Primitive recursion for higher-order abstract syntax. Theoretical Computer Science 266, 1–57 (2001)

25. Schürmann, C., Poswolsky, A., Sarnat, J.: The $\triangledown$-calculus. Functional programming with higher-order encodings. In: Urzyczyn, P. (ed.) TLCA 2005. LNCS, vol. 3461, pp. 339–353. Springer, Heidelberg (2005)

26. Stump, A.: POPLmark 1a with named bound variables. Presented at the Progress on Poplmark Workshop (January 2006)

27. Werner, B.: Méta-théorie du Calcul des Constructions Inductives. PhD thesis, Université Paris 7 (1994)

Deciding Equality in the Constructor Theory[*]

Pierre Corbineau

Institute for Computing and Information Science
Radboud University Nijmegen, Postbus 9010
6500GL Nijmegen, The Netherlands
pierre.corbineau@cs.ru.nl

Abstract. We give a decision procedure for the satisfiability of finite sets of ground equations and disequations in the *constructor theory*: the terms used may contain both uninterpreted and constructor function symbols. Constructor function symbols are by definition injective and terms built with distinct constructors are themselves distinct. This corresponds to properties of (co-)inductive type constructors in inductive type theory. We do this in a framework where function symbols can be partially applied and equations between functions are allowed. We describe our algorithm as an extension of congruence-closure and give correctness, completeness and termination arguments. We then proceed to discuss its limits and extension possibilities by describing its implementation in the Coq proof assistant.

Among problems in equational reasoning, a crucial one is the word problem: does a set of equations entail another one? In 1947, Post and Markov [15,7] showed that this is undecidable. What is decidable is whether an equation between *closed* terms is the consequence of a *finite* conjunction of equations between *closed* terms. This problem is called congruence-closure and its decidability was established by Nelson and Oppen [11], and Downey, Sethi, and Tarjan [6]. The algorithms that are described have a quasi-linear complexity. Nelson and Oppen, and also Shostak, described methods to extend the scope of this closed case with some specific equational theories: lists, tuples ...

In general, semi-decision procedures over equations use syntactic methods (replacement, rewriting ...). Hence the algorithm used for congruence-closure is an exception since its principle is to directly build a set of term equivalence classes. The link between congruence-closure and syntactic methods was established by the concept of abstract congruence closure [2,1].

Because of its simplicity and efficiency, this algorithm makes a good candidate for an implementation of a Coq tactic. Moreover, we wish to extend the decision procedure in order to cover specific properties of constructors for inductive and co-inductive types in Coq:

- Injectivity : $C\,x_1 \ldots x_n = C\,y_1 \ldots y_n \Rightarrow x_1 = y_1 \wedge \cdots \wedge x_n = y_n$
- Discrimination : $C\,x_1 \ldots x_n \neq D\,y_1 \ldots y_p$

---

[*] This work was done at project ProVal, Pole Commun de Recherche en Informatique du plateau de Saclay: CNRS, École Polytechnique, INRIA, Universit Paris-Sud.

The Coq proof assistant already contained, on the one hand, the `injection` and `discriminate` tactics, that deal with immediate consequences of an equation between constructor-based terms, and on the other hand the `autorewrite` tactic that allows us to normalize a term with respect to a set of oriented equations. Those tactics cannot cooperate with each other, which we could do by designing a tactic for congruence-closure with the constructor theory.

In 2001, I [4] proved in Coq the correctness of the congruence-closure algorithm and designed a tagging system for the data structure allowing us to extract proofs of the equations. That method was implemented in the Coq system in 2002 as a tactic named `congruence`. The proof production method used there is similar to the one discovered and studied in detail by Nieuwenhuis and Oliveras [12].

In 2003, I implemented an extended version of the `congruence` tactic that included the constructor theory. No theoretical ground was given to support that extension, so this is the purpose of this article. Our aim is thus to describe a decision procedure that can decide the combination of the theories of constructors and uninterpreted symbols, by extending the congruence-closure algorithm. This is done by proving that the decision problem can be reduced to a finite set of terms. Moreover, the higher-order logic of the Coq system advocates for solving the problem for simply typed terms, allowing equations in functional types.

In this article, we will only manipulate closed terms (without variables), that we will abusively call *terms* for clarity purposes. We will also talk about *term algebra* instead of *closed term algebra*.

1   The Congruence Closure Problem

The language for the theory of equality uses a distinguished binary relation symbol $\approx$, which is supposed to be well-sorted (i.e. $\approx$-related terms have the same sort). Hence this polymorphic relation symbol is a notation for a set of monomorphic predicates $(\approx_s)_{s \in \mathcal{S}}$ where $\mathcal{S}$ is the set of sorts.

In order to define the congruence-closure problem, we will first explain what a congruence is.

Definition 1 (Congruence). *A well-sorted relation $\approx$ over a first-order term algebra is called a congruence if, and only if, the following conditions hold :*

1. $\approx$ is an equivalence relation, i.e. satisfies these three rules :

$$\frac{}{t \approx t}\ \text{REFL} \qquad \frac{s \approx t}{t \approx s}\ \text{SYM} \qquad \frac{s \approx u \qquad u \approx t}{s \approx t}\ \text{TRANS}$$

2. Any n-ary function symbol f in the signature is a $\approx$-morphism, i.e. it satisfies the following congruence rule :

$$\frac{s_1 \approx t_1 \qquad \cdots \qquad s_n \approx t_n}{f(s_1, \ldots, s_n) \approx f(t_1, \ldots, t_n)}\ \text{CONGR}_f$$

An interesting property is that congruence relations are stable under arbitrary intersection, thus for any well-sorted binary relation R over a term algebra, there is a unique minimal congruence $\approx_R$ such that $R \subseteq \approx_R$: $\approx_R$ is the intersection of all congruences coarser than R (there is at least one, which relates all terms of the same sort).

The congruence-closure problem can be stated in two equivalent ways:

- Is the finite set of propositions $\{s_1 \approx t_1, \ldots, s_n \approx t_n, u \not\approx v\}$ satisfiable in the theory of equality ?
- Is there a congruence that satisfies both $s_1 \approx t_1, \ldots, s_n \approx t_n$ and $u \not\approx v$?

The problem can be generalized to any number of inequalities $u \not\approx v$, the issue being to determine if one among them contradicts the equations $s \approx t$.

The equivalence between the two questions relies on two arguments: first, if the set of propositions is satisfiable, by an interpretation I, then the $\approx$ relation defined by $s \approx t \Leftrightarrow I(s) = I(t)$ is a congruence such that $s_i \approx t_i$ for any i and $u \not\approx v$. Second, if we have a congruence $\approx$ such that $s_i \approx t_i$ for any i and $u \not\approx v$, then we can build an interpretation mapping any term to its $\approx$ equivalence class. The interpretation is well-defined since $\approx$ satisfies the congruence rule, and it satisfies the equations and the inequality.

The key fact in the proof of congruence-closure decidability is : if T is the set of terms and subterms appearing in the problem instance, then any proof derivation of the equation $u \approx v$ from the axioms using the rules REFL, SYM, TRANS and CONGR, can be turned into a proof where only terms in T are used. From the dual point of view, any interpretation satisfying locally the set of axioms can be extended to the set of all terms.

Nelson and Oppen [10,11] and Downey, Sethi, Tarjan [6], start from this property to build algorithms representing equivalence classes by a forest of trees (UNION-FIND structure), in order to obtain an optimal complexity.

1.1 Simply Typed Term Algebra

Here, we wish to deal with simply typed terms, so we will use an encoding corresponding to curryfied terms. This representation allows to represent partial function application. This is done by restricting our first-order signatures to a small subset of simply typed signatures.

Definition 2 (Simply typed term). *A simply typed term algebra is defined by giving a signature $(\mathcal{S}, \Sigma)$ such that:*

- $\mathcal{S} = \bigcup_{n \in \mathbb{N}} \mathcal{S}_n$, *where* $\mathcal{S}_{n+1} = S_n \cup \{s{\to}s' | (s, s') \in \mathcal{S}_n \times \mathcal{S}_n\}$ *and* $\mathcal{S}_0$ *is a set of base sorts* $\alpha_1, \ldots, \alpha_n$. *The sorts in* $\mathcal{S} \setminus \mathcal{S}_0$ *are called* functional sorts.
- Σ *contains only constants and a set of binary function symbols* $@_{s,s'}$ *called* application symbols *which take one argument* $s{\to}s'$, *the second in s and yields a result in s'.*

In practice, only a finite number of the sorts and application symbols will be used since only a finite number of sorts will occur in a given problem. As an abuse of notation, we will often omit the indices of application symbols, and we will shorten $@(@(f,x),y)$ as $(f\,x\,y)$. Any first-order signature can be turned into a simply typed signature by applying a curryfication mapping as follows:

Definition 3 (Curryfied signature). *Let $(\mathcal{S}, \Sigma)$ be a first-order signature. The corresponding curryfied signature is $(\hat{\mathcal{S}}, \hat{\Sigma})$, where $\hat{\mathcal{S}}$ is the set of simple sorts generated by using $\mathcal{S}$ as the set of base sorts, and $\hat{\Sigma}$ contains, for each function symbol $f : (s_1, \ldots, s_n) \to s \in \Sigma$, a constant symbol $\hat{f} : s_1 \to (\ldots \to (s_n \to s))$, along with the n application symbols required to apply $\hat{f} : @_{s_1, s_2 \to \ldots \to s}, \ldots, @_{s_n, s}$.*

From now on, we suppose the symbol $\to$ is right-associative and will omit unnecessary brackets. Every term t with sort s in the first-order term algebra Σ can be mapped to a curryfied term $\hat{t}$ with the same sort in the term algebra generated by $\hat{\Sigma}$, by following the next example :

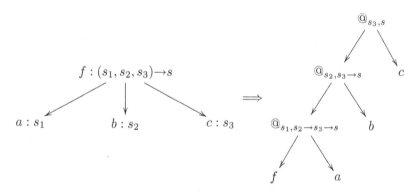

The reverse statement, however, is only true for terms with a base sort — note that many simply typed signatures are not curryfied signature, e.g. if they have symbols with sort $(\mathbb{N} \to \mathbb{N}) \to \mathbb{N}$ — and this gives us a bijection between first-order terms and curryfied terms.

Example 1. The first-order signature for Peano arithmetic has $\mathcal{S}_0 = \{\mathbb{N}\}$ as sorts and the following symbols:
$$\{\mathbf{0} : () \to \mathbb{N}\,;$$
$$S : (\mathbb{N}) \to \mathbb{N}\,;$$
$$+ : (\mathbb{N}, \mathbb{N}) \to \mathbb{N}\,;$$
$$\times : (\mathbb{N}, \mathbb{N}) \to \mathbb{N}\}$$

The sorts used in the corresponding curryfied signature are
$$\mathcal{S} = \{\mathbb{N}\,;\mathbb{N} \to \mathbb{N}\,;\mathbb{N} \to \mathbb{N} \to \mathbb{N}\,\}$$

and the set of curryfied symbols is :

$$\{ \quad \hat{\mathbf{0}} \quad : N;$$
$$\hat{S} \quad : N \to N;$$
$$\hat{+} \quad : N \to N \to N;$$
$$\hat{\times} \quad : N \to N \to N;$$
$$@_{N,N} \quad : (N \to N, N) \to N;$$
$$@_{N,N \to N} : (N \to N \to N, N) \to N \to N\}$$

Please note that we can express functional equations in the curryfied signature, such as: $@(\hat{+}, @(\hat{S}, \hat{\mathbf{0}})) \approx \hat{S}$.

We might think that allowing to express more intermediate properties (higher-order equations) in the curryfied framework would allow us to prove more properties in the theory of equality, but this is not the case:

Lemma 1 (Conservativity of curryfication). *The curryfication is conservative with respect to congruence-closure, i.e. any derivation of $\hat{u} \approx \hat{v}$ from $\hat{s}_1 \approx \hat{t}_1, \ldots, \hat{s}_n \approx \hat{t}_n$ in the curryfied system can be turned into a derivation in the first-order system. The converse also holds.*

Proof. By explicit tree transformation, only termination is tricky. Please note that by construction, curryfied terms are fully applied.

2   Extension to Constructor Theory

We now wish to extend the historical algorithm in order to deal with datatype constructors. Those constructors are injective function symbols, similarly to tuples constructors. They also have a discrimination property: two different constructors in the same sort produce distinct objects. These properties are built into the system Coq's type theory: they hold for constructors of both inductive and co-inductive type.

We do not claim here that those two properties are sufficient to characterize those constructors. For example, inductive type constructors also have acyclicity properties in the style of Peano's axiom about the successor function S, which satisfies $\forall x, x \not\approx S\, x$. These acyclicity properties have been studied in [14].

Other kinds of deductions can be based on the assumption that datatypes are totally generated by their constructors: considering a finite type such as the booleans, with two constant constructors T and F, we can deduce that the set of propositions $\{g\, T \approx a, g\, F \approx a, g\, c \not\approx a\}$ in unsatisfiable. This kind of property is not included in our notion of constructor theory, which would be closer to Objective Caml [13] sum types when taking in account the possibility of raising exceptions.

Definition 4 (Signature with constructors). *A signature with constructors is a simply typed signature with some distinguished constants called constructors. A constructor with shape $C_i^\alpha : \tau_1 \to (\ldots \to (\tau_n \to \alpha))$ where alpha is a base sort is called a n-ary constructor of the sort α.*

By convention, we will now use lowercase letters for regular function symbols and uppercase letters for constructor symbols. The notation C_i^α will refer to some constructor of the base sort α, with the convention that identical constructors will always have identical indices. We now have to explain a congruence satisfying the constructor theory.

Definition 5 (Constructor theory). *A congruence $\approx$ satisfies the constructor theory over the signature Σ if, and only if, the following conditions hold:*

— *$\approx$ satisfies all instances of the following injectivity rule :*

$$\frac{C_i^\alpha s_1 \ldots s_n \approx C_i^\alpha t_1 \ldots t_n}{s_i \approx t_i} \; \text{INJ}_i \; \text{if} \begin{cases} C_i^\alpha s_1 \ldots s_n : \alpha \\ C_i^\alpha t_1 \ldots t_n : \alpha \end{cases}$$

— *For any pair of terms $C s_1 \ldots s_n$ and $D t_1 \ldots t_p$ in sort α, such that C and D are distinct constructors, $C s_1 \ldots s_n \not\approx D t_1 \ldots t_p$.*

Since we allow equations between terms in any functional sort, we need to give a semantic notion of constructor-generated term relatively to a congruence $\approx$. We will call such terms *inductive term*.

Definition 6 (Inductive term). *A term t is inductive with respect to $\approx$ if, and only if, $t \approx C$ for some constructor C or $t \approx (fx)$ for some term x and inductive term f.*

We can immediately remark that if s is inductive w.r.t. $\approx$ and $s \approx t$, then t is also inductive. This allows us to use the notion of *inductive equivalence class*, which contains (only) inductive terms.

The problem we now wish to solve is the satisfiability of a finite set E of ground equations and inequalities of terms inside a constructor theory, which is equivalent to finding a congruence satisfying both the constructor theory and E.

Similarly to basic congruence-closure, we can see that congruences satisfying the *injectivity* of constructors are closed under arbitrary intersection. We can conclude that there exists a smallest congruence satisfying the equations of a given problem and the injectivity rule. If there is a congruence satisfying this problem in the constructor theory, then this smallest congruence also does since it satisfies less inequalities.

In order to use those facts to design our algorithm, we must check that we are able to reduce the decision problem to a finite set of terms, as is done for the congruence-closure problems. Unfortunately, the set of terms and subterms from the problem is not sufficient to decide our problem, as shown by the following example.

Example 2. Let α, β be two base sorts, f, C two symbols in sort $\beta \rightarrow \alpha$, D in sort α, b in sort β, among which C and D are (distinct) constructors of the sort α. Then $\{C \approx f; f b \approx D\}$ is unsatisfiable, since it allows to prove that $C b \approx D$, but the proof uses the term $C b$ which does not appear in the initial problem.

$$\frac{\dfrac{C \approx f \quad \overline{b \approx b} \; \text{REFL}}{C b \approx f b} \; \text{CONGR@} \quad f b \approx D}{C b \approx D} \; \text{TRANS}$$

In order to cope with this, we define application-closed sets of terms. As we deal with closed terms, a sort can be *empty* in a signature Σ. This is the case if no closed term can be built from symbols in Σ, e.g. if $\Sigma = \{f : \beta \rightarrow \beta\}$ then β is an empty sort. The following definition takes this remark into account.

Definition 7 (Application-closed set). *A set T of simply typed terms with constructors is application-closed with respect to the congruence $\approx$ if for each inductive equivalence class $\mathcal{C}$ in $T/_\approx$ with a functional sort $\tau \rightarrow \tau'$, if the sort τ is non-empty then there exist terms $t \in \mathcal{C}$ and $u \in T$ such that $(tu) \in T$.*

Now, we can define a local version of the constructor theory which uses a secondary relation $\rightsquigarrow$, in order to keep the term set considered as small as possible. The $\rightsquigarrow$ is just an asymmetric and external version of $\approx$ that allows its left-hand side to be a term with a constructor in the head position that is not mentioned in the problem. It is actually this *local theory* that will be decided by the algorithm.

Definition 8 (Local constructor theory). *Let T be a subterm-closed set of terms and $\approx$ a well-sorted relation on T. $\approx$ is said to be a* local congruence *on T if it is an equivalence relation on T satisfying every instance of the* CONGR@ *rule where terms occuring are in T.*

Furthermore, a local congruence $\approx$ satisfies locally the constructor theory if, and only if, there exists a relation $\rightsquigarrow$ satisfying the following rules :

$$\frac{C_i^\alpha \approx f}{C_i^\alpha \rightsquigarrow f} \text{ PROMOTION} \qquad \frac{C_i^\alpha\, t_1 \ldots t_k \rightsquigarrow u \quad u\, t_{k+1} \approx v}{C_i^\alpha\, t_1 \ldots t_k\, t_{k+1} \rightsquigarrow v} \text{ COMPOSE}$$

$$\frac{C_i^\alpha\, s_1 \ldots s_n \rightsquigarrow u \quad C_i^\alpha\, t_1 \ldots t_n \rightsquigarrow u}{s_i \approx t_i} \text{ INJECTION}$$

and such that there is no term $t \in T$ such that $C_i^\alpha\, s_1 \ldots s_n \rightsquigarrow t$ and $C_j^\alpha\, t_1 \ldots t_p \rightsquigarrow t$, where $i \neq j$ and n and p are the respective arities of C_i^α and C_j^α.

The theorem for completeness of the original congruence-closure algorithm states that any local congruence is the restriction of a congruence on all terms. Now we have this similar local notion for constructor theory, that we use to formulate the theorem that restricts our decision problem to a finite set of terms.

Theorem 1 (Finite set restriction). *Let E be a finite set of equations and inequalities and T be a subterm-closed set of terms containing all the terms occuring in E.*

Let $\approx$ be a local congruence on T satisfying E and satisfying locally the theory of constructors, such that T is application-closed for $\approx$.

Then there exists a congruence $\approx^*$ over the whole term algebra, which, on the one hand, satisfies E and the theory of constructors, and on the other hand, is a conservative extension of $\approx$, i.e. for all terms u, v in T, $u \approx^* v \Leftrightarrow u \approx v$.

Proof. In order to prove this, we explicitly construct a binary tree model for all terms and define the relation $\approx^*$ as relating terms having the same interpretation in the model. We build this model so that $\approx$ is the restriction of $\approx^*$ to T.

Let S be the set of binary trees with leaves labelled by the set $T/_{\approx} \cup \{e_\tau\}_{\tau in S}$ where $T/_{\approx}$ is the set of equivalence classes modulo $\approx$ and and e_τ is an arbitrary object in sort τ, which will be the representation for unconstrained terms.

We also restrict S to contain only *well-sorted trees*, where every leaf and node has a additional sort label, and for any node with sort τ, there is a sort τ' such that the left son of the tree has sort $\tau' \to \tau$ and the right son has sort τ'. We also require that e_τ is labelled by the sort τ and that any equivalence class is labelled by the sort of the terms it contains. Let S_τ be the subset of trees whose root has sort τ.

The notation (s, t) stands for the well-sorted tree with s and t as (well-sorted) subtrees. A tree is said to be *inductive* if it is an equivalence class containing an inductive term (for $\approx$), or if it is not just a leaf.

We can then define the $App_{\tau,\tau'}$ function from $S_{\tau \to \tau'} \times S_\tau$ to $S_{\tau'}$ as follows: suppose we have $a \in S_{\tau \to \tau'}$ and $b \in S_\tau$.

– if $a = e_{\tau \to \tau'}$, then $App_{\tau,\tau'}(e_{\tau \to \tau'}, b) = e_{\tau'}$
– if a is not a leaf then $App_{\tau,\tau'}(a, b) = (a, b)$
– if $a \in T/_{\approx}$ then :
  • if $b \in T/_{\approx}$ and there is $t_a \in a$ and $t_b \in b$ such that $(t_a\, t_b) \in T$, then let c be the equivalence class of $(t_a\, t_b)$, we put $App_{\tau,\tau'}(a, b) = c$. This definition is not ambiguous since $\approx$ is a congruence.
  • otherwise
    * if a is inductive, then $App_{\tau,\tau'}(a, b) = (a, b)$
    * else $App_{\tau,\tau'}(a, b) = e_{\tau'}$.

Our interpretation domain is the set S. The interpretation of constants and constructors is their equivalence class in $T/_{\approx}$: $I(f) = \bar{f}$, and we set $I(@_{\tau,\tau'}) = App_{\tau,\tau'}$. Then $\approx^*$ is defined by $s \approx^* t \Leftrightarrow I(s) = I(t)$.

Now first, $\approx^*$ is a congruence relation since it is by construction an equivalence relation, it is well-sorted since the S_τ sets are disjoint, and since $I(@(f, x)) = App(I(f), I(x))$, $\approx$ satisfies the congruence rule.

Again by construction, terms in T are interpreted into their equivalence class for $\approx$, so $\approx^*$ is conservative with respect to $\approx$, and satisfies E.

Then we show that $\approx^*$ satisfies the constructor theory, which is a bit technical, and too long to be detailed here, see [5], Theorem 2.13, pages 42–45 for the full proof. This concludes the proof.   □

Now that we have reduced our problem to a more manageable one, we will give a description of the algorithm itself.

3   The Decision Algorithm

3.1   Description

We now give an algorithm to decide problems with ground equations and inequalities and a constructor theory. This algorithm is very similar to the one in [6], except for the part concerning constructors. This algorithm is given by a set of transition rules (read from top to bottom) between quintuples (T, E, M, U, S) made of :

T: Set of **T**erms not up to date in the signature table and for marking

E: Set of unprocessed **E**quations ($\approx$-constraints)

M: Set of unprocessed **M**arks ($\rightsquigarrow$-constraints)

U: **U**NION-FIND structure (map from terms to equivalence classes)

S: **S**ignature table

An equivalence class is a quadruple (r, L, R, C) made of:

r: the **r**epresentative term for the class

L: Set of **L**eft fathers of the class

R: Set of **R**ight fathers of the class

C: Set of **C**onstructor marks of the equivalence class, i.e. set of terms $\rightsquigarrow$-related to terms in the class.

We write $U(u) = (\bar{u}, L(u), R(u), C(u))$ to simplify notation. A *left* (resp. *right*) *father* of a class c is a term whose left (resp. right) subterm is in the class c.

The UNION-FIND structure is best described as a forest of trees in which all nodes contain a term and point to their father node, and the roots contain an additional class quadruple. The class searching operation is done by following the path from a node to the root. The union(U, s, t) operation merges two equivalence classes by making $\bar{s}$ an immediate child of $\bar{t}$, thus making $\bar{t}$ the representative term for all terms in the former class of s, the sets of left and right fathers are then merged, but $\bar{t}$ keeps its original set C.

The *signature table* acts as a cache. It maps a pair of representative terms $(\bar{u}, \bar{v})$ to a term whose left (resp. right) immediate subterms are in the class of u (resp. v), if such a term is registered, and otherwise it yields $\bot$. The suppr(S, R) operation removes bindings to terms in R from the table.

We start with a set $\mathcal{A}$ of equations and disequalities. The set $\mathcal{A}{\downarrow}$ is the set of terms and subterms occuring in $\mathcal{A}$. The set $\mathcal{A}^+$ (resp. $\mathcal{A}^-$) is the set of pairs of terms appearing in an equation (resp. inequality) in $\mathcal{A}$.

The transition rules for the algorithm are given in Figure 1. The INIT rule builds the initial structure, where all terms are apart, all equations and all constructor marks are pending in E and M.

Three issues are possible for the algorithm: the REFUTE rule detects if an inequality in $\mathcal{A}$ has been proved false, it answers that the problem is unsatisfiable. The CONFLICT rule concludes the computation if two terms made of distinct fully applied constructors are proven equal. If there is no constraint left to propagate and none of the above rules apply, then the problem is declared satisfiable using the OK rule.

The MERGE and NO-MERGE rules are used to deal with pending equality constraints, the NO-MERGE rule avoids unnecessary operations. The MERGE rule has to expunge all outdated information from the tables: remove fathers of merged terms from the signature table and put them in T to re-process them later.

initialization rule

$$\frac{Sat(\mathcal{A})?}{T_0, E, M, U, \emptyset} \text{ INIT} \begin{cases} T_0 = \{@(u,v) \in \mathcal{A}\downarrow\} \\ E = \mathcal{A}^+ \\ M = \{(C_i^\alpha, C_i^\alpha) | C_i^\alpha \in \mathcal{A}\downarrow\} \\ U = u \mapsto (u, \{@(u,v) \in \mathcal{A}\downarrow\}, \{@(v,u) \in \mathcal{A}\downarrow\}, \emptyset) \end{cases}$$

conclusion rules

$$\frac{\emptyset, \emptyset, \emptyset, U, S}{SAT} \text{ OK} \{\forall(u,v) \in \mathcal{A}^-, \bar{u} \neq \bar{v}$$

$$\frac{T, E, M, U, S}{NOSAT} \text{ REFUTE} \{\exists(u,v) \in \mathcal{A}^-, \bar{u} = \bar{v}$$

$$\frac{T, E, M \cup \{(t, C_i^\alpha\, \boldsymbol{u})\}, U, S}{NOSAT} \text{ CONFLICT} \begin{cases} C_i^\alpha \boldsymbol{u} : \alpha \text{ (fully applied)} \\ C_j^\alpha \boldsymbol{v} \in C(t) \\ i \neq j \end{cases}$$

deduction rules

$$\frac{T, E \cup \{(s,t)\}, M, U, S}{\substack{T \cup L(s) \cup R(s), E, M \cup \{(t,c)|c \in C(s)\}, \\ \text{union}(U, s, t), \text{suppr}(S, L(s) \cup R(s))}} \text{ MERGE} \{\bar{s} \neq \bar{t}$$

$$\frac{T, E \cup \{(s,t)\}, M, U, S}{T, E, M, U, S} \text{ NO-MERGE} \{\bar{s} = \bar{t}$$

$$\frac{T, E, M \cup \{(t, C_i^\alpha\, \boldsymbol{u})\}, U, S}{T, E \cup \{(\boldsymbol{u}, \boldsymbol{v})\}, M, U, S} \text{ MATCH} \begin{cases} C_i^\alpha \boldsymbol{u} : \alpha \text{ (fully applied)} \\ C_i^\alpha \boldsymbol{v} \in C(t) \end{cases}$$

$$\frac{T, E, M \cup \{(t, C_i^\alpha\, \boldsymbol{u})\}, U, S}{T \cup L(t), E, M, U\{C(t) \leftarrow C(t) \cup C_i^\alpha\, \boldsymbol{u}\}, S} \text{ MARK} \begin{cases} C_i^\alpha \boldsymbol{u} : \tau \to \tau' \text{ (partially applied)} \\ \text{or} \\ C(t) = \emptyset \end{cases}$$

$$\frac{T \cup \{t\}, E, M, U, S}{T, E \cup \{(t,s)\}, M \cup \{(t, F\, v)|F \in \mathcal{C}\}, U, S} \text{ UPDATE}_1 \begin{cases} t = @(u,v) \\ S(\bar{u}, \bar{v}) = s \\ C(u) = \mathcal{C} \end{cases}$$

$$\frac{T \cup \{t\}, E, M, U, S}{T, E, M \cup \{(t, F\, v)|F \in \mathcal{C}\}, U, S \cup \{(\bar{u}, \bar{v}) \mapsto t\}} \text{ UPDATE}_2 \begin{cases} t = @(u,v) \\ S(\bar{u}, \bar{v}) = \bot \\ C(u) = \mathcal{C} \end{cases}$$

Fig. 1. Rules of our algorithm for the problem $\mathcal{A}$

The MATCH and MARK rules allow to mark an equivalence class with the inductive terms that belong to it. The MATCH rule propagates the consequences of the injectivity rule to E. The UPDATE$_{1/2}$ rules are used to update the signature

$$\frac{\dfrac{SAT(f\,a \approx a, f(f\,a) \not\approx a)?}{\{f\,a, f(f\,a)\}, \{(f\,a, a)\}, \emptyset, \emptyset, \emptyset}\ \text{INIT}}{\dfrac{\{f\,a, f(f\,a)\}, \emptyset, \emptyset, \{f\,a \mapsto a\}, \emptyset}{\dfrac{\{f(f\,a)\}, \emptyset, \emptyset, \{f\,a \mapsto a\}, \{(f, a) \mapsto f\,a\}}{\dfrac{\emptyset, \{(f(f\,a), f\,a)\}, \emptyset, \{f\,a \mapsto a\}, \{(f, a) \mapsto f\,a\}}{\dfrac{\emptyset, \emptyset, \emptyset, \left\{\begin{array}{l} f\,a\ \mapsto a \\ f(f\,a) \mapsto a \end{array}\right\}, \{(f, a) \mapsto f\,a\}}{NOSAT}\ \text{REFUTE}\quad \overline{f(f\,a)} = a = \bar{a}}}\ \text{MERGE}}\ \text{UPDATE}_1\quad S(\bar{f}, \overline{f\,a}) = f\,a}}{}\ \text{UPDATE}_2}\ \text{MERGE}}$$

Fig. 2. Satisfiability computation for $\{f\,a \approx a, f(f\,a) \not\approx a\}$

table, propagate the consequences of the congruence rule to E, and propagate marks from subterms to their left fathers.

Example 3. This first example is a very simple one used to illustrate how the signature table works. In the U set, the notation $s \mapsto t$ stands for $\bar{s} = t$, and by default $\bar{u} = u$. Similarly, $C(t) = \emptyset$ for all representative terms not mentioned.

Consider the signature $S_0 = \{s\}, \Sigma = \{a : s, f : s \rightarrow s\}$ and the following problem : $\{f\,a \approx a, f(f\,a) \not\approx a\}$. A possible execution of the algorithm is given in Figure 2.

Example 4. In order to illustrate the use of marks for constructors on equivalence classes, we now give an example with both injection and discrimination. Consider the following signature :

$$S_0 = \{\text{bool}, \text{box}\},$$
$$\Sigma = \{g : \text{bool} \rightarrow \text{box}, B : \text{bool} \rightarrow \text{box}, T : \text{bool}, F : \text{bool}\}$$

The satisfiability problem $\{B \approx g, g\,T \approx g\,F\}$ is solved by our procedure in Figure 3.

3.2   Properties

Theorem 2 (Termination). *Any successive application of the rules of the algorithm reaches SAT or NOSAT in a finite number of steps.*

Proof. By checking the rules, we remark that there is a rule that we can apply in every situation and we can check that the tuples decrease for the lexicographic ordering made of:

1. the number of equivalence classes in the UNION-FIND structure.
2. the union $T \cup M$, ordered by the multiset extension of the size of term sorts, considering that identical sorts are bigger in T than in M.
3. the size of E

$$SAT(B \approx g, g\,T \approx g\,F)?$$

$$\frac{}{\{g\,T, g\,F\}, \{(B,g), (g\,T, g\,F)\}, \{(T,T), (F,F), (B,B)\}, \emptyset, \emptyset} \text{ INIT}$$

$$\frac{}{\{g\,T, g\,F\}, \{(g\,T, g\,F)\}, \{(T,T), (F,F), (B,B)\}, \{B \mapsto g\}, \emptyset} \text{ MERGE}$$

$$\frac{}{\{g\,T, g\,F\}, \emptyset, \{(T,T), (F,F), (B,B)\}, \left\{ \begin{array}{l} g\,T \mapsto g\,F \\ B \mapsto g \end{array} \right\}, \emptyset} \text{ MERGE}$$

$$\frac{}{\{g\,T, g\,F\}, \emptyset, \{(F,F), (B,B)\}, \left\{ \begin{array}{l} g\,T \mapsto g\,F \;\; C(T) = \{T\} \\ B \mapsto g \end{array} \right\}, \emptyset} \text{ MARK}$$

$$\frac{}{\{g\,T, g\,F\}, \emptyset, \{(B,B)\}, \left\{ \begin{array}{ll} g\,T \mapsto g\,F & C(T) = \{T\} \\ B \mapsto g & C(F) = \{F\} \end{array} \right\}, \emptyset} \text{ MARK}$$

$$\frac{}{\{g\,T, g\,F\}, \emptyset, \emptyset, \left\{ \begin{array}{ll} g\,T \mapsto g\,F & C(T) = \{T\} \\ B \mapsto g & C(F) = \{F\} \\ & C(g) = \{B\} \end{array} \right\}, \emptyset} \text{ MARK}$$

$$\frac{}{\{g\,F\}, \emptyset, \{(g\,T, B\,T)\}, \left\{ \begin{array}{ll} g\,T \mapsto g\,F & C(T) = \{T\} \\ B \mapsto g & C(F) = \{F\} \\ & C(g) = \{B\} \end{array} \right\}, \emptyset} \text{ UPDATE}_2$$

$$\frac{}{\{g\,F\}, \emptyset, \emptyset, \left\{ \begin{array}{ll} g\,T \mapsto g\,F & C(T) = \{T\} \\ B \mapsto g & C(F) = \{F\} \\ C(g\,F) = \{B\,T\} & C(g) = \{B\} \end{array} \right\}, \emptyset} \text{ MARK}$$

$$\frac{}{\emptyset, \emptyset, \{(g\,F, B\,F)\}, \left\{ \begin{array}{ll} g\,T \mapsto g\,F & C(T) = \{T\} \\ B \mapsto g & C(F) = \{F\} \\ C(g\,F) = \{B\,T\} & C(g) = \{B\} \end{array} \right\}, \emptyset} \text{ UPDATE}_2$$

$$\frac{}{\emptyset, \{(F,T)\}, \emptyset, \left\{ \begin{array}{ll} g\,T \mapsto g\,F & C(T) = \{T\} \\ B \mapsto g & C(F) = \{F\} \\ C(g\,F) = \{B\,T\} & C(g) = \{B\} \end{array} \right\}, \emptyset} \text{ MATCH } B\,F \approx B\,T$$

$$\frac{}{\{g\,F\}, \emptyset, \{(F,F)\}, \left\{ \begin{array}{ll} g\,T \mapsto g\,F & C(T) = \{T\} \\ B \mapsto g & C(g) = \{B\} \\ F \mapsto T & C(g\,F) = \{B\,T\} \end{array} \right\}, \emptyset} \text{ MERGE}$$

$$\frac{}{NOSAT} \text{ CONFLICT } F \approx T$$

Fig. 3. Computation for $\{B \approx g, g\,T \approx g\,F\}$

Theorem 3 (Correctness). *The algorithm is correct, i.e. if it answers NOSAT then the problem is unsatisfiable.*

Proof. The proof is made by showing that the rules preserve key invariants and that the side-conditions for final transitions ensure correctness. The detailed proof can be found in [5], section 2.3.4, p. 51.

Completeness is proven in two steps: first, we prove that if the algorithm reaches SAT then the equivalence relation induced by the UNION-FIND structures, together with the marking relation, gives a congruence that satisfies locally the constructor theory.

Lemma 2 (Local completeness). *If the algorithm reaches SAT, then the relation $\approx$ defined by $u \approx v \Leftrightarrow \bar{u} = \bar{v}$ is a local congruence in $\mathcal{A}{\downarrow}$ which satisfies locally the constructor theory.*

To establish global completeness, we need a hypothesis: the set $\mathcal{A}{\downarrow}$ has to be application-closed for $\approx$. This is not true in general. However the algorithm can be adapted to detect if this occurs, and we can try to add terms to our term set in order to have the terms we want fully applied. The incompleteness detection is done in the following way :

- Initially, every class has *UNKNOWN* status.
- When a class with *UNKNOWN* status is marked with a partially-applied constructor, it goes to *INCOMPLETE* status.
- When an UPDATE rule is applied to a term $(f\,x)$ and the class of f has *INCOMPLETE* status, it goes to *COMPLETE* status.
- When classes are merged, they get the most advanced status of the two merged classes, considering *COMPLETE* most advanced, *INCOMPLETE* less advanced and *UNKNOWN* least advanced.

That way, when the algorithm reaches SAT, the final set is application-closed if, and only if, no class is left with *INCOMPLETE* status. This said, global completeness can be stated as follows:

Theorem 4 (Completeness). *Let $\mathcal{A}$ a non-empty finite set of axioms, if the algorithm reaches SAT from $\mathcal{A}$ and the final set is application-closed for $\approx$, then $\mathcal{A}$ is satisfiable.*

Proof. By using the previous lemma together with Theorem 1.

If all sorts are inhabited, we can add to the final state of the algorithm the partially applied inductive terms, applied to arbitrary terms in the correct sort. Once this is done, a re-run of the algorithm from this state will yield an application-closed set of terms. Thus there is no need to carry on this process more than once. Nevertheless, the first run is necessary to determine how to complete the set of terms.

4   The Congruence Tactic

A previous version of the algorithm has been implemented in the Coq proof assistant (version 8.0). This version is incomplete but the upcoming version 8.1 will allow you to benefit from a corrected version. The **congruence** tactic extracts equations and inequalities from the context and adds the negated conclusion if it is an equality. Otherwise, we add to the problem the inequalities $H \not\approx C$, H which is a hypotheses which is not an equation and C the conclusion. We also add $H \not\approx H'$ for every pair of hypotheses $h : H$ and $h' : \neg H'$ where neither H nor H' are equalities.

Then the algorithm is executed with a fixed strategy. If the result is SAT it fails to prove the goal. Otherwise it is instrumented to produce a constructive proof of the equation that was proven to dissatisfy the problem. The method used is similar to that explained in [4] and [9] and more recently in [12].

Moreover, the congruence tactic detects if the final set of terms is application-closed and if not, it can complete it by adding extra terms with meaningless constants, so that if the completed problem is unsatisfiable, the procedure can ask the user for a replacement to these constants. We exemplify this with the type of triples in the Coq system (type **nat**) :

```
Inductive Cube:Set :=
| Triple: nat -> nat -> nat -> Cube.
```

We now wish to prove :

```
Theorem incomplete : ∀ a b c d : nat,
   Triple a = Triple b → Triple d c = Triple d b → a = c.
```

After having introduced the hypotheses by using the **intros** command, we are able to extract the following set of equations :

$$\{\mathtt{Triple\,a} \approx \mathtt{Triple\,b}, \mathtt{Triple\,d\,c} \approx \mathtt{Triple\,d\,b}, \mathtt{a} \not\approx \mathtt{c}\}$$

The algorithm finds out that the set of equations is unsatisfiable, but the set of terms had to be completed to become application-closed, so it fails to produce the proof and explains it to the user :

```
Goal is solvable by congruence but some arguments are missing.
Try "congruence with (Triple a ?₁ ?₂) (Triple d c ?₃)",
replacing metavariables by arbitrary terms.
```

The user may then give manually the additional terms to solve the goal :

```
congruence with (Triple a 0 0) (Triple d c 0).
```

5   Conclusion

The **congruence** tactic is a fast tactic which solves a precise problem, so it can be used as a goal filter, to eliminate trivial subgoals. This is very useful when doing multiple nested case-analysis of which most cases are failure cases for example when doing the proofs for the reflexion schemes used in [3].

The work lacks both a proper complexity analysis and some benchmarks, but since it is quite specific it is difficult to compare with other tools.

An obvious improvement would be to add λ-abstraction to the term algebra. This would require us to keep track of all possible β-reductions. Then, it would be interesting to see what can be done with dependently-typed terms, since the conclusion of the CONGR@ rule is ill-typed in that case. An exploration of the relation between heterogeneous equality [8] and standard homogeneous equality is needed for closed terms. Whichever extension will be done will have to keep to the original aim: design a specific procedure which can solve a problem or fail to solve it very fast so it can be used widely.

References

1. Bachmair, L., Tiwari, A., Vigneron, L.: Abstract congruence closure. Journal of Automated Reasoning 31(2), 129–168 (2003)
2. Bachmair, L., Tiwari, A.: Abstract congruence closure and specializations. In: McAllester, D. (ed.) CADE-17. LNCS, vol. 1831, pp. 64–78. Springer, Heidelberg (2000)
3. Contejean, E., Corbineau, P.: Reflecting proofs in first-order logic with equality. In: Nieuwenhuis, R. (ed.) CADE-20. LNCS (LNAI), vol. 3632, pp. 7–22. Springer, Heidelberg (2005)
4. Corbineau, P.: Autour de la clôture de congruence avec Coq. Master's thesis, Université Paris 7 (2001), http://www.cs.ru.nl/~corbinea/ftp/publis/mem-dea.ps
5. Corbineau, P.: Démonstration automatique en Théorie des Types. PhD thesis, Université Paris 11 (2005), http://www.cs.ru.nl/~corbinea/ftp/publis/Corbineau-these.pdf
6. Downey, P.J., Sethi, R., Tarjan, R.E.: Variations on the common subexpressions problem. Journal of the ACM 27(4), 771–785 (1980)
7. Markov, A.A.: On the impossibility of certain algorithms in the theory of associative systems. Doklady Akademii Nauk SSSR 55(7), 587–590 (1947) (in Russian, English translation in C.R. Acad. Sci. URSS, 55, 533–586)
8. McBride, C.: Dependently Typed Functional Programs and their Proofs. PhD thesis, University of Edinburgh (1999), http://www.lfcs.informatics.ed.ac.uk/reports/00/-LFCS-00-419/
9. Necula, G.C.: Compiling with Proofs. PhD thesis, Carnegie-Mellon University, available as Technical Report CMU-CS-98-154 (1998)
10. Nelson, G., Oppen, D.C.: Simplification by cooperating decision procedures. ACM Trans. on Programming, Languages and Systems 1(2), 245–257 (1979)
11. Nelson, G., Oppen, D.C.: Fast decision procedures based on congruence closure. Journal of the ACM 27, 356–364 (1980)
12. Nieuwenhuis, R., Oliveras, A.: Proof-Producing Congruence Closure. In: Giesl, J. (ed.) RTA 2005. LNCS, vol. 3467, pp. 453–468. Springer, Heidelberg (2005)
13. The Objective Caml language, http://www.ocaml.org/
14. Oppen, D.C.: Reasoning about recursively defined data structures. Journal of the ACM (1978)
15. Post, E.L.: Recursive unsolvability of a problem of Thue. Journal of Symbolic Logic 13, 1–11 (1947)

A Formalisation of a
Dependently Typed Language as an
Inductive-Recursive Family

Nils Anders Danielsson

Chalmers University of Technology

Abstract. It is demonstrated how a dependently typed lambda calculus (a logical framework) can be formalised inside a language with inductive-recursive families. The formalisation does not use raw terms; the well-typed terms are defined directly. It is hence impossible to create ill-typed terms.

As an example of programming with strong invariants, and to show that the formalisation is usable, normalisation is proved. Moreover, this proof seems to be the first formal account of normalisation by evaluation for a dependently typed language.

1 Introduction

Programs can be verified in many different ways. One difference lies in how invariants are handled. Consider a type checker, for instance. The typing rules of the language being type checked are important invariants of the resulting abstract syntax. In the *external* approach to handling invariants the type checker works with raw terms. Only later, when verifying the soundness of the type checker, is it necessary to verify that the resulting, supposedly well-typed terms satisfy the invariants (typing rules). In the *internal* approach the typing rules are instead represented directly in the abstract syntax data types, and soundness thus follows automatically from the type of the type checking function, possibly at the cost of extra work in the implementation. For complicated invariants the internal approach requires strong forms of data types, such as inductive families or generalised algebraic data types.

Various aspects of many different essentially simply typed programming languages have been formalised using the internal approach [CD97, AR99, Coq02, XCC03, PL04, MM04, AC06, MW06, McBb]. Little work has been done on formalising *dependently* typed languages using this approach, though; Dybjer's work [Dyb96] on formalising so-called categories with families, which can be seen as the basic framework of dependent types, seems to be the only exception. The present work attempts to fill this gap.

This paper describes a formalisation of the type system, and a proof of normalisation, for a dependently typed lambda calculus (basically the logical framework of Martin-Löf's monomorphic type theory [NPS90] with explicit substitutions). Moreover, the proof of normalisation seems to be the first formal implementation

T. Altenkirch and C. McBride (Eds.): TYPES 2006, LNCS 4502, pp. 93–109, 2007.

of normalisation by evaluation [ML75, BS91] for a dependently typed language. The ultimate goal of this work is to have a formalised implementation of the core type checker for a full-scale implementation of type theory.

To summarise, the contributions of this work are as follows:

- A fully typed representation of a dependently typed language (Sect. 3).
- A proof of normalisation (Sect. 5). This proof seems to be the first account of a formal implementation of normalisation by evaluation for a dependently typed language.
- Everything is implemented and type checked in the proof checker AgdaLight [Nor07]. The code can be downloaded from the author's web page [Dan07].

2   Meta Language

Let us begin by introducing the meta language in which the formalisation has been carried out, AgdaLight [Nor07], a prototype of a dependently typed programming language. It is in many respects similar to Haskell, but, naturally, deviates in some ways.

One difference is that AgdaLight lacks polymorphism, but has *hidden arguments*, which in combination with dependent types compensate for this loss. For instance, the ordinary list function *map* could be given the following type signature:

$$map : \{\, a, b : Set \,\} \; \rightarrow \; (a \; \rightarrow \; b) \; \rightarrow \; List \; a \; \rightarrow \; List \; b$$

Here *Set* is the type of types from the first universe. Arguments within $\{\ldots\}$ are hidden, and need not be given explicitly, if the type checker can infer their values from the context in some way. If the hidden arguments cannot be inferred, then they can be given explicitly by enclosing them within $\{\ldots\}$:

$$map \, \{\, Integer \,\} \, \{\, Bool \,\} : (Integer \; \rightarrow \; Bool) \; \rightarrow \; List \; Integer \; \rightarrow \; List \; Bool$$

AgdaLight also has inductive-recursive families [DS06], illustrated by the following example (which is not recursive, just inductive). Data types are introduced by listing the constructors and giving their types; natural numbers can for instance be defined as follows:

data *Nat* : *Set* **where**
  zero : *Nat*
  suc : *Nat* $\rightarrow$ *Nat*

Vectors, lists of a given fixed length, may be more interesting:

data *Vec* (*a* : *Set*) : *Nat* $\rightarrow$ *Set* **where**
  nil : *Vec a zero*
  cons : $\{\, n : Nat \,\}$ $\rightarrow$ *a* $\rightarrow$ *Vec a n* $\rightarrow$ *Vec a* (*suc n*)

Note how the *index* (the natural number introduced after the last : in the definition of *Vec*) is allowed to vary between the constructors. *Vec a* is a *family* of types, with one type for every index n.

To illustrate the kind of pattern matching AgdaLight allows for an inductive family, let us define the tail function:

$$tail : \{\, a : Set \,\} \;\rightarrow\; \{\, n : Nat \,\} \;\rightarrow\; Vec\ a\ (suc\ n) \;\rightarrow\; Vec\ a\ n$$
$$tail\ (cons\ x\ xs) = xs$$

We can and need only pattern match on *cons*, since the type of *nil* does not match the type *Vec a* (*suc n*) given in the type signature for *tail*. As another example, consider the definition of the append function:

$$(+\!\!+) : Vec\ a\ n_1 \;\rightarrow\; Vec\ a\ n_2 \;\rightarrow\; Vec\ a\ (n_1 + n_2)$$
$$nil \qquad +\!\!+\ ys = ys$$
$$cons\ x\ xs +\!\!+\ ys = cons\ x\ (xs +\!\!+ ys)$$

In the *nil* case the variable n_1 in the type signature is unified with *zero*, transforming the result type into *Vec a* n_2, allowing us to give *ys* as the right-hand side. (This assumes that *zero* + n_2 evaluates to n_2.) The *cons* case can be explained in a similar way.

Note that the hidden arguments of $(+\!\!+)$ were not declared in its type signature. This is not allowed by AgdaLight, but often done in the paper to reduce notational noise. Some other details of the formalisation are also ignored, to make the paper easier to follow. The actual code can be downloaded for inspection [Dan07].

Note also that some of the inductive-recursive families in this formalisation do not quite meet the requirements of [DS06]; see Sects. 3.2 and 5.2. Furthermore [DS06] only deals with functions defined using elimination rules. The functions in this paper are defined using pattern matching and structural recursion.

AgdaLight currently lacks (working) facilities for checking that the code is terminating and that all pattern matching definitions are exhaustive. However, for the formalisation presented here this has been verified manually. Unless some mistake has been made all data types are strictly positive (with the exception of *Val*; see Sect. 5.2), all definitions are exhaustive, and every function uses structural recursion of the kind accepted by the termination checker foetus [AA02].

3   Object Language

The object language that is formalised is a simple dependently typed lambda calculus with explicit substitutions. Its type system is sketched in Fig. 1. The labels on the rules correspond to constructors introduced in the formalisation. Note that $\Gamma \Rightarrow \Delta$ is the type of substitutions taking terms with variables in Γ to terms with variables in Δ, and that the symbol $=_\star$ stands for $\beta\eta$-equality between types. Some things are worth noting about the language:

Contexts

$$\frac{}{\varepsilon \ context} \ (\varepsilon) \qquad \frac{\Gamma \ context \qquad \Gamma \vdash \tau \ type}{\Gamma, x : \tau \ context} \ (\triangleright)$$

Types

$$\frac{}{\Gamma \vdash \star \ type} \ (\star) \qquad \frac{\Gamma \vdash \tau_1 \ type \qquad \Gamma, x : \tau_1 \vdash \tau_2 \ type}{\Gamma \vdash \Pi(x : \tau_1) \, \tau_2 \ type} \ (\Pi) \qquad \frac{\Gamma \vdash t : \star}{\Gamma \vdash El \ t \ type} \ (El)$$

Terms

$$\frac{(x, \tau) \in \Gamma}{\Gamma \vdash x : \tau} \ (var) \qquad \frac{\Gamma, x : \tau_1 \vdash t : \tau_2}{\Gamma \vdash \lambda x : \tau_1 . t : \Pi(x : \tau_1) \, \tau_2} \ (\lambda) \qquad \frac{\Gamma \vdash t : \tau \qquad \rho : \Gamma \Rightarrow \Delta}{\Delta \vdash t \, \rho : \tau \, \rho} \ (/\!\!\vdash)$$

$$\frac{\Gamma \vdash t_1 : \Pi(x : \tau_1) \, \tau_2 \qquad \Gamma \vdash t_2 : \tau_1}{\Gamma \vdash t_1 \, t_2 : \tau_2 \, [x \mapsto t_2]} \ (@) \qquad \frac{\Gamma \vdash t : \tau_1 \qquad \tau_1 =_\star \tau_2}{\Gamma \vdash t : \tau_2} \ (::\!\stackrel{\equiv}{\vdash})$$

Substitutions

$$\frac{\Gamma \vdash t : \tau}{[x \mapsto t] \ : \ \Gamma, x : \tau \Rightarrow \Gamma} \ (sub) \qquad \frac{}{wk \ x \ \tau \ : \ \Gamma \Rightarrow \Gamma, x : \tau} \ (wk) \qquad \frac{}{id \ \Gamma \ : \ \Gamma \Rightarrow \Gamma} \ (id)$$

$$\frac{\rho \ : \ \Gamma \Rightarrow \Delta}{\rho \uparrow_x \tau \ : \ \Gamma, x : \tau \Rightarrow \Delta, x : \tau \, \rho} \ (\uparrow) \qquad \frac{\rho_1 \ : \ \Gamma \Rightarrow \Delta \qquad \rho_2 \ : \ \Delta \Rightarrow X}{\rho_1 \, \rho_2 \ : \ \Gamma \Rightarrow X} \ (\odot)$$

Fig. 1. Sketch of the type system that is formalised. If a rule mentions $\Gamma \vdash t : \tau$, then it is implicitly assumed that $\Gamma \ context$ and $\Gamma \vdash \tau \ type$; similar assumptions apply to the other judgements as well. All freshness side conditions have been omitted.

- It has explicit substitutions in the sense that the application of a substitution to a term is an explicit construction in the language. However, the application of a substitution to a *type* is an implicit operation.
- There does not seem to be a "standard" choice of basic substitutions. The set chosen here is the following:
  - $[x \mapsto t]$ is the substitution mapping x to t and every other variable to itself.
  - $wk \ x \ \tau$ extends the context with a new, unused variable.
  - $id \ \Gamma$ is the identity substitution on Γ.
  - $\rho \uparrow_x \tau$ is a lifting; variable x is mapped to itself, and the other variables are mapped by ρ.
  - $\rho_1 \, \rho_2$ is composition of substitutions.
- Heterogeneous equality is used. Two types can be equal ($\tau_1 =_\star \tau_2$) even though their contexts are not definitionally equal in the meta-theory. Contexts of equal types are always provably equal in the object-theory, though (see Sect. 3.6).

The following subsections describe the various parts of the formalisation: contexts, types, terms, variables, substitutions and equalities. Section 3.7 discusses some of the design choices made. The table below summarises the types defined;

the concept being defined, typical variable names used for elements of the type, and the type name (fully applied):

Contexts	Γ, Δ, X	$Ctxt$
Types	τ, σ	$Ty\ \Gamma$
Terms	t	$\Gamma \vdash \tau$
Variables	v	$\Gamma \ni \tau$
Substitutions	ρ	$\Gamma \Rightarrow \Delta$
Equalities	eq	$\Gamma_1 =_{Ctxt} \Gamma_2,\ \tau_1 =_\star \tau_2, \ldots$

Note that all the types in this section are part of the same mutually recursive definition, together with the function $(/)$ (see Sect. 3.2).

3.1 Contexts

Contexts are represented in a straight-forward way. The empty context is written ε, and $\Gamma \rhd \tau$ is the context Γ extended with the type τ. Variables are represented using de Bruijn indices, so there is no need to mention variables here:

data $Ctxt : Set$ **where**
$\quad \varepsilon \quad : Ctxt$
$\quad (\rhd) : (\Gamma : Ctxt) \ \to \ Ty\ \Gamma \ \to \ Ctxt$

$Ty\ \Gamma$ is the type, introduced below, of object-language types with variables in Γ.

3.2 Types

The definition of the type family Ty of object-level types follows the type system sketch in Fig. 1:

data $Ty : Ctxt \ \to \ Set$ **where**
$\quad \star \ : \{\Gamma : Ctxt\} \ \to \ Ty\ \Gamma$
$\quad \Pi \ : (\tau : Ty\ \Gamma) \ \to \ Ty\ (\Gamma \rhd \tau) \ \to \ Ty\ \Gamma$
$\quad El : \Gamma \vdash \star \ \to \ Ty\ \Gamma$

The type $\Gamma \vdash \tau$ stands for terms of type τ with variables in Γ, so terms can only be viewed as types if they have type $\star$.

Note that types are indexed on the context to which their variables belong, and similarly terms are indexed on both contexts and types ($\Gamma \vdash \tau$). The meta-theory behind indexing a type by a type family defined in the *same* mutually recursive definition has not been worked out properly yet. It is, however, crucial to this formalisation.

Let us now define the function $(/)$, which applies a substitution to a type (note that postfix application is used). The type $\Gamma \Rightarrow \Delta$ stands for a substitution which, when applied to something in context Γ (a type, for instance), transforms this into something in context Δ:

$$(/) : Ty\ \Gamma\ \to\ \Gamma \Rightarrow \Delta\ \to\ Ty\ \Delta$$
$$\star \qquad /\ \rho = \star$$
$$\Pi\ \tau_1\ \tau_2\ /\ \rho\ =\ \Pi\ (\tau_1\ /\ \rho)\ (\tau_2\ /\ \rho \uparrow \tau_1)$$
$$El\ t \quad /\ \rho\ =\ El\ (t \not\vdash \rho)$$

The constructor $(\not\vdash)$ is the analogue of $(/)$ for terms (see Sect. 3.3). The substitution transformer $(\uparrow)$ is used when going under binders; $\rho \uparrow \tau_1$ behaves as ρ, except that the new variable zero in the original context is mapped to the new variable zero in the resulting context:

$$(\uparrow) : (\rho : \Gamma \Rightarrow \Delta)\ \to\ (\sigma : Ty\ \Gamma)\ \to\ \Gamma \triangleright \sigma \Rightarrow \Delta \triangleright (\sigma\ /\ \rho)$$

Substitutions are defined in Sect. 3.5.

3.3 Terms

The types $\Gamma \vdash \tau$ and $\Gamma \ni \tau$ stand for terms and variables, respectively, of type τ in context Γ. Note that what is customarily written $\Gamma \vdash t : \tau$, like in Fig. 1, is now written $t : \Gamma \vdash \tau$. There are five kinds of terms: variables (var), abstractions (λ), applications (@), casts ($::\overset{\equiv}{\vdash}$) and substitution applications ($\not\vdash$):

$$\textbf{data}\ (\vdash) : (\Gamma : Ctxt)\ \to\ Ty\ \Gamma\ \to\ Set\ \textbf{where}$$

var	$: \Gamma \ni \tau$		$\to\ \Gamma \vdash \tau$
λ	$: \Gamma \triangleright \tau_1 \vdash \tau_2$		$\to\ \Gamma \vdash \Pi\ \tau_1\ \tau_2$
$(@)$	$: \Gamma \vdash \Pi\ \tau_1\ \tau_2\ \to$	$(t_2 : \Gamma \vdash \tau_1)\ \to$	$\Gamma \vdash \tau_2\ /\ sub\ t_2$
$(::\overset{\equiv}{\vdash})$	$: \Gamma \vdash \tau_1$	$\to\ \tau_1 =_\star \tau_2\ \to$	$\Gamma \vdash \tau_2$
$(\not\vdash)$	$: \Gamma \vdash \tau$	$\to\ (\rho : \Gamma \Rightarrow \Delta)\ \to$	$\Delta \vdash \tau\ /\ \rho$

Notice the similarity to the rules in Fig. 1. The substitution $sub\ t_2$ used in the definition of (@) replaces vz with t_2, and lowers the index of all other variables by one:

$$sub : \Gamma \vdash \tau\ \to\ \Gamma \triangleright \tau \Rightarrow \Gamma$$

The conversion rule defined here ($::\overset{\equiv}{\vdash}$) requires the two contexts to be definitionally equal in the meta-theory. A more general version of the rule would lead to increased complexity when functions that pattern match on terms are defined. However, we can *prove* a general version of the conversion rule, so no generality is lost:

$$(::\vdash) : \Gamma_1 \vdash \tau_1\ \to\ \tau_1 =_\star \tau_2\ \to\ \Gamma_2 \vdash \tau_2$$

In this formalisation, whenever a cast constructor named ($::\overset{\equiv}{\bullet}$) is introduced (where $\bullet$ can be $\vdash$ or $\ni$, for instance), a corresponding generalised variant ($::\bullet$) is always proved.

Before moving on to variables, note that all typing information is present in a term, including casts (the conversion rule). Hence this type family actually represents typing derivations.

3.4   Variables

Variables are represented using de Bruijn indices (the notation ($\ni$) is taken from [McBb]):

$$\textbf{data } (\ni) : (\Gamma : Ctxt) \rightarrow Ty\ \Gamma \rightarrow Set\ \textbf{where}$$
$$vz \quad : \qquad\qquad\qquad \{\sigma : Ty\ \Gamma\} \rightarrow \Gamma \triangleright \sigma \ni \sigma\,/\,wk\ \sigma$$
$$vs \quad : \Gamma \ni \tau \rightarrow \quad \{\sigma : Ty\ \Gamma\} \rightarrow \Gamma \triangleright \sigma \ni \tau\,/\,wk\ \sigma$$
$$(::\overset{\equiv}{\ni}) : \Gamma \ni \tau_1 \rightarrow \tau_1 =_\star \tau_2 \quad \rightarrow \Gamma \ni \tau_2$$

The rightmost variable in the context is denoted by vz ("variable zero"), and $vs\ v$ is the variable to the left of v. The substitution $wk\ \sigma$ is a weakening, taking something in context Γ to the context $\Gamma \triangleright \sigma$:

$$wk : (\sigma : Ty\ \Gamma) \rightarrow \Gamma \Rightarrow \Gamma \triangleright \sigma$$

The use of weakening is necessary since, for instance, σ is a type in Γ, whereas vz creates a variable in $\Gamma \triangleright \sigma$.

The constructor $(::\overset{\equiv}{\ni})$ is a variant of the conversion rule for variables. It might seem strange that the conversion rule is introduced twice, once for variables and once for terms. However, note that $var\ v ::\overset{\equiv}{\vdash} eq$ is a term and not a variable, so if the conversion rule is needed to show that a variable has a certain type, then $(::\overset{\equiv}{\vdash})$ cannot be used.

3.5   Substitutions

Substitutions are defined as follows:

$$\textbf{data } (\Rightarrow) : Ctxt \rightarrow Ctxt \rightarrow Set\ \textbf{where}$$
$$sub : \Gamma \vdash \tau \qquad\qquad\qquad \rightarrow \Gamma \triangleright \tau \Rightarrow \Gamma$$
$$wk \ : (\sigma : Ty\ \Gamma) \qquad\qquad\quad \rightarrow \Gamma \Rightarrow \Gamma \triangleright \sigma$$
$$(\uparrow) \ : (\rho : \Gamma \Rightarrow \Delta) \rightarrow (\sigma : Ty\ \Gamma) \rightarrow \Gamma \triangleright \sigma \Rightarrow \Delta \triangleright (\sigma\,/\,\rho)$$
$$id \ : \qquad\qquad\qquad\qquad\qquad \Gamma \Rightarrow \Gamma$$
$$(\odot) : \Gamma \Rightarrow \Delta \qquad \rightarrow \Delta \Rightarrow X \quad \rightarrow \Gamma \Rightarrow X$$

Single-term substitutions (sub), weakenings (wk) and liftings ($\uparrow$) have been introduced above. The remaining constructors denote the identity substitution (id) and composition of substitutions ($\odot$). The reasons for using this particular definition of ($\Rightarrow$) are outlined in Sect. 3.7.

3.6   Equality

The following equalities are defined:

$$(=_{Ctxt}) : Ctxt \qquad \rightarrow Ctxt \qquad \rightarrow Set$$
$$(=_\star) \quad : Ty\ \Gamma_1 \qquad \rightarrow Ty\ \Gamma_2 \quad \rightarrow Set$$
$$(=_\vdash) \quad : \Gamma_1 \vdash \tau_1 \quad \rightarrow \Gamma_2 \vdash \tau_2 \quad \rightarrow Set$$

$$(=_\ni) \quad : \Gamma_1 \ni \tau_1 \quad \to \quad \Gamma_2 \ni \tau_2 \quad \to \quad Set$$
$$(=_\Rightarrow) \quad : \Gamma_1 \Rightarrow \Delta_1 \quad \to \quad \Gamma_2 \Rightarrow \Delta_2 \quad \to \quad Set$$

As mentioned above heterogeneous equality is used. As a sanity check every equality is associated with one or more lemmas like the following one, which states that equal terms have equal types:

$$eq\llcorner eq_\star : \{\, t_1 : \Gamma_1 \vdash \tau_1 \,\} \; \to \; \{\, t_2 : \Gamma_2 \vdash \tau_2 \,\} \; \to \; t_1 =_\vdash t_2 \; \to \; \tau_1 =_\star \tau_2$$

The context and type equalities are the obvious congruences. For instance, type equality is defined as follows:

data $(=_\star) : Ty\ \Gamma_1 \; \to \; Ty\ \Gamma_2 \; \to \; Set$ **where**
$\quad \star_{Cong} \; : \Gamma_1 =_{Ctxt} \Gamma_2 \; \to \; \star\{\Gamma_1\} =_\star \star\{\Gamma_2\}$
$\quad \Pi_{Cong} : \tau_{11} =_\star \tau_{12} \; \to \; \tau_{21} =_\star \tau_{22} \; \to \; \Pi\ \tau_{11}\ \tau_{21} =_\star \Pi\ \tau_{12}\ \tau_{22}$
$\quad El_{Cong} : t_1 =_\vdash t_2 \; \to \; El\ t_1 =_\star El\ t_2$

In many presentations of type theory it is also postulated that type equality is an equivalence relation. This introduces an unnecessary amount of constructors into the data type; when proving something about a data type one typically needs to pattern match on all its constructors. Instead I have chosen to *prove* that every equality (except $(=_\vdash)$) is an equivalence relation:

$$refl_\star \; : \tau =_\star \tau$$
$$sym_\star \; : \tau_1 =_\star \tau_2 \; \to \; \tau_2 =_\star \tau_1$$
$$trans_\star : \tau_1 =_\star \tau_2 \; \to \; \tau_2 =_\star \tau_3 \; \to \; \tau_1 =_\star \tau_3$$

(And so on for the other equalities.)

The semantics of a variable should not change if a cast is added, so the variable equality is a little different. In order to still be able to prove that the relation is an equivalence the following definition is used:

data $(=_\ni) : \Gamma_1 \ni \tau_1 \; \to \; \Gamma_2 \ni \tau_2 \; \to \; Set$ **where**
$\quad vz_{Cong} \quad : \qquad\qquad\qquad\qquad \sigma_1 =_\star \sigma_2 \; \to \; vz\ \ \{\sigma_1\} =_\ni vz\ \ \{\sigma_2\}$
$\quad vs_{Cong} \quad : v_1 =_\ni v_2 \; \to \; \sigma_1 =_\star \sigma_2 \; \to \; vs\ v_1\ \{\sigma_1\} =_\ni vs\ v_2\ \{\sigma_2\}$
$\quad castEq_\ni^\ell : v_1 =_\ni v_2 \qquad\qquad\qquad\qquad \to \; v_1 ::_{\overline{\ni}}^\equiv eq \; =_\ni v_2$
$\quad castEq_\ni^r : v_1 =_\ni v_2 \qquad\qquad\qquad\qquad \to \; v_1 \qquad\qquad =_\ni v_2 ::_{\overline{\ni}}^\equiv eq$

For substitutions extensional equality is used:

data $(=_\Rightarrow)\ (\rho_1 : \Gamma_1 \Rightarrow \Delta_1)\ (\rho_2 : \Gamma_2 \Rightarrow \Delta_2) : Set$ **where**
$\quad extEq \; : \; \Gamma_1 =_{Ctxt} \Gamma_2 \; \to \; \Delta_1 =_{Ctxt} \Delta_2$
$\quad\qquad \to \; (\forall v_1\ v_2.\ v_1 =_\ni v_2 \; \to \; var\ v_1 \not\llcorner \rho_1 =_\vdash var\ v_2 \not\llcorner \rho_2)$
$\quad\qquad \to \; \rho_1 =_\Rightarrow \rho_2$

Note that this data type contains negative occurrences of Ty, $(\ni)$ and $(=_\ni)$, which are defined in the same mutually recursive definition as $(=_\Rightarrow)$. In order to keep this definition strictly positive a first-order variant of $(=_\Rightarrow)$ is used, which simulates the higher-order version by explicitly enumerating all the variables. The first-order variant is later proved equivalent to the definition given here.

Term equality is handled in another way than the other equalities. The presence of the β and η laws makes it hard to prove that $(=_\vdash)$ is an equivalence relation, and hence this is postulated:

data $(=_\vdash) : \Gamma_1 \vdash \tau_1 \to \Gamma_2 \vdash \tau_2 \to Set$ **where**
    -- Equivalence.
$refl_\vdash$ $: (t : \Gamma \vdash \tau) \to t =_\vdash t$
$sym_\vdash$ $: t_1 =_\vdash t_2 \to t_2 =_\vdash t_1$
$trans_\vdash : t_1 =_\vdash t_2 \to t_2 =_\vdash t_3 \to t_1 =_\vdash t_3$
    -- Congruence.
var_{Cong} $: v_1 =_\ni v_2 \to var\ v_1 =_\vdash var\ v_2$
λ_{Cong} $: t_1 =_\vdash t_2 \to \lambda\ t_1 =_\vdash \lambda\ t_2$
$(@_{Cong}) : t_{11} =_\vdash t_{12} \to t_{21} =_\vdash t_{22} \to t_{11}@t_{21} =_\vdash t_{12}@t_{22}$
$(/\!\!\vdash_{Cong}) : t_1 =_\vdash t_2 \to \rho_1 =_\Rightarrow \rho_2 \to t_1 /\!\!\vdash \rho_1 =_\vdash t_2 /\!\!\vdash \rho_2$
    -- Cast, β and η equality.
$castEq_\vdash : t ::_{\overline{\vdash}}^{\equiv} eq =_\vdash t$
β $: (\lambda\ t_1)@t_2 =_\vdash t_1 /\!\!\vdash sub\ t_2$
η $: \{t : \Gamma \vdash \Pi\ \tau_1\ \tau_2\} \to \lambda\ ((t /\!\!\vdash wk\ \tau_1)@var\ vz) =_\vdash t$
    -- Substitution application axioms.
    . . .

The η law basically states that, if x is not free in t, and t is of function type, then $\lambda x.t\ x = t$. The first precondition on t is handled by explicitly weakening t, though.

The behaviour of $(/\!\!\vdash)$ also needs to be postulated. The abstraction and application cases are structural; the *id* case returns the term unchanged, and the $(\odot)$ case is handled by applying the two substitutions one after the other; a variable is weakened by applying *vs* to it; substituting t for variable zero results in t, and otherwise the variable's index is lowered by one; and finally lifted substitutions need to be handled appropriately:

data $(=_\vdash) : \Gamma_1 \vdash \tau_1 \to \Gamma_2 \vdash \tau_2 \to Set$ **where**
    . . .
    -- Substitution application axioms.

$substLam$	$: \lambda\ t$	$/\!\!\vdash \rho$	$=_\vdash \lambda\ (t /\!\!\vdash \rho \uparrow \tau_1)$
$substApp$	$: (t_1@t_2)$	$/\!\!\vdash \rho$	$=_\vdash (t_1 /\!\!\vdash \rho)@(t_2 /\!\!\vdash \rho)$
$idVanishesTm$	$: t$	$/\!\!\vdash id$	$=_\vdash t$
$compSplitsTm$	$: t$	$/\!\!\vdash (\rho_1 \odot \rho_2) =_\vdash t /\!\!\vdash \rho_1 /\!\!\vdash \rho_2$	
$substWk$	$: var\ v$	$/\!\!\vdash wk\ \sigma$	$=_\vdash var\ (vs\ v)$
$substVzSub$	$: var\ vz$	$/\!\!\vdash sub\ t$	$=_\vdash t$
$substVsSub$	$: var\ (vs\ v) /\!\!\vdash sub\ t$	$=_\vdash var\ v$	
$substVzLift$	$: var\ vz$	$/\!\!\vdash (\rho \uparrow \sigma)$	$=_\vdash var\ vz$
$substVsLift$	$: var\ (vs\ v) /\!\!\vdash (\rho \uparrow \sigma)$	$=_\vdash var\ v /\!\!\vdash \rho /\!\!\vdash wk\ (\sigma\ /\ \rho)$	

3.7   Design Choices

Initially I tried to formalise a language with implicit substitutions, i.e. I tried to implement $(/\!\!\vdash)$ as a function instead of as a constructor. This turned out to be difficult, since when $(/\!\!\vdash)$ is defined as a function many substitution lemmas need to be proved in the initial mutually recursive definition containing all the type families above, and when the mutual dependencies become too complicated it is hard to prove that the code is terminating.

As an example of why substitution lemmas are needed, take the axiom $substApp$ above. If $substApp$ is considered as a pattern matching equation in the definition of $(/\!\!\vdash)$, then it needs to be modified in order to type check:

$$(t_1 @ t_2) /\!\!\vdash \rho = (t_1 /\!\!\vdash \rho)@(t_2 /\!\!\vdash \rho) ::\stackrel{\equiv}{\vdash} subCommutes_\star$$

Here $subCommutes_\star$ states that, in certain situations, sub commutes with other substitutions:

$$subCommutes_\star : \tau \,/\, (\rho \uparrow \sigma) \,/\, sub\ (t /\!\!\vdash \rho) =_\star \tau \,/\, sub\ t \,/\, \rho$$

Avoidance of substitution lemmas is also the reason for making the equalities heterogeneous. It would be possible to enforce directly that, for instance, two terms are only equal if their respective types are equal. It suffices to add the type equality as an index to the term equality:

$$(=_\vdash) : \{\tau_1 =_\star \tau_2\} \rightarrow \Gamma_1 \vdash \tau_1 \rightarrow \Gamma_2 \vdash \tau_2 \rightarrow Set$$

However, in this case $substApp$ could not be defined without a lemma like $subCommutes_\star$. Furthermore this definition of $(=_\vdash)$ easily leads to a situation where two equality *proofs* need to be proved equal. These problems are avoided by, instead of enforcing equality directly, proving that term equality implies type equality ($eq_\vdash eq_\star$) and so on. These results also require lemmas like $subCommutes_\star$, but the lemmas can be proved after the first, mutually recursive definition.

The problems described above could be avoided in another way, by postulating the substitution lemmas needed, i.e. adding them as type equality constructors. This approach has not been pursued, as I have tried to minimise the amount of "unnecessary" postulates and definitions.

The postulate $substApp$ discussed above also provides motivation for defining $(/)$ as a function, even though $(/\!\!\vdash)$ is a constructor: if $(/)$ were a constructor then $t_1 /\!\!\vdash \rho$ would not have a Π type as required by $(@)$ (the type would be $\Pi\ \tau_1\ \tau_2 \,/\, \rho$), and hence a cast would be required in the definition of $substApp$. I have not examined this approach in detail, but I suspect that it would be harder to work with.

Another important design choice is the basic set of substitutions. The following definition is a natural candidate for this set:

data $(\stackrel{\Rightarrow}{\cdot}) : Ctxt \rightarrow Ctxt \rightarrow Set$ **where**
$\quad \emptyset\quad : \varepsilon \stackrel{\Rightarrow}{\cdot} \Delta$
$\quad (\blacktriangleright) : (\rho : \Gamma \stackrel{\Rightarrow}{\cdot} \Delta) \rightarrow \Delta \vdash \tau \,/\, \rho \rightarrow \Gamma \triangleright \tau \stackrel{\Rightarrow}{\cdot} \Delta$

This type family encodes simultaneous (parallel) substitutions; for every variable in the original context a term in the resulting context is given. So far so good, but the substitutions used in the type signatures above (*sub* and *wk*, for instance) need to be implemented in terms of $\emptyset$ and ($\blacktriangleright$), and these implementations seem to require various substitution lemmas, again leading to the problems described above.

Note that, even though ($\Rightarrow$) is not used to define what a substitution is, the substitutions $\emptyset$ and ($\blacktriangleright$) can be defined in terms of the other substitutions, and they are used in Sect. 5.3 when value environments are defined.

4   Removing Explicit Substitutions

In Sect. 5 a normalisation proof for the lambda calculus introduced in Sect. 3 is presented. The normalisation function defined there requires terms without explicit substitutions ("implicit terms"). This section defines a data type Tm^- representing such terms.

The type Tm^- provides a view of the ($\vdash$) terms (the "explicit terms"). Other views will be introduced later, for instance normal forms (Sect. 5.1), and they will all follow the general scheme employed by Tm^-, with minor variations.

Implicit terms are indexed on explicit terms to which they are, in a sense, $\beta\eta$-equal; the function tm^-ToTm converts an implicit term to the corresponding explicit term, and $tm^-ToTm\ t^- =_\vdash t$ for every implicit term $t^- : Tm^-\ t$:

data $Tm^- : \Gamma \vdash \tau \rightarrow Set$ **where**
$\quad var^- : (v : \Gamma \ni \tau) \qquad\qquad \rightarrow Tm^-\ (var\ v)$
$\quad \lambda^- \quad : Tm^-\ t \qquad\qquad\quad\ \rightarrow Tm^-\ (\lambda\ t)$
$\quad (@^-) : Tm^-\ t_1 \rightarrow Tm^-\ t_2 \rightarrow Tm^-\ (t_1 @ t_2)$
$\quad (::^{\equiv}_{\vdash-}) : Tm^-\ t_1 \rightarrow t_1 =_\vdash t_2 \rightarrow Tm^-\ t_2$

$tm^-ToTm : \{t : \Gamma \vdash \tau\} \rightarrow Tm^-\ t \rightarrow \Gamma \vdash \tau$
$tm^-ToTm\ (var^-\ v) \quad = var\ v$
$tm^-ToTm\ (\lambda^-\ t^-) \quad = \lambda\ (tm^-ToTm\ t^-)$
$tm^-ToTm\ (t_1^-\ @^-\ t_2^-) = (tm^-ToTm\ t_1^-) @ (tm^-ToTm\ t_2^-) ::^{\equiv}_{\vdash} \ldots$
$tm^-ToTm\ (t^-\ ::^{\equiv}_{\vdash-}\ eq) = tm^-ToTm\ t^- ::^{\equiv}_{\vdash}\ eq_\vdash eq_\star\ eq$

(The ellipsis stands for uninteresting code that has been omitted.)

It would be possible to index implicit terms on types instead. However, by indexing on explicit terms soundness results are easily expressed in the types of functions constructing implicit terms. For instance, the function $tmToTm^-$ which converts explicit terms to implicit terms has the type $(t : \Gamma \vdash \tau) \rightarrow Tm^-\ t$, which guarantees that the result is $\beta\eta$-equal to t. The key to making this work is the cast constructor $(::^{\equiv}_{\vdash-})$, which makes it possible to include equality proofs in an implicit term; without $(::^{\equiv}_{\vdash-})$ no implicit term could be indexed on $t \not\vdash \rho$, for instance.

Explicit terms are converted to implicit terms using techniques similar to those in [McBb]. Due to lack of space this conversion is not discussed further here.

5   Normalisation Proof

This section proves that every explicit term has a normal form. The proof uses normalisation by evaluation (NBE). Type-based NBE proceeds as follows:

- First terms (in this case implicit terms) are evaluated by a function $[\![\cdot]\!]$ (Sect. 5.4), resulting in "values" (Sect. 5.2). Termination issues are avoided by representing function types using the function space of the meta-language.
- Then these values are converted to normal forms by using two functions, often called *reify* and *reflect*, defined by recursion on the (spines of the) types of their arguments (Sect. 5.5).

5.1   Normal Forms

Let us begin by defining what a normal form is. Normal forms (actually long $\beta\eta$-normal forms) and atomic forms are defined simultaneously. Both type families are indexed on a $\beta\eta$-equivalent term, just like Tm^- (see Sect. 4):

data $Atom : \Gamma \vdash \tau \to Set$ **where**
$\quad var_{At} : (v : \Gamma \ni \tau) \to Atom\ (var\ v)$
$\quad (@_{At}) : Atom\ t_1 \to NF\ t_2 \to Atom\ (t_1 @ t_2)$
$\quad (::_{At}^{\overline{\overline{\equiv}}}) : Atom\ t_1 \to t_1 =_\vdash t_2 \to Atom\ t_2$
data $NF : \Gamma \vdash \tau \to Set$ **where**
$\quad atom_{NF}^\star : \{t : \Gamma \vdash \star\} \quad\ \to Atom\ t \to NF\ t$
$\quad atom_{NF}^{El} : \{t : \Gamma \vdash El\ t'\} \to Atom\ t \to NF\ t$
$\quad \lambda_{NF} \quad\ : NF\ t \to NF\ (\lambda\ t)$
$\quad (::_{NF}^{\overline{\overline{\equiv}}}) \quad : NF\ t_1 \to t_1 =_\vdash t_2 \to NF\ t_2$

The two $atom_{NF}$ constructors ensure that the only normal forms of type $\Pi\ \tau_1\ \tau_2$ are lambdas and casts; this is how long η-normality is ensured.

A consequence of the inclusion of the cast constructors $(::_{At}^{\overline{\overline{\equiv}}})$ and $(::_{NF}^{\overline{\overline{\equiv}}})$ is that normal forms are not unique. However, the equality on normal and atomic forms (congruence plus postulates stating that casts can be removed freely) ensures that equality can be decided by erasing all casts and annotations and then checking syntactic equality.

A normal form can be converted to a term in the obvious way, and the resulting term is $\beta\eta$-equal to the index (cf. tm^-ToTm in Sect. 4):

$nfToTm \quad : \{t : \Gamma \vdash \tau\} \to NF\ t \to \Gamma \vdash \tau$
$nfToTmEq : (nf : NF\ t) \to nfToTm\ nf =_\vdash t$

Similar functions are defined for atomic forms.

We also need to weaken normal and atomic forms. In fact, multiple weakenings will be performed at once. In order to express these multi-weakenings context extensions are introduced. The type $Ctxt^+\ \Gamma$ stands for context extensions which can be put "to the right of" the context Γ by using $(+\!\!+)$:

data $Ctxt^+$ $(\Gamma : Ctxt) : Set$ **where**

$\quad \varepsilon^+ \quad : Ctxt^+ \ \Gamma$

$\quad (\triangleright^+) : (\Gamma' : Ctxt^+ \ \Gamma) \to Ty \ (\Gamma \mathbin{+\!\!+} \Gamma') \to Ctxt^+ \ \Gamma$

$\quad (\mathbin{+\!\!+}) : (\Gamma : Ctxt) \to Ctxt^+ \ \Gamma \to Ctxt$

$\quad \Gamma \mathbin{+\!\!+} \varepsilon^+ \qquad = \Gamma$

$\quad \Gamma \mathbin{+\!\!+} (\Gamma' \triangleright^+ \tau) = (\Gamma \mathbin{+\!\!+} \Gamma') \triangleright \tau$

Now the following type signatures can be understood:

$$wk^* \ : (\Gamma' : Ctxt^+ \ \Gamma) \to \Gamma \Rightarrow \Gamma \mathbin{+\!\!+} \Gamma'$$

$$wk^*_{At} : Atom \ t \to (\Gamma' : Ctxt^+ \ \Gamma) \to Atom \ (t \not\vdash wk^* \ \Gamma')$$

5.2 Values

Values are represented using one constructor for each type constructor, plus a case for casts (along the lines of previously introduced types indexed on terms). Values of function type are represented using meta-language functions:

data $Val : \Gamma \vdash \tau \to Set$ **where**

$\quad (::_{Val}) : Val \ t_1 \to t_1 =_\vdash t_2 \to Val \ t_2$

$\quad \star_{Val} \quad : \{t : \Gamma \vdash \star\} \qquad \to Atom \ t \to Val \ t$

$\quad El_{Val} : \{t : \Gamma \vdash El \ t'\} \to Atom \ t \to Val \ t$

$\quad \Pi_{Val} \quad : \quad \{t_1 : \Gamma \vdash \Pi \ \tau_1 \ \tau_2\}$

$\qquad\qquad \to (f \ : \ (\Gamma' : Ctxt^+ \ \Gamma)$

$\qquad\qquad\qquad \to \{t_2 : \Gamma \mathbin{+\!\!+} \Gamma' \vdash \tau_1 \ / \ wk^* \ \Gamma'\}$

$\qquad\qquad\qquad \to (v : Val \ t_2)$

$\qquad\qquad\qquad \to Val \ ((t_1 \not\vdash wk^* \ \Gamma')@t_2))$

$\qquad\qquad \to Val \ t_1$

The function f given to $\Pi_{Val} \ \{t_1\}$ essentially takes an argument value and evaluates t_1 applied to this argument. For technical reasons, however, we need to be able to weaken t_1 (see *reify* in Sect. 5.5). This makes Val look suspiciously like a Kripke model [MM91] (suitably generalised to a dependently typed setting); this has not been verified in detail, though. The application operation of this supposed model is defined as follows. Notice that the function component of Π_{Val} is applied to an empty $Ctxt^+$ here:

$\quad (@_{Val}) : Val \ t_1 \to Val \ t_2 \to Val \ (t_1@t_2)$

$\quad \Pi_{Val} \ f \qquad @_{Val} \ v_2 = f \ \varepsilon^+ \ (v_2 ::_{Val} \ ...) ::_{Val} \ ...$

$\quad (v_1 ::_{Val} \ eq) @_{Val} \ v_2 = (v_1 @_{Val} \ (v_2 ::_{Val} \ ...)) ::_{Val} \ ...$

The transition function of the model weakens values:

$$wk^*_{Val} : Val \ t \to (\Gamma' : Ctxt^+ \ \Gamma) \to Val \ (t \not\vdash wk^* \ \Gamma')$$

Note that Val is not a positive data type, due to the negative occurrence of Val inside of Π_{Val}, so this data type is not part of the treatment in [DS06]. In practise this should not be problematic, since the type index of that occurrence, $\tau_1 \,/\, wk^*\, \Gamma'$, is smaller than the type index of $\Pi_{Val}\, f$, which is $\Pi\; \tau_1\; \tau_2$. Here we count just the *spine* of the type, ignoring the contents of *El*, so that τ and $\tau\,/\,\rho$ have the same size, and two equal types also have the same size. In fact, by supplying a spine argument explicitly it should not be difficult to define *Val* as a structurally recursive function instead of as a data type.

5.3   Environments

The function $\llbracket \cdot \rrbracket$, defined in Sect. 5.4, makes use of environments, which are basically substitutions containing values instead of terms:

> **data** $Env : \Gamma \Rightarrow \Delta \;\rightarrow\; Set$ **where**
> $\emptyset_{Env}$ $\;\;: Env\; \emptyset$
> $(\blacktriangleright_{Env}) : Env\; \rho \;\rightarrow\; Val\; t \;\rightarrow\; Env\; (\rho \blacktriangleright t)$
> $(::\overline{\overline{\equiv}}_{Env}) : Env\; \rho_1 \;\rightarrow\; \rho_1 =_{\Rightarrow} \rho_2 \;\rightarrow\; Env\; \rho_2$

Note that the substitutions $\emptyset$ and $(\blacktriangleright)$ from Sect. 3.7 are used as indices here.

It is straight-forward to define functions for looking up a variable in an environment and weakening an environment:

> $lookup : (v : \Gamma \ni \tau) \;\rightarrow\; Env\; \rho \;\rightarrow\; Val\; (var\; v \,\mid\!\!\!-\, \rho)$
> $wk^{\star}_{Env} : Env\; \rho \;\rightarrow\; (\Delta' : Ctxt^+\, \Delta) \;\rightarrow\; Env\; (\rho \odot wk^*\, \Delta')$

5.4   Evaluating Terms

Now we can evaluate an implicit term, i.e. convert it to a value. The most interesting case is $\lambda^-\; t_1^-$, where t_1^- is evaluated in an extended, weakened environment:

> $\llbracket \cdot \rrbracket : Tm^-\; t \;\rightarrow\; Env\; \rho \;\rightarrow\; Val\; (t \,\mid\!\!\!-\, \rho)$
> $\llbracket var^-\; v \rrbracket\, \gamma\;\;\; = lookup\; v\; \gamma$
> $\llbracket t_1^-\; @^-\; t_2^- \rrbracket\, \gamma = (\llbracket t_1^- \rrbracket\, \gamma\; @_{Val}\; \llbracket t_2^- \rrbracket\, \gamma)\; ::_{Val}\; \cdots$
> $\llbracket t^- \;::\overline{\overline{\equiv}}_{\mid\!-}\; eq \rrbracket\, \gamma = \llbracket t^- \rrbracket\, \gamma\; ::_{Val}\; \cdots$
> $\llbracket \lambda^-\; t_1^- \rrbracket\, \gamma\;\;\; =$
> $\quad \Pi_{Val}\, (\backslash \Delta'\; v_2 \;\rightarrow\; \llbracket t_1^- \rrbracket\, (wk^{\star}_{Env}\; \gamma\; \Delta'\; \blacktriangleright_{Env}\; (v_2\; ::_{Val}\; \cdots))\; ::_{Val}\; \cdots\, \beta \cdots)$

(The notation $\backslash x \rightarrow \cdots$ is lambda abstraction in the meta-language.) It would probably be straightforward to evaluate explicit terms directly, without going through implicit terms (cf. [Coq02]). Here I have chosen to separate these two steps, though.

5.5   *Reify* and *Reflect*

Let us now define *reify* and *reflect*. These functions are implemented by recursion over spines (see Sect. 5.2), in order to make them structurally recursive, but to avoid clutter the spine arguments are not written out below.

The interesting cases correspond to function types, for which *reify* and *reflect* use each other recursively. Notice how *reify* applies the function component of Π_{Val} to a singleton $Ctxt^+$, to enable using the reflection of variable zero, which has a weakened type; this is the reason for including weakening in the definition of Π_{Val}:

$$reify : (\tau : Ty\ \Gamma) \rightarrow \{t : \Gamma \vdash \tau\} \rightarrow Val\ t \rightarrow NF\ t$$
$$reify\ (\Pi\ \tau_1\ \tau_2)\ (\Pi_{Val}\ f) =$$
$$\lambda_{NF}\ (reify\ (\tau_2\ /\ \_\ /\ \_)$$
$$(f\ (\varepsilon^+ \rhd^+ \tau_1)\ (reflect\ (\tau_1\ /\ \_)\ (var_{At}\ vz)\ ::_{Val}\ ...)))$$
$$::_{NF}\ ...\ \eta\ ...$$
$$reflect : (\tau : Ty\ \Gamma) \rightarrow \{t : \Gamma \vdash \tau\} \rightarrow Atom\ t \rightarrow Val\ t$$
$$reflect\ (\Pi\ \tau_1\ \tau_2)\ at = \Pi_{Val}\ (\backslash\Gamma'\ v \rightarrow$$
$$reflect\ (\tau_2\ /\ \_\ /\ \_)\ (wk^*_{At}\ at\ \Gamma'\ @_{At}\ reify\ (\tau_1\ /\ \_)\ v))$$

Above underscores (_) have been used instead of giving non-hidden arguments which can be inferred automatically by the AgdaLight type checker, and some simple and boring cases have been omitted to save space.

5.6 Normalisation

After having defined $[\![\cdot]\!]$ and *reify* it is very easy to normalise a term. First we build an identity environment by applying *reflect* to all the variables in the context:

$$id_{Env} : (\Gamma : Ctxt) \rightarrow Env\ (id\ \Gamma)$$

Then an explicit term can be normalised by converting it to an implicit term, evaluating the result in the identity environment, and then reifying:

$$normalise : (t : \Gamma \vdash \tau) \rightarrow NF\ t$$
$$normalise\ t = reify\ \_\ ([\![tmToTm^- \ t]\!]\ (id_{Env}\ \_)\ ::_{Val}\ ...)$$

Since a normal form is indexed on an equivalent term it is easy to show that *normalise* is sound:

$$normaliseEq : (t : \Gamma \vdash \tau) \rightarrow nfToTm\ (normalise\ t) =_\vdash t$$
$$normaliseEq\ t = nfToTmEq\ (normalise\ t)$$

If this normalising function is to be really useful (as part of a type checker, for instance) it should also be proved, for the normal form equality ($=_{NF}$), that $t_1 =_\vdash t_2$ implies that $normalise\ t_1 =_{NF} normalise\ t_2$. This is left for future work, though.

6 Related Work

As stated in the introduction Dybjer's formalisation of categories with families [Dyb96] seems to be the only prior example of a formalisation of a dependently typed language done using the internal approach to handle the type system invariants. Other formalisations of dependently typed languages, such as

McKinna/Pollack [MP99] and Barras/Werner [BW97], have used the external approach. There is also an example, due to Adams [Ada04], of a hybrid approach which handles some invariants internally, but not the type system.

Normalisation by evaluation (NBE) seems to have been discovered independently by Martin-Löf (for a version of his type theory) [ML75] and Berger and Schwichtenberg (for simply typed lambda calculus) [BS91]. Martin-Löf has also defined an NBE algorithm for his logical framework [ML04], and recently Dybjer, Abel and Aehlig have done the same for Martin-Löf type theory with one universe [AAD07].

NBE has been formalised, using the internal approach, by T. Coquand and Dybjer, who treated a combinatory version of Gödel's System T [CD97]. C. Coquand has formalised normalisation for a simply typed lambda calculus with explicit substitutions, also using the internal approach [Coq02]. Her normalisation proof uses NBE and Kripke models, and in that respect it bears much resemblance to this one. McBride has implemented NBE for the untyped lambda calculus [McBa]. His implementation uses an internal approach (nested types in Haskell) to ensure that terms are well-scoped, and that aspect of his code is similar to mine.

My work seems to be the first formalised NBE algorithm for a dependently typed language.

7   Discussion

I have presented a formalisation of a dependently typed lambda calculus, including a proof of normalisation, using the internal approach to handle typing rules. This formalisation demonstrates that, at least in this case, it is feasible to use the internal approach when programming with invariants strong enough to encode the typing rules of a dependently typed language. How this method compares to other approaches is a more difficult question, which I do not attempt to answer here.

Acknowledgements

I am grateful to Ulf Norell, who has taught me a lot about dependently typed programming, discussed many aspects of this work, and fixed many of the bugs in AgdaLight that I have reported. I would also like to thank Thierry Coquand and Peter Dybjer for interesting discussions and useful pointers, and Patrik Jansson and the anonymous referees for helping me with the presentation.

References

[AA02]   Abel, A., Altenkirch, T.: A predicative analysis of structural recursion. Journal of Functional Programming 12(1), 1–41 (2002)

[AAD07]   Abel, A., Aehlig, K., Dybjer, P.: Normalization by evaluation for Martin-Löf type theory with one universe. Submitted for publication (2007)

[AC06]   Altenkirch, T., Chapman, J.: Tait in one big step. In: MSFP 2006 (2006)
[Ada04]   Adams, R.: Formalized metatheory with terms represented by an indexed family of types. In: TYPES 2004. LNCS, vol. 3839, pp. 1–16. Springer, Heidelberg (2004)
[AR99]   Altenkirch, T., Reus, B.: Monadic presentations of lambda terms using generalized inductive types. In: Flum, J., Rodríguez-Artalejo, M. (eds.) CSL 1999. LNCS, vol. 1683, pp. 453–468. Springer, Heidelberg (1999)
[BS91]   Berger, U., Schwichtenberg, H.: An inverse of the evaluation functional for typed λ-calculus. In: LICS '91, pp. 203–211 (1991)
[BW97]   Barras, B., Werner, B.: Coq in Coq. Unpublished (1997)
[CD97]   Coquand, T., Dybjer, P.: Intuitionistic model constructions and normalization proofs. Mathematical Structures in Computer Science 7(1), 75–94 (1997)
[Coq02]   Coquand, C.: A formalised proof of the soundness and completeness of a simply typed lambda-calculus with explicit substitutions. Higher-Order and Symbolic Computation 15, 57–90 (2002)
[Dan07]   Danielsson, N.A.: Personal web page. Available (2007) at http://www.cs.chalmers.se/~nad/
[DS06]   Dybjer, P., Setzer, A.: Indexed induction-recursion. Journal of Logic and Algebraic Programming 66(1), 1–49 (2006)
[Dyb96]   Dybjer, P.: Internal type theory. In: Berardi, S., Coppo, M. (eds.) TYPES 1995. LNCS, vol. 1158, pp. 120–134. Springer, Heidelberg (1996)
[McBa]   McBride, C.: Beta-normalization for untyped lambda-calculus (unpublished program)
[McBb]   McBride, C.: Type-preserving renaming and substitution (unpublished)
[ML75]   Martin-Löf, P.: An intuitionistic theory of types: Predicative part. In: Logic Colloquium '73, pp. 73–118. North-Holland, Amsterdam (1975)
[ML04]   Martin-Löf, P.: Normalization by evaluation and by the method of computability. Lecture series given at Logikseminariet Stockholm–Uppsala (2004)
[MM91]   Mitchell, J.C., Moggi, E.: Kripke-style models for typed lambda calculus. Annals of Pure and Applied Logic 51, 99–124 (1991)
[MM04]   McBride, C., McKinna, J.: The view from the left. Journal of Functional Programming 14(1), 69–111 (2004)
[MP99]   McKinna, J., Pollack, R.: Some lambda calculus and type theory formalized. Journal of Automated Reasoning 23(3), 373–409 (1999)
[MW06]   McKinna, J., Wright, J.: A type-correct, stack-safe, provably correct expression compiler in Epigram. Accepted for publication in the Journal of Functional Programming (2006)
[Nor07]   Norell, U.: AgdaLight home page. Available (2007) at http://www.cs.chalmers.se/~ulfn/agdaLight/
[NPS90]   Nordström, B., Petersson, K., Smith, J.M.: Programming in Martin-Löf's Type Theory, An Introduction. Oxford University Press, Oxford (1990)
[PL04]   Pašalić, E., Linger, N.: Meta-programming with typed object-language representations. In: Karsai, G., Visser, E. (eds.) GPCE 2004. LNCS, vol. 3286, pp. 136–167. Springer, Heidelberg (2004)
[XCC03]   Xi, H., Chen, C., Chen, G.: Guarded recursive datatype constructors. In: POPL '03, pp. 224–235 (2003)

Truth Values Algebras and Proof Normalization

Gilles Dowek

École polytechnique and INRIA,
LIX, École polytechnique, 91128 Palaiseau Cedex, France
Gilles.Dowek@polytechnique.edu
http://lix.polytechnique.fr/~dowek/

Abstract. We extend the notion of Heyting algebra to a notion of *truth values algebra* and prove that a theory is consistent if and only if it has a $\mathcal{B}$-valued model for some non trivial truth values algebra $\mathcal{B}$. A theory that has a $\mathcal{B}$-valued model for all truth values algebras $\mathcal{B}$ is said to be *super-consistent*. We prove that super-consistency is a model-theoretic sufficient condition for strong normalization.

1 Introduction

Proving that a theory has the cut elimination property has some similarities with proving that it has a model. These similarities appear, for instance, in the model theoretic proofs of cut elimination, where cut elimination is obtained as a corollary of a strengthening of the completeness theorem, expressing that if a formula is valid in all models of a theory, then it has a *cut free* proof in this theory. Such a method has been used, for instance, by Schütte, Kanger, Beth, Hintikka and Smullyan. It has then been used by Tait [15], Prawitz [13], Takahashi [17] and Andrews [1] to prove cut elimination for simple type theory. It has been generalized, more recently, by De Marco and Lipton [2] to prove cut elimination for an intuitionistic variant of simple type theory, by Hermant [8,10] to prove cut elimination for classical and intuitionistic theories in deduction modulo and by Okada [11] to prove cut elimination for intuitionistic linear logic.

An alternative method to prove cut elimination is to prove that all proofs strongly normalize. Following Tait [16] and Girard [7], this is proved by assigning a set of proofs, called a *reducibility candidate*, to each formula. Here also, the proofs have some similarities with the construction of models, except that, in these models, the truth values 0 and 1 are replaced by reducibility candidates. This analogy has been exploited in a joint work with Werner [5], where we have defined a notion of reducibility candidate valued models, called *pre-models*, and proved that if a theory in deduction modulo has such a model, then it has the strong normalization property.

The fact that both cut elimination proofs and strong normalization proofs proceed by building models raises the problem of the difference between cut elimination and strong normalization. It is well-known that strong normalization implies cut elimination, but what about the converse? This problem can be precisely stated in deduction modulo, where instead of using an *ad hoc* notion

T. Altenkirch and C. McBride (Eds.): TYPES 2006, LNCS 4502, pp. 110–124, 2007.

of cut for each theory of interest, we can formulate a general notion of cut for a large class of theories, that subsumes the usual *ad hoc* notions. This problem has been solved by Hermant [9] and surprisingly the answer is negative: there are theories that have the cut elimination property, but not the strong normalization property and even not the weak normalization property. Thus, although the model theoretic cut elimination proofs and the strong normalization proofs both proceed by building models, these methods apply to different theories.

In this paper, we focus on the model theoretic characterization of theories in deduction modulo that have the strong normalization property. It has been proved in [5] that a theory has the strong normalization property if it has a reducibility candidate valued model. However, the usual model constructions use very little of the properties of reducibility candidates. In particular, these constructions seem to work independently of the chosen variant of the closure conditions defining reducibility candidates. This suggests that this notion of reducibility candidate valued model can be further generalized, by considering an abstract notion of reducibility candidate.

Abstracting this way on the notion of reducibility candidate leads to introduce a class of algebras, called *truth values algebras*, that also generalize Heyting algebras. However there is an important difference between truth values algebras and Heyting algebras: in a Heyting algebra valued model the formula $P \Leftrightarrow Q$ is valid if and only if the formulae P and Q have the same denotation. In particular, all theorems have the same denotation. This is not necessarily the case in truth values algebra valued models where two theorems may have different denotation. Thus, truth values algebra valued models are more "intentional" than Heyting algebra valued models. In particular, it is possible to distinguish in the model between the computational equivalence of formulae (the congruence of deduction modulo, or the definitional equality of Martin-Löf's type theory) and the provable equivalence: the denotations of two computationally equivalent formulae are the same, but not necessarily those of two logically equivalent formulae. Thus, independently of normalization, this generalization of Heyting algebras seems to be of interest for the model theory of deduction modulo and type theory.

We shall first introduce the notion of truth values algebra and compare it with the notion of Heyting algebra. Then, we shall consider plain predicate logic, define a notion of model based on these truth values algebras and prove a soundness and a completeness theorem for this notion of model. We shall then show that this notion of model extends to deduction modulo. Finally, we shall strengthen the notion of consistency into a notion of *super-consistency* and prove that all super-consistent theories have the strong normalization property. We refer to the long version of the paper for the proofs omitted in this abstract.

2 Truth Values Algebras

2.1 Definition

Definition 1 (Truth values algebra). *Let $\mathcal{B}$ be a set, whose elements are called* truth values, *$\mathcal{B}^+$ be a subset of $\mathcal{B}$, whose elements are called* positive *truth* *values,* $\mathcal{A}$ *and* $\mathcal{E}$ *be subsets of* $\wp(\mathcal{B})$, $\tilde{\top}$ *and* $\tilde{\bot}$ *be elements of* $\mathcal{B}$, $\tilde{\Rightarrow}$, $\tilde{\wedge}$, *and* $\tilde{\vee}$ *be functions from* $\mathcal{B} \times \mathcal{B}$ *to* $\mathcal{B}$, $\tilde{\forall}$ *be a function from* $\mathcal{A}$ *to* $\mathcal{B}$ *and* $\tilde{\exists}$ *be a function from* $\mathcal{E}$ *to* $\mathcal{B}$. *The structure* $\mathcal{B} = \langle \mathcal{B}, \mathcal{B}^+, \mathcal{A}, \mathcal{E}, \tilde{\top}, \tilde{\bot}, \tilde{\Rightarrow}, \tilde{\wedge}, \tilde{\vee}, \tilde{\forall}, \tilde{\exists} \rangle$ *is said to be a truth value algebra if the set* $\mathcal{B}^+$ *is* closed by the intuitionistic deduction rules i.e. *if for all* a, b, c *in* $\mathcal{B}$, A *in* $\mathcal{A}$ *and* E *in* $\mathcal{E}$,

1. *if* $a \tilde{\Rightarrow} b \in \mathcal{B}^+$ *and* $a \in \mathcal{B}^+$ *then* $b \in \mathcal{B}^+$,
2. $a \tilde{\Rightarrow} b \tilde{\Rightarrow} a \in \mathcal{B}^+$,
3. $(a \tilde{\Rightarrow} b \tilde{\Rightarrow} c) \tilde{\Rightarrow} (a \tilde{\Rightarrow} b) \tilde{\Rightarrow} a \tilde{\Rightarrow} c \in \mathcal{B}^+$,
4. $\tilde{\top} \in \mathcal{B}^+$,
5. $\tilde{\bot} \tilde{\Rightarrow} a \in \mathcal{B}^+$,
6. $a \tilde{\Rightarrow} b \tilde{\Rightarrow} (a \tilde{\wedge} b) \in \mathcal{B}^+$,
7. $(a \tilde{\wedge} b) \tilde{\Rightarrow} a \in \mathcal{B}^+$,
8. $(a \tilde{\wedge} b) \tilde{\Rightarrow} b \in \mathcal{B}^+$,
9. $a \tilde{\Rightarrow} (a \tilde{\vee} b) \in \mathcal{B}^+$,
10. $b \tilde{\Rightarrow} (a \tilde{\vee} b) \in \mathcal{B}^+$,
11. $(a \tilde{\vee} b) \tilde{\Rightarrow} (a \tilde{\Rightarrow} c) \tilde{\Rightarrow} (b \tilde{\Rightarrow} c) \tilde{\Rightarrow} c \in \mathcal{B}^+$,
12. *the set* $a \tilde{\Rightarrow} A = \{a \tilde{\Rightarrow} e \mid e \in A\}$ *is in* $\mathcal{A}$ *and the set* $E \tilde{\Rightarrow} a = \{e \tilde{\Rightarrow} a \mid e \in E\}$ *is in* $\mathcal{A}$,
13. *if all elements of* A *are in* $\mathcal{B}^+$ *then* $\tilde{\forall} A \in \mathcal{B}^+$,
14. $\tilde{\forall} (a \tilde{\Rightarrow} A) \tilde{\Rightarrow} a \tilde{\Rightarrow} (\tilde{\forall} A) \in \mathcal{B}^+$,
15. *if* $a \in A$, *then* $(\tilde{\forall} A) \tilde{\Rightarrow} a \in \mathcal{B}^+$,
16. *if* $a \in E$, *then* $a \tilde{\Rightarrow} (\tilde{\exists} E) \in \mathcal{B}^+$,
17. $(\tilde{\exists} E) \tilde{\Rightarrow} \tilde{\forall} (E \tilde{\Rightarrow} a) \tilde{\Rightarrow} a \in \mathcal{B}^+$.

Definition 2 (Full). *A truth values algebra is said to be* full *if* $\mathcal{A} = \mathcal{E} = \wp(\mathcal{B})$, i.e. *if* $\tilde{\forall} A$ *and* $\tilde{\exists} A$ *exist for all subsets* A *of* $\mathcal{B}$.

Definition 3 (Trivial). *A truth values algebra is said to be* trivial *if* $\mathcal{B}^+ = \mathcal{B}$.

Example 1. Let $\mathcal{B} = \{0, 1\}$. Let $\mathcal{B}^+ = \{1\}$, $\mathcal{A} = \mathcal{E} = \wp(\mathcal{B})$, $\tilde{\top} = 1$, $\tilde{\bot} = 0$, $\tilde{\Rightarrow}$, $\tilde{\wedge}$, $\tilde{\vee}$ be the usual boolean operations, $\tilde{\forall}$ be the function mapping the sets $\{0\}$ and $\{0, 1\}$ to 0 and $\varnothing$ and $\{1\}$ to 1 and $\tilde{\exists}$ be the function mapping the sets $\varnothing$ and $\{0\}$ to 0 and $\{1\}$ and $\{0, 1\}$ to 1. Then $\langle \mathcal{B}, \mathcal{B}^+, \mathcal{A}, \mathcal{E}, \tilde{\top}, \tilde{\bot}, \tilde{\Rightarrow}, \tilde{\wedge}, \tilde{\vee}, \tilde{\forall}, \tilde{\exists} \rangle$ is a truth value algebra.

Example 2. Let $\mathcal{B}$ be an arbitrary set, $\mathcal{B}^+ = \mathcal{B}$, $\mathcal{A} = \mathcal{E} = \wp(\mathcal{B})$ and $\tilde{\top}$, $\tilde{\bot}$, $\tilde{\Rightarrow}$, $\tilde{\wedge}$, $\tilde{\vee}$, $\tilde{\forall}$ and $\tilde{\exists}$ be arbitrary operations. Then $\langle \mathcal{B}, \mathcal{B}^+, \mathcal{A}, \mathcal{E}, \tilde{\top}, \tilde{\bot}, \tilde{\Rightarrow}, \tilde{\wedge}, \tilde{\vee}, \tilde{\forall}, \tilde{\exists} \rangle$ is a trivial truth value algebra.

2.2   Pseudo-heyting Algebras

In this section, we show that truth values algebras can alternatively be characterized as pseudo-Heyting algebras.

Definition 4 (Pseudo-Heyting algebra)
Let $\mathcal{B}$ *be a set,* $\leq$ *be a relation on* $\mathcal{B}$, $\mathcal{A}$ *and* $\mathcal{E}$ *be subsets of* $\wp(\mathcal{B})$, $\tilde{\top}$ *and* $\tilde{\bot}$ *be elements of* $\mathcal{B}$, $\tilde{\Rightarrow}$, $\tilde{\wedge}$, *and* $\tilde{\vee}$ *be functions from* $\mathcal{B} \times \mathcal{B}$ *to* $\mathcal{B}$, $\tilde{\forall}$ *be a function from* $\mathcal{A}$ *to* $\mathcal{B}$ *and* $\tilde{\exists}$ *be a function from* $\mathcal{E}$ *to* $\mathcal{B}$, *the structure* $\mathcal{B} = \langle \mathcal{B}, \leq, \mathcal{A}, \mathcal{E}, \tilde{\top}, \tilde{\bot}, \tilde{\Rightarrow},$

$\tilde{\wedge}, \tilde{\vee}, \tilde{\forall}, \tilde{\exists}\rangle$ *is said to be a* pseudo-Heyting algebra *if for all a, b, c in $\mathcal{B}$, A in $\mathcal{A}$ and E in $\mathcal{E}$, (the relation $\leq$ is a pre-order)*

- $a \leq a$,
- *if $a \leq b$ and $b \leq c$ then $a \leq c$,*

($\tilde{\top}$ and $\tilde{\bot}$ are maximum and minimum elements (notice that these need not be unique))

- $a \leq \tilde{\top}$,
- $\tilde{\bot} \leq a$,

(a $\tilde{\wedge}$ b is a greatest lower bound of a and b and and a $\tilde{\vee}$ b is a least upper bound of a and b (again, these need not be unique))

- $a \tilde{\wedge} b \leq a$,
- $a \tilde{\wedge} b \leq b$,
- *if $c \leq a$ and $c \leq b$ then $c \leq a \tilde{\wedge} b$,*
- $a \leq a \tilde{\vee} b$,
- $b \leq a \tilde{\vee} b$,
- *if $a \leq c$ and $b \leq c$ then $a \tilde{\vee} b \leq c$,*

(the set $\mathcal{A}$ and $\mathcal{E}$ have closure conditions)

- $a \tilde{\Rightarrow} A$ *and* $E \tilde{\Rightarrow} a$ *are in $\mathcal{A}$,*

($\tilde{\forall}$ and $\tilde{\exists}$ are infinite greatest lower bound and least upper bound)

- *if $a \in A$ then $\tilde{\forall} A \leq a$,*
- *if for all a in A, $b \leq a$ then $b \leq \tilde{\forall} A$,*
- *if $a \in E$ then $a \leq \tilde{\exists} E$,*
- *if for all a in E, $a \leq b$ then $\tilde{\exists} E \leq b$,*

and

- $a \leq b \tilde{\Rightarrow} c$ *if and only if $a \tilde{\wedge} b \leq c$.*

Proposition 1. *Consider a truth values algebra $\langle \mathcal{B}, \mathcal{B}^+, \mathcal{A}, \mathcal{E}, \tilde{\top}, \tilde{\bot}, \tilde{\Rightarrow}, \tilde{\wedge}, \tilde{\vee}, \tilde{\forall}, \tilde{\exists}\rangle$ then the algebra $\langle \mathcal{B}, \leq, \mathcal{A}, \mathcal{E}, \tilde{\top}, \tilde{\bot}, \tilde{\Rightarrow}, \tilde{\wedge}, \tilde{\vee}, \tilde{\forall}, \tilde{\exists}\rangle$ where the relation $\leq$ is defined by $a \leq b$ if and only if $a \tilde{\Rightarrow} b \in \mathcal{B}^+$ is a pseudo-Heyting algebra.*

Conversely, consider a pseudo-Heyting algebra $\langle \mathcal{B}, \leq, \mathcal{A}, \mathcal{E}, \tilde{\top}, \tilde{\bot}, \tilde{\Rightarrow}, \tilde{\wedge}, \tilde{\vee}, \tilde{\forall}, \tilde{\exists}\rangle$, then the algebra $\langle \mathcal{B}, \mathcal{B}^+, \mathcal{A}, \mathcal{E}, \tilde{\top}, \tilde{\bot}, \tilde{\Rightarrow}, \tilde{\wedge}, \tilde{\vee}, \tilde{\forall}, \tilde{\exists}\rangle$, where $\mathcal{B}^+ = \{x \mid \tilde{\top} \leq x\}$ is a truth values algebra.

Definition 5 (Heyting algebra). *A pseudo-Heyting algebra is said to be a Heyting algebra if the relation $\leq$ is antisymmetric*

- $x \leq y \Rightarrow y \leq x \Rightarrow x = y$.

Remark. If the pseudo-Heyting algebra $\langle \mathcal{B}, \leq, \mathcal{A}, \mathcal{E}, \tilde{\top}, \tilde{\bot}, \tilde{\Rightarrow}, \tilde{\wedge}, \tilde{\vee}, \tilde{\forall}, \tilde{\exists}\rangle$ is a Heyting algebra, then the set $\mathcal{B}^+ = \{x \mid \tilde{\top} \leq x\}$ is the singleton $\{\tilde{\top}\}$. Indeed, if $a \in \mathcal{B}^+$ then $\tilde{\top} \leq a$ and $a \leq \tilde{\top}$. Hence $a = \tilde{\top}$.

Definition 6. *A function F from a truth value algebra $\mathcal{B}_1$ to a truth value algebra $\mathcal{B}_2$ is said to be* a morphism of truth values algebras *if*

- $x \in \mathcal{B}_1^+$ *if and only if* $F(x) \in \mathcal{B}_2^+$,
- *if* $A \in \mathcal{A}_1$ *then* $F(A) \in \mathcal{A}_2$, *if* $E \in \mathcal{E}_1$ *then* $F(E) \in \mathcal{E}_2$,
- $F(\tilde{\top}_1) = \tilde{\top}_2$, $F(\tilde{\bot}_1) = \tilde{\bot}_2$, $F(a \; \tilde{\Rightarrow}_1 \; b) = F(a) \; \tilde{\Rightarrow}_2 \; F(b)$, $F(a \; \tilde{\wedge}_1 \; b) = F(a) \; \tilde{\wedge}_2 \; F(b)$, $F(a \; \tilde{\vee}_1 \; b) = F(a) \; \tilde{\vee}_2 \; F(b)$, $F(\tilde{\forall}_1 \; A) = \tilde{\forall}_2 \; F(A)$, $F(\tilde{\exists}_1 \; E) = \tilde{\exists}_2 \; F(E)$.

Morphisms of pseudo-Heyting algebras *are defined in a similar way except that the first condition is replaced by*

- $x \leq_1 y$ *if and only if* $F(x) \leq_2 F(y)$.

Proposition 2. *Let $\mathcal{B}$ be a pseudo-Heyting algebra, then there exists a pseudo-Heyting algebra $\mathcal{B}/\mathcal{B}^+$ that is a Heyting algebra and a morphism of pseudo-Heyting algebras Φ from $\mathcal{B}$ to $\mathcal{B}/\mathcal{B}^+$.*

Remark. We have proved that, in the definition of Heyting algebras, the antisymmetry is useless and can be dropped. The equivalence of truth values algebras and pseudo-Heyting algebras shows that antisymmetry is the only property that can be dropped and that truth values algebras are, in some sense, the best possible generalization of Heyting algebras, as we cannot require less than closure by intuitionistic deduction rules.

2.3   Examples of Truth Values Algebras

We have seen that the algebra $\{0, 1\}$ is a truth value algebra and more generally that all Heyting algebras are truth values algebras. We give in this section two examples of truth values algebras that are not Heyting algebras.

Example 3. The truth value algebra $\mathcal{T}_1$ is defined as follows. The set $\mathcal{T}_1$ is $\{0, I, 1\}$ and the set $\mathcal{T}_1^+$ is $\{I, 1\}$. The sets $\mathcal{A}$ and $\mathcal{E}$ are $\wp(\mathcal{T}_1)$. The functions $\tilde{\top}$, $\tilde{\bot}$, $\tilde{\wedge}$, $\tilde{\vee}$, $\tilde{\forall}$ and $\tilde{\exists}$ are the same as in the algebra $\{0, 1\}$, except that their value on I is the same as their value on 1. For instance the table of the operation $\tilde{\vee}$ is

	0	I	1
0	0	1	1
I	1	1	1
1	1	1	1

The function $\tilde{\Rightarrow}$ is defined by the table

	0	I	1
0	1	1	1
I	0	1	1
1	0	I	I

Notice that as $I \; \tilde{\Rightarrow} \; 1$ and $1 \; \tilde{\Rightarrow} \; I$ are both in $\mathcal{T}_1^+$ we have $I \leq 1$ and $1 \leq I$. Hence the relation $\leq$ is not antisymmetric and the truth value algebra $\mathcal{T}_1$ is not a Heyting algebra.

Example 4. The truth value algebra $\mathcal{T}_2$ is similar to $\mathcal{T}_1$, except that the function $\tilde{\Rightarrow}$ is defined by the table

	0	I	1
0	1	1	I
I	0	1	I
1	0	1	I

2.4   Ordered Truth Values Algebras

We consider truth values algebras extended with an order relation $\sqsubseteq$ on $\mathcal{B}$. This order relation extends to sets of truth values in a trivial way: $A \sqsubseteq B$ if for all x in A there exists a y in B such that $x \sqsubseteq y$.

Definition 7 (Ordered truth values algebra). *An ordered truth values algebra is a truth values algebra together with a relation $\sqsubseteq$ on $\mathcal{B}$ such that*

- $\sqsubseteq$ *is an order relation,*
- $\mathcal{B}^+$ *is upward closed,*
- $\tilde{\top}$ *is a maximal element,* $\tilde{\bot}$ *a minimal element,*
- $\tilde{\wedge}$, $\tilde{\vee}$, $\tilde{\forall}$ *and* $\tilde{\exists}$ *are monotonous,* $\tilde{\Rightarrow}$ *is left anti-monotonous and right monotonous.*

Definition 8 (Complete ordered truth values algebra). *A ordered truth values algebra is said to be* complete *if every subset of $\mathcal{B}$ has a greatest lower bound for $\sqsubseteq$. Notice that this implies that every subset also has a least upper bound. We write $glb(a, b)$ and $lub(a, b)$ the greatest lower bound and the least upper bound of a and b for the order $\sqsubseteq$.*

Example 5. The algebra $\mathcal{T}_1$ ordered by $0 \sqsubseteq I \sqsubseteq 1$ is complete.

Example 6. The algebra $\mathcal{T}_2$ cannot be extended to a complete ordered algebra. Indeed the set $\{I, 1\}$ would need to have a least upper bound. This least upper bound cannot be 0 because $\mathcal{T}_2^+$ would then not be upward closed. If it were 1 then we would have $I \sqsubseteq 1$ and thus $1 \tilde{\Rightarrow} I \sqsubseteq 1 \tilde{\Rightarrow} 1$, *i.e.* $1 \sqsubseteq I$. Thus the relation $\sqsubseteq$ would not be antisymmetric. If it were I then we would have $1 \sqsubseteq I$ and thus $1 \tilde{\Rightarrow} 1 \sqsubseteq 1 \tilde{\Rightarrow} I$, *i.e.* $I \sqsubseteq 1$. Thus the relation $\sqsubseteq$ would not be antisymmetric.

Proposition 3. *In a Heyting algebra, $\leq$ and $\sqsubseteq$ are extensionally equal, i.e. $a \sqsubseteq b$ if and only if $a \leq b$.*

2.5   Completion

We now want to prove that for any truth value algebra $\mathcal{B}$, there is another truth value algebra $\mathcal{B}_C$ that is full, ordered and complete and a morphism Φ from $\mathcal{B}$ to $\mathcal{B}_C$. Notice that we do not require the morphism Φ to be injective.

There are two ways to prove this, the first is to use Proposition 2 in a first step to build a truth value algebra $\mathcal{B}/\mathcal{B}^+$ that is a Heyting algebra and a morphism for

$\mathcal{B}$ to $\mathcal{B}/\mathcal{B}^+$ and then apply in a second step MacNeille completion to the algebra $\mathcal{B}/\mathcal{B}^+$ to embed it into a full Heyting algebra. Together with its natural order, this algebra is a full, ordered and complete truth value algebra. The second is to apply MacNeille completion directly to $\mathcal{B}$ noticing that antisymmetry is not used in MacNeille completion, except to prove the injectivity of the morphism. The proof is detailed in the long version of the paper.

Example 7. The algebra $\mathcal{T}_2$ cannot be extended to a complete ordered algebra, but it can be embedded with a non injective morphism in the full ordered and complete algebra $\{0, 1\}$.

3   Predicate Logic

3.1   Models

Definition 9 ($\mathcal{B}$-valued structure). *Let $\mathcal{L} = \langle f_i, P_j \rangle$ be a language in predicate logic and $\mathcal{B}$ be a truth values algebra, a $\mathcal{B}$-valued structure for the language $\mathcal{L}$, $\mathcal{M} = \langle M, \mathcal{B}, \hat{f}_i, \hat{P}_j \rangle$ is a structure such that $\hat{f}_i$ is a function from M^n to M where n is the arity of the symbol f_i and $\hat{P}_j$ is a function from M^n to $\mathcal{B}$ where n is the arity of the symbol P_i.*

This definition extends trivially to many-sorted languages.

Definition 10 (Denotation). *Let $\mathcal{B}$ be a truth values algebra, $\mathcal{M}$ be a $\mathcal{B}$-valued structure and ϕ be an assignment. The denotation $[\![A]\!]_\phi$ of a formula A in $\mathcal{M}$ is defined as follows*

- $[\![x]\!]_\phi = \phi(x)$,
- $[\![f(t_1, ..., t_n)]\!]_\phi = \hat{f}([\![t_1]\!]_\phi, ..., [\![t_n]\!]_\phi)$,
- $[\![P(t_1, ..., t_n)]\!]_\phi = \hat{P}([\![t_1]\!]_\phi, ..., [\![t_n]\!]_\phi)$,
- $[\![\top]\!]_\phi = \tilde{\top}$,
- $[\![\bot]\!]_\phi = \tilde{\bot}$,
- $[\![A \Rightarrow B]\!]_\phi = [\![A]\!]_\phi \tilde{\Rightarrow} [\![B]\!]_\phi$,
- $[\![A \wedge B]\!]_\phi = [\![A]\!]_\phi \tilde{\wedge} [\![B]\!]_\phi$,
- $[\![A \vee B]\!]_\phi = [\![A]\!]_\phi \tilde{\vee} [\![B]\!]_\phi$,
- $[\![\forall x\, A]\!]_\phi = \tilde{\forall} \{[\![A]\!]_{\phi + \langle x, e \rangle} \mid e \in \mathcal{M}\}$,
- $[\![\exists x\, A]\!]_\phi = \tilde{\exists} \{[\![A]\!]_{\phi + \langle x, e \rangle} \mid e \in \mathcal{M}\}$.

Notice that the denotation of a formula containing quantifiers may be undefined, but it is always defined if the truth value algebra is full.

Definition 11 (Model). *A formula A is said to be* valid *in a $\mathcal{B}$-valued structure $\mathcal{M}$, and the $\mathcal{B}$-valued structure $\mathcal{M}$ is said to be* a model *of A, $\mathcal{M} \models A$, if for all assignments ϕ, $[\![A]\!]_\phi$ is defined and is a positive truth value.*

The $\mathcal{B}$-valued structure $\mathcal{M}$ is said to be a model *of a theory $\mathcal{T}$ if it is a model of all the axioms of $\mathcal{T}$.*

3.2   Soundness and Completeness

As the notion of truth values algebra extends that of Heyting algebra, the completeness theorem for the notion of model introduced above is a simple corollary of the completeness theorem for the notion of model based on Heyting algebras. But, it has a simpler direct proof. It is well-known that completeness proofs for boolean algebra valued models and Heyting algebra valued models are simpler than for $\{0, 1\}$-valued models. For truth values algebra valued models, it is even simpler. We want to prove that if A is valid in all models of $\mathcal{T}$ where it has a denotation then $\mathcal{T} \vdash A$. To do so, we consider a theory $\mathcal{T}$ and we construct a model of $\mathcal{T}$ such that the formulae valid in this model are the intuitionistic theorems of $\mathcal{T}$.

Definition 12 (Lindenbaum model). *Let $\mathcal{T}$ be a theory in a language $\mathcal{L}$. Let S be an infinite set of constants and $\mathcal{L}' = \mathcal{L} \cup S$. Let $\mathcal{M}$ be the set of closed terms of $\mathcal{L}'$ and $\mathcal{B}_\mathcal{T}$ be the set of closed formulae of $\mathcal{L}'$. Let $\mathcal{B}_\mathcal{T}^+$ be the set of elements A of $\mathcal{B}_\mathcal{T}$, such that the sequent $\mathcal{T} \vdash A$ is provable. Let $\mathcal{A} = \mathcal{E}$ be the set of subsets of $\mathcal{B}_\mathcal{T}$ of the form $\{(t/x)A \mid t \in \mathcal{M}\}$ for some A. Notice that, in this case, the formula A is unique.*

The operations $\tilde{\top}$, $\tilde{\bot}$, $\tilde{\Rightarrow}$, $\tilde{\wedge}$ and $\tilde{\vee}$ are $\top$, $\bot$, $\Rightarrow$, $\wedge$ and $\vee$. The operations $\tilde{\forall}$ and $\tilde{\exists}$ are defined as follows

- $\tilde{\forall} \{(t/x)A \mid t \in \mathcal{M}\} = (\forall x\ A)$,
- $\tilde{\exists} \{(t/x)A \mid t \in \mathcal{M}\} = (\exists x\ A)$.

If f is a function symbol, we let $\hat{f}$ be the function mapping $t_1, ..., t_n$ to $f(t_1, ..., t_n)$. If P is a predicate symbol, we let $\hat{P}$ be the function mapping $t_1, ..., t_n$ to $P(t_1, ..., t_n)$.

Proposition 4. *The algebra $\mathcal{B}_\mathcal{T}$ is a truth values algebra.*

Proposition 5 (Completeness). *If A is valid in all the models of $\mathcal{T}$ where it is defined, then $\mathcal{T} \vdash A$.*

Using completion, we can strengthen this completeness theorem.

Proposition 6. *If A is valid in all the models of $\mathcal{T}$ where the truth values algebra is full, ordered and complete then $\mathcal{T} \vdash A$.*

The converse is a simple induction over proof structure.

Proposition 7 (Soundness). *If $\mathcal{T} \vdash A$ then A is valid in all the models of $\mathcal{T}$ where the truth value algebra is full, ordered and complete.*

We finally get the following theorem.

Theorem 1. *$\mathcal{T} \vdash A$ if and only if A is valid in all the models of $\mathcal{T}$ where the truth values algebra is full, ordered and complete.*

3.3 Consistency

Definition 13. *A theory is said to be* consistent *if there exists a non provable formula in this theory.*

In the completeness theorem above, we did not assume the theory $\mathcal{T}$ to be consistent. If it is not, then the algebra of the Lindenbaum model is trivial, *i.e.* all truth values are positive and every formula is valid. But we have the following theorem.

Proposition 8. *The theory $\mathcal{T}$ is consistent if and only if it has a $\mathcal{B}$-valued model, for some non trivial full, ordered and complete truth values algebra $\mathcal{B}$.*

4 Deduction Modulo

4.1 Deduction Modulo

In Deduction modulo [3,5], a theory is defined by a set of axioms $\mathcal{T}$ and a congruence $\equiv$ defined by a confluent rewrite system rewriting terms to terms and atomic formulae to formulae. The deduction rules are modified to take the congruence $\equiv$ into account. For instance, the *modus ponens* rule is not stated as usual

$$\frac{\Gamma \vdash_\equiv A \Rightarrow B \quad \Gamma \vdash_\equiv A}{\Gamma \vdash_\equiv B}$$

but

$$\frac{\Gamma \vdash_\equiv C \quad \Gamma \vdash_\equiv A}{\Gamma \vdash_\equiv B} \, C \equiv A \Rightarrow B$$

In deduction modulo, some proofs, in some theories, do not normalize. For instance, in the theory formed with the rewrite rule $P \longrightarrow (P \Rightarrow Q)$, the proof

$$\cfrac{\cfrac{\cfrac{\overline{P \vdash_\equiv P \Rightarrow Q}\text{ axiom} \quad \overline{P \vdash_\equiv P}\text{ axiom}}{P \vdash_\equiv Q}\Rightarrow\text{-elim}}{\vdash_\equiv P \Rightarrow Q}\Rightarrow\text{-intro} \quad \cfrac{\overline{P \vdash_\equiv P \Rightarrow Q}\text{ axiom} \quad \overline{P \vdash_\equiv P}\text{ axiom}}{\cfrac{P \vdash_\equiv Q}{\vdash_\equiv P}\Rightarrow\text{-intro}}\Rightarrow\text{-elim}}{\vdash_\equiv Q}\Rightarrow\text{-elim}$$

does not normalize and, moreover, the formula Q has no normal proof.

But, as we shall see, in some other theories, such as the theory formed with the rewrite rule $P \longrightarrow (Q \Rightarrow P)$, all proofs strongly normalize.

In deduction modulo, like in predicate logic, normal proofs of a sequent of the form $\vdash_\equiv A$ always end with an introduction rule. Thus, when a theory can be expressed in deduction modulo with rewrite rules only, *i.e.* with no axioms, in such a way that proofs modulo these rewrite rules strongly normalize, then the theory is consistent, it has the disjunction property and the witness property, various proof search methods for this theory are complete, ...

Many theories can be expressed this way in deduction modulo, in particular arithmetic [6] and simple type theory [4] and the notion of cut of deduction modulo subsumes the *ad hoc* notions of cut defined for these theories.

4.2   Models

Definition 14 (Model). *Let* $\mathcal{T}, \equiv$ *be a theory in deduction modulo. The* $\mathcal{B}$-*valued structure* $\mathcal{M}$ *is said to be* a model of *the theory* $\mathcal{T}, \equiv$ *if all axioms of* $\mathcal{T}$ *are valid in* $\mathcal{M}$ *and for all terms or formulae* A *and* B *such that* $A \equiv B$ *and assignment* ϕ, $[\![A]\!]_\phi$ *and* $[\![B]\!]_\phi$ *are defined and* $[\![A]\!]_\phi = [\![B]\!]_\phi$.

Example 8. Let $\mathcal{B}$ be an arbitrary truth value algebra, then the theory $P \longrightarrow$ $(Q \Rightarrow R)$ has a $\mathcal{B}$-valued model. We take $\hat{P} = (\tilde{\top} \tilde{\Rightarrow} \tilde{\top})$ and $\hat{Q} = \hat{R} = \tilde{\top}$.

Example 9. Let $\mathcal{B}$ be an arbitrary full, ordered and complete truth value algebra, then the theory $P \longrightarrow (Q \Rightarrow P)$ has a $\mathcal{B}$-valued model. The function $a \mapsto$ $(\tilde{\bot} \tilde{\Rightarrow} a)$ is monotonous for the order $\sqsubseteq$ and this order is complete. Hence, it has a fixed point b. We define a $\mathcal{B}$-valued model of this theory by $\hat{P} = b$ and $\hat{Q} = \tilde{\bot}$.

In the same way, if $\mathcal{B}$ be an arbitrary full, ordered and complete truth value algebra, then the theory $P \longrightarrow (\bot \Rightarrow P)$ has a $\mathcal{B}$-valued model.

Example 10. The theory $P \longrightarrow (P \Rightarrow Q)$ has a $\{0, 1\}$-valued model ($\hat{P} = \hat{Q} = 1$), but no $\mathcal{T}_1$-valued model. Indeed there is no 0 in the line 0 of the table of the function $\tilde{\Rightarrow}$ of $\mathcal{T}_1$, no I in the line I and no 1 in the line 1.

4.3   Soundness and Completeness

To extend the completeness and the soundness theorem to deduction modulo, we replace terms by classes of congruent terms and formulae by classes of congruent formulae.

Definition 15 (Lindenbaum model). *Let* $\mathcal{T}, \equiv$ *be a theory in a language* $\mathcal{L}$. *Let* S *be an infinite set of constants and* $\mathcal{L}' = \mathcal{L} \cup S$. *Let* $\mathcal{M}$ *be the set of* $\equiv$-*classes of closed terms of* $\mathcal{L}'$ *and* $\mathcal{B}$ *be the set of* $\equiv$-*classes of closed formulae of* $\mathcal{L}'$. *Let* $\mathcal{B}^+$ *be the set of elements of* $\mathcal{B}$ *containing a formula* A *such that the sequent* $\mathcal{T} \vdash_\equiv A$ *is provable. Let* $\mathcal{A} = \mathcal{E}$ *be the set of subsets of* $\mathcal{B}$ *of the form* $\{(t/x)A/ \equiv \;|\; t \in \mathcal{M}\}$ *for some* A.

The operations $\tilde{\top}$, $\tilde{\bot}$, $\tilde{\Rightarrow}$, $\tilde{\wedge}$ *and* $\tilde{\vee}$ *are* $\top$, $\bot$, $\Rightarrow$, $\wedge$ *and* $\vee$ *extended to* $\equiv$-*classes. To define the operations* $\tilde{\forall}$ *and* $\tilde{\exists}$, *we choose for each element* a *of* $\mathcal{A}$ *and* $\mathcal{E}$ *a formula* A *such that* $a = \{(t/x)A/ \equiv \;|\; t \in \mathcal{M}\}$ *and we let*

- $\tilde{\forall} a = (\forall x\ A)$,
- $\tilde{\exists} a = (\exists x\ A)$.

If f *is a function symbol, we let* $\hat{f}$ *be the function mapping the classes of* $t_1, ..., t_n$ *to that of* $f(t_1, ..., t_n)$. *If* P *is a predicate symbol, we let* $\hat{P}$ *be the function mapping the classes of* $t_1, ..., t_n$ *to that of* $P(t_1, ..., t_n)$.

However, this introduces some additional complexity as we lose the property that the class of formulae A is uniquely determined by the set of classes of formulae $\{(t/x)A \mid t \in \mathcal{M}\}$. Indeed, consider, for instance, the congruence generated by the rewrite rule $f(f(x)) \longrightarrow x$. Then the formulae $P(x)$ and $P(f(x))$ have the same set of instances (the instance t in one formula corresponds to the instance $f(t)$ in the other). Thus, we need first the following definition and proposition.

Definition 16. *Two sets A and B of (classes of) formulae are said to be* equivalent modulo $\equiv$ *if for each formula P in A there exists a formula Q in B such that $P \Leftrightarrow Q$ is provable in deduction modulo $\equiv$ and vice versa.*

Proposition 9. *Let $\equiv$ be a congruence defined by a confluent rewrite system rewriting terms to terms and atomic formulae to formulae. Let P and Q be two formulae such that the sets $\{(t/x)P/ \equiv \; | \; t \in \mathcal{M}\}$ and $\{(t/x)Q/ \equiv \; | \; t \in \mathcal{M}\}$ are equivalent modulo $\equiv$, then $(\forall x\ P) \Rightarrow (\forall x\ Q)$ is provable in deduction modulo the congruence $\equiv$.*

Proposition 10. *The algebra $\mathcal{B}$ is a truth values algebra.*

Proposition 11 (Completeness). *If A is valid in all the models of $T, \equiv$ where it is defined, then $T \vdash_{\equiv} A$.*

Using completion, we can strengthen this completeness theorem.

Proposition 12. *If A is valid in all the models of $T, \equiv$ where the truth values algebra is full, ordered and complete then $T \vdash_{\equiv} A$.*

The converse is a simple induction over proof structure.

Proposition 13 (Soundness). *If $T \vdash_{\equiv} A$ then A is valid in all the models of $T, \equiv$ where the truth value algebra is full, ordered and complete.*

We finally get the following theorem.

Theorem 2. *$T \vdash_{\equiv} A$ if and only if A is valid in all the models of $T, \equiv$ where the truth values algebra is full, ordered and complete.*

4.4   Consistency

Proposition 14. *The theory $T, \equiv$ is consistent if and only if it has a $\mathcal{B}$-valued model, for some non trivial full, ordered and complete truth values algebra $\mathcal{B}$.*

5   Super-Consistency

5.1   Definition

By Proposition 14, a theory is consistent if it has a $\mathcal{B}$-valued model for some non trivial full, ordered and complete truth values algebra. We now strengthen this condition and require that the theory has a $\mathcal{B}$-valued model for all full, ordered and complete truth values algebras $\mathcal{B}$.

Definition 17 (Super-consistent). *A theory $T, \equiv$ in deduction modulo is* super-consistent *if it has a $\mathcal{B}$-valued model for all full, ordered and complete truth values algebras $\mathcal{B}$.*

Notice that, as there exists non trivial full, ordered and complete truth values algebras (*e.g.* $\{0, 1\}$), super-consistent theories are consistent.

5.2   Examples of Super-Consistent Theories

We have seen that the theories $P \longrightarrow (Q \Rightarrow R)$ and $P \longrightarrow (Q \Rightarrow P)$ are super-consistent, but that the theory $P \longrightarrow (P \Rightarrow Q)$ is not. We give other examples of super-consistent theory. In particular, we show that all the theories that have been proved to have the strong normalization property in [5,6] *i.e.* arithmetic, simple type theory, the theories defined by a confluent, terminating and quantifier free rewrite system, the theories defined by a confluent, terminating and positive rewrite systems and the theories defined by a positive rewrite systems such that each atomic formula has at most one one-step reduct are super-consistent. In this abstract, we detail only the case of simple type theory.

Definition 18 (Simple type theory). *Simple type theory is a many-sorted theory defined as follows. The sorts are inductively defined by ι and o are sorts and if T and U are sorts then $T \to U$ is a sort. The language contains the constants $S_{T,U,V}$ of sort $(T \to U \to V) \to (T \to U) \to T \to V$, $K_{T,U}$ of sort $T \to U \to T$, $\dot{\top}$ of sort o and $\dot{\bot}$ of sort o, $\dot{\Rightarrow}$, $\dot{\wedge}$ and $\dot{\vee}$ of sort $o \to o \to o$, $\dot{\forall}_T$ and $\dot{\exists}_T$ of sort $(T \to o) \to o$, the function symbols $\alpha_{T,U}$ of rank $\langle T \to U, T, U \rangle$ and the predicate symbol ε of rank $\langle o \rangle$. The rules are*

$$\alpha(\alpha(\alpha(S_{T,U,V}, x), y), z) \longrightarrow \alpha(\alpha(x, z), \alpha(y, z))$$
$$\alpha(\alpha(K_{T,U}, x), y) \longrightarrow x$$
$$\varepsilon(\dot{\top}) \longrightarrow \top$$
$$\varepsilon(\dot{\bot}) \longrightarrow \bot$$
$$\varepsilon(\alpha(\alpha(\dot{\Rightarrow}, x), y)) \longrightarrow \varepsilon(x) \Rightarrow \varepsilon(y)$$
$$\varepsilon(\alpha(\alpha(\dot{\wedge}, x), y)) \longrightarrow \varepsilon(x) \wedge \varepsilon(y)$$
$$\varepsilon(\alpha(\alpha(\dot{\vee}, x), y)) \longrightarrow \varepsilon(x) \vee \varepsilon(y)$$
$$\varepsilon(\alpha(\dot{\forall}, x)) \longrightarrow \forall y \; \varepsilon(\alpha(x, y))$$
$$\varepsilon(\alpha(\dot{\exists}, x)) \longrightarrow \exists y \; \varepsilon(\alpha(x, y))$$

Proposition 15. *Simple type theory is super-consistent.*

Proof. Let $\mathcal{B}$ be a full truth values algebra. The model $\mathcal{M}_\iota = \{0\}$, $\mathcal{M}_o = \mathcal{B}$, $\mathcal{M}_{T \to U} = \mathcal{M}_U^{\mathcal{M}_T}$, $\hat{S}_{T,U,V} = a \mapsto (b \mapsto (c \mapsto a(c)(b(c))))$, $\hat{K}_{T,U} = a \mapsto (b \mapsto a)$, $\hat{\alpha}(a, b) = a(b)$, $\hat{\varepsilon}(a) = a$, $\hat{\dot{\top}} = \tilde{\top}$, $\hat{\dot{\bot}} = \tilde{\bot}$, $\hat{\dot{\Rightarrow}} = \tilde{\Rightarrow}$, $\hat{\dot{\wedge}} = \tilde{\wedge}$, $\hat{\dot{\vee}} = \tilde{\vee}$, $\hat{\dot{\forall}}_T = a \mapsto \tilde{\forall}(Range(a))$, $\hat{\dot{\exists}}_T = a \mapsto \tilde{\exists}(Range(a))$ where $Range(a)$ is the range of the function a, is a $\mathcal{B}$-valued model of simple type theory.

5.3   Normalization

We have seen that the theory $P \longrightarrow (P \Rightarrow Q)$, that does not have the strong normalization property, is consistent but not super-consistent, *i.e.* it has $\mathcal{B}$-valued models for some non trivial, full, ordered and complete truth values algebras $\mathcal{B}$, but not all. We prove now that, in contrast, all super-consistent theories have the strong normalization property. To prove this, we build a particular full, ordered and complete truth values algebra: the algebra of reducibility candidates.

We refer, for instance, to [5] for the definition of proof-terms, neutral proof-terms and of proof-term reduction $\triangleright$ and we define the following operations on sets of proofs.

Definition 19

- *The set $\tilde{\top}$ is the set of strongly normalizing proof-terms.*
- *The set $\tilde{\perp}$ is the set of strongly normalizing proof-terms.*
- *If a and b are two sets of proofs-terms, then $a \tilde{\Rightarrow} b$ is the set of strongly normalizing proof-terms π such that if π reduces to $\lambda \alpha\ \pi_1$ then for every π' in a, $(\pi'/\alpha)\pi_1$ is in b.*
- *If a and b are two sets of proof-terms, then then $a \tilde{\wedge} b$ is the set of strongly normalizing proof-terms π such that if π reduces to $\langle \pi_1, \pi_2 \rangle$ then π_1 is in a and π_2 is in b.*
- *If a and b are two sets of proof-terms, then $a \tilde{\vee} b$ is the set of strongly normalizing proof-terms π such that if π reduces to $i(\pi_1)$ (resp. $j(\pi_2)$) then π_1 (resp. π_2) is in a (resp. b).*
- *If A is a set of sets of proof-terms, then $\tilde{\forall} A$ is the set of strongly normalizing proof-terms π such that if π reduces to $\lambda x\ \pi_1$ then for every term t and every element a of A, $(t/x)\pi_1$ is in a.*
- *If A is a set of sets of proof-terms, then $\tilde{\exists} A$ is the set of strongly normalizing proof-terms π such that if π reduces to $\langle t, \pi_1 \rangle$, there exists an element a of A such that π_1 is an element of a.*

Definition 20 (Reducibility candidate). *A set R of proof-terms is a reducibility candidate if*

- *if $\pi \in R$, then π is strongly normalizable,*
- *if $\pi \in R$ and $\pi \triangleright^* \pi'$ then $\pi' \in R$,*
- *if π is neutral and if for every π' such that $\pi \triangleright^1 \pi'$, $\pi' \in R$ then $\pi \in R$.*

Proposition 16. *The set of reducibility candidates is closed by the operations of Definition 19.*

Definition 21 (The algebra of reducibility candidates). *The set $\mathcal{B}$ is the set of reducibility candidates. The set $\mathcal{B}^+$ may be any set closed by intuitionistic deduction rules, e.g. the set of all candidates. The sets $\mathcal{A}$ and $\mathcal{E}$ are $\wp(\mathcal{B})$. The operations are those of definition 19. The order $\sqsubseteq$ is inclusion.*

Theorem 3 (Normalization). *If the theory $\mathcal{T}, \equiv$ is super-consistent, then all proofs strongly normalize in $\mathcal{T}, \equiv$.*

Proof. Consider the full, ordered and complete truth values algebra $\mathcal{B}$ of reducibility candidates. As it is super-consistent, the theory $\mathcal{T}, \equiv$ has a $\mathcal{B}$-valued model. This model is a reducibility candidate valued model of $\equiv$ [5], called *pre-models* there. Hence all proofs strongly normalize in $\mathcal{T}, \equiv$.

An alternative would be to define the set of candidates directly as the smallest set of sets of proofs closed by the operations of definition 19 and arbitrary intersections, like [12].

Notice that the pre-order $\leq$ is trivial and thus not antisymmetric. Hence, the truth values algebra of reducibility candidates is not a Heyting algebra. The fact that the choice of the set $\mathcal{B}^+$ is immaterial is due to the fact that $\mathcal{B}^+$ matters for the interpretation of axioms but not for that of the congruence and cut elimination is a property of the congruence of a theory, not of its axioms.

6   Conclusion

We have generalized the notion of Heyting algebra into a notion of *truth values algebra* and proved that a theory is consistent if and only if it has a $\mathcal{B}$-valued model for some non trivial full, ordered and complete truth values algebra $\mathcal{B}$. Unlike Heyting algebra valued models, truth values algebra valued models allow to distinguish computational equivalence from provable equivalence.

When a theory has a $\mathcal{B}$-valued model for all full, ordered and complete truth values algebras, it is said to be super-consistent and all proofs strongly normalize in this theory. Proving strong normalization by proving super-consistency is easier than proving strong normalization directly. For instance the proof that simple type theory is super-consistent (Proposition 15) takes only a few lines. All the technicalities related to the notion of reducibility candidate are now hidden in the proof that super-consistency implies strong normalization and are not used in the proof that the theory of interest is super-consistent.

The notion of super-consistency is a model theoretic sufficient condition for strong normalization. It remains to understand if it also a necessary condition or if some theories have the strong normalization property without being super-consistent. To prove that strong normalization implies super-consistency, we might need to restrict further the notion of super-consistency. For instance, we have already restricted it by considering only ordered and complete truth values algebras. Indeed, without such a completeness property, we could not use the fixed point theorem to prove that the theory $P \longrightarrow (\perp \Rightarrow P)$ had a $\mathcal{B}$-valued model for all $\mathcal{B}$, and indeed, this theory does not have a $\mathcal{T}_2$-valued model. Thus, the fact that the algebra of reducibility candidates, ordered by inclusion, is complete seems to be an essential property that needs to be kept when abstracting on reducibility candidates. It remains to understand if there are other essential properties of candidates that need to be kept this way, so that strong normalization may imply super-consistency.

Acknowledgments

Thierry Coquand suggested to characterize truth values algebras as pseudo-Heyting algebras. Lisa Allali, Olivier Hermant and Milly Maietti provided many helpful comments on a previous version of the paper.

References

1. Andrews, P.B.: Resolution in type theory. The Journal of Symbolic Logic 36(3), 414–432 (1971)
2. De Marco, M., Lipton, J.: Completeness and cut-elimination in the intuitionistic theory of types. Journal of Logic and Computation 15, 821–854 (2005)
3. Dowek, G., Hardin, T., Kirchner, C.: Theorem proving modulo. Journal of Automated Reasoning 31, 32–72 (2003)
4. Dowek, G., Hardin, T., Kirchner, C.: HOL-lambda-sigma: an intentional first-order expression of higher-order logic. Mathematical Structures in Computer Science 11, 1–25 (2001)
5. Dowek, G., Werner, B.: Proof normalization modulo. The Journal of Symbolic Logic 68(4), 1289–1316 (2003)
6. Dowek, G., Werner, B.: Arithmetic as a theory modulo. In: Giesl, J. (ed.) RTA 2005. LNCS, vol. 3467, pp. 423–437. Springer, Heidelberg (2005)
7. Girard, J.-Y.: Une extension de l'interprétation de Gödel à l'analyse, et son application à l'élimination des coupures dans l'analyse et la théorie des types. In: Fenstad, J. (ed.) 2nd Scandinavian Logic Symposium, pp. 63–92. North Holland (1971)
8. Hermant, O.: A model based cut elimination proof. In: 2nd St-Petersbourg Days in Logic and Computability (2003)
9. Hermant, O.: Méthodes sémantiques en déduction modulo. Doctoral Thesis. Université de Paris 7 (2005)
10. Hermant, O.: Semantic cut elimination in the intuitionistic sequent calculus. In: Urzyczyn, P. (ed.) TLCA 2005. LNCS, vol. 3461, pp. 221–233. Springer, Heidelberg (2005)
11. Okada, M.: A uniform semantic proof for cut elimination and completeness of various first and higher order logics. Theoretical Computer Science 281, 471–498 (2002)
12. Parigot, M.: Strong normalization for the second orclassical natural deduction. Logic in Computer Science, pp. 39–46 (1993)
13. Prawitz, D.: Hauptsatz for higher order logic. The Journal of Symbolic Logic 33, 452–457 (1968)
14. Rasiowa, H., Sikorski, R.: The mathematics of metamathematics. Polish Scientific Publishers (1963)
15. Tait, W.W.: A non constructive proof of Gentzen's Hauptsatz for second order predicate logic. Bulletin of the American Mathematical Society 72, 980–983 (1966)
16. Tait, W.W.: Intentional interpretations of functionals of finite type I. The Journal of Symbolic Logic 32, 198–212 (1967)
17. Takahashi, M.o.: A proof of cut-elimination theorem in simple type theory. Journal of the Mathematical Society of Japan 19, 399–410 (1967)

Curry-Style Types for Nominal Terms[*]

Maribel Fernández[1] and Murdoch J. Gabbay[2]

[1] King's College London
Dept. Computer Science
Strand, London WC2R 2LS, UK
maribel.fernandez@kcl.ac.uk
[2] Heriot-Watt University
Dept. Mathematics and Computer Science
Riccarton, Edinburgh EH14 4AS, UK
murdoch.gabbay@gmail.com

Abstract. We define a rank 1 polymorphic type system for nominal terms, where typing environments type atoms, variables and function symbols. The interaction between type assumptions for atoms and substitution for variables is subtle: substitution does not avoid capture and so can move an atom into multiple different typing contexts. We give typing rules such that principal types exist and are decidable for a fixed typing environment. α-equivalent nominal terms have the same types; a non-trivial result because nominal terms include explicit constructs for renaming atoms. We investigate rule formats to guarantee subject reduction. Our system is in a convenient Curry-style, so the user has no need to explicitly type abstracted atoms.

Keywords: binding, polymorphism, type inference, rewriting.

1 Introduction

Nominal terms are used to specify and reason about formal languages with binding, such as logics or programming languages. Consider denumerable sets of **atoms** $a, b, c, \ldots$, **variables** $X, Y, Z, \ldots$, and **term-formers** or **function symbols** $f, g, \ldots$. Following previous work [10,21], **nominal terms** t are defined by:

$$s, t ::= a \mid [a]t \mid ft \mid (t_1, \ldots, t_n) \mid \pi{\cdot}X \qquad \pi ::= \mathbf{Id} \mid (a\ b) \circ \pi$$

and called respectively **atoms**, **abstractions**, **function applications**, **tuples**, and **moderated variables** (or just **variables**). We call π a **permutation** and read $(a\ b){\cdot}X$ as '**swap** a **and** b **in** X'. We say that permutations **suspend** on variables. X is not a term but $\mathbf{Id}{\cdot}X$ is and we usually write it just as X. Similarly we omit the final $\mathbf{Id}$ in π, writing $(a\ c){\cdot}X$ instead of $((a\ c) \circ \mathbf{Id}){\cdot}X$.

For example, suppose term-formers lam and app. Then the nominal terms $lam[a]a$ and $app(lam[a]a, a)$ represent λ-terms $\lambda x.x$ and $(\lambda x.x)x$, and $lam[a]X$ and $app(lam[a]X, X)$ represent λ-term 'contexts' $\lambda x.\text{-}$ and $(\lambda x.\text{-})\text{-}$. Note how X

---

[*] Research partially supported by the EPSRC (EP/D501016/1 "CANS").

T. Altenkirch and C. McBride (Eds.): TYPES 2006, LNCS 4502, pp. 125–139, 2007.
© Springer-Verlag Berlin Heidelberg 2007

occurs under different abstractions. Substitution for X is grafting of syntax trees, it does *not* avoid capture of atoms by abstractions; we may call it **instantiation**.

Nominal terms differ from other treatments of syntax-with-binding because they support a capturing substitution, and the notation, although formal, is close to standard informal practice; for example β-reduction may be represented simply but explicitly as $app(lam[a]X, Y) \to sub([a]X, Y)$ where $sub([a]X, Y)$ is a term which may be given the behaviour of 'non-capturing substitution' (once we instantiate X and Y) by rewrite rules [10,12].

Now consider *static* semantics, i.e. types like $\tau ::= \mathbb{N} \mid \tau \to \tau$ where we read $\mathbb{N}$ as **numbers** and $\tau \to \tau$ as **function types**. Assigning types to terms partitions the language into 'numbers', or 'functions between numbers', and so on. Java [16], ML [9], and System F [15] demonstrate how this is commercially useful and theoretically interesting.

Existing static semantics for nominal terms type atoms with a special type of atoms $\mathbb{A}$ [21,20]. But when we write $lam[a]X$ or $lam[a]a$, our intention is $\lambda x.$- or $\lambda x.x$ and we do not expect a to be forbidden from having any type other than $\mathbb{A}$. We can use explicit casting function symbols to inject $\mathbb{A}$ into other types; however the a in $lam[a]X$ still has type $\mathbb{A}$, so we cannot infer more about a until X is instantiated. This notion of typing can only usefully type terms without variables and in the presence of polymorphism such strategies break down entirely.

We now present a Curry-style system with rank 1 polymorphism (*ML-style polymorphism* or *Hindley-Milner types* [9]). Atoms can inhabit *any* type. We can write $lam[a]X$, or $fix[f]X$, or $forall[a]X$, or $app(lam[a]X, lam[b]X)$, and so on, and expect the type system to make sense of these terms, even though these terms explicitly feature context holes representing unknown terms *and* abstractions over those holes. Different occurrences of X may be under different numbers of abstractions, and for different atoms. This means that, when we instantiate X with t, the atoms in t may move into different type contexts and so receive different types. At the same time, the entire type system is consistent with a functional intuition, so X of type $\mathbb{N}$ manages to simultaneously behave like an 'unknown number' and an 'unknown term'.

We give syntax-directed typing rules and show that every typable term has a *principal type* (one which subsumes all others in a suitable sense) in a given environment. Type inference is decidable and types are compatible with α-equivalence on nominal terms. We give a notion of *typable rewrite rule* such that rewriting preserves types. In future, we plan to extend the system with intersection types, to derive a system with principal typings (a type and a type *environment* which subsume all others). With this system we will study normalisation of nominal terms.

2   Background

We continue the notation from the Introduction. Write $V(t)$ for the variables in t and $A(t)$ for the atoms in t (e.g., $a, b, c \in A([a][b](a\ c) \cdot X)$, $X \in V([a][b](a\ c) \cdot X)$). We define $\pi \cdot t$, the **permutation action** of π on t, inductively by:

Id$\cdot t \equiv t$ and $((a\ b) \circ \pi)\cdot t \equiv (a\ b)\cdot(\pi\cdot t)$, where

$$(a\ b)\cdot a \equiv b \quad (a\ b)\cdot b \equiv a \quad (a\ b)\cdot c \equiv c \quad (c \not\equiv a, b)$$
$$(a\ b)\cdot(\pi\cdot X) \equiv ((a\ b) \circ \pi)\cdot X \quad (a\ b)\cdot ft \equiv f(a\ b)\cdot t$$
$$(a\ b)\cdot[n]t \equiv [(a\ b)\cdot n](a\ b)\cdot t \quad (a\ b)\cdot(t_1, \ldots, t_n) \equiv ((a\ b)\cdot t_1, \ldots, (a\ b)\cdot t_n)$$

For example $(a\ b)\cdot lam[a](a, b, X) \equiv lam[b](b, a, (a\ b)\cdot X)$.
Define $t[X \mapsto s]$, the **(term-)substitution of X for s in t**, by:

$$(ft)[X \mapsto s] \equiv f(t[X \mapsto s]) \quad ([a]t)[X \mapsto s] \equiv [a](t[X \mapsto s]) \quad (\pi\cdot X)[X \mapsto s] \equiv \pi\cdot s$$
$$a[X \mapsto s] \equiv a \quad (t_1, \ldots)[X \mapsto s] \equiv (t_1[X \mapsto s], \ldots) \quad (\pi\cdot Y)[X \mapsto s] \equiv \pi\cdot Y \quad (X \not\equiv Y)$$

Term-substitutions are defined by $\theta ::= \textbf{Id} \mid [X \mapsto s]\theta$ and have an action given by $t\textbf{Id} \equiv t$ and $t([X \mapsto s]\theta) \equiv (t[X \mapsto s])\theta$. We write substitutions postfix, and write $\circ$ for composition of substitutions: $t(\theta \circ \theta') \equiv (t\theta)\theta'$.

Nominal syntax represents systems with binding, closely following informal notation. See [21,10,11] for examples and discussions of nominal terms and nominal rewriting. It has the same applications as higher-order systems such as Klop's CRSs, Khasidashvili's ERSs, and Nipkow's HRSs [18,17,19]. Intuitively, the distinctive features of nominal syntax are:

- X is an 'unknown term'; the **substitution action** $t[X \mapsto s]$ which does not avoid capture, makes this formal. Therefore X behaves differently from 'free variables' of systems such as HRSs [19] or meta-variables of CRSs [18].
- $[a]X$ denotes 'X with a abstracted in X'. We do *not* work modulo α-equivalence and $[a]X$ and $[b]X$ are not equal in any sense; for example $([a]X)[X \mapsto a] \equiv [a]a$ and $([b]X)[X \mapsto a] \equiv [b]a$, and we certainly expect from the intuition '$\lambda x.x$' and '$\lambda x.y$' that $[a]a$ and $[b]a$ should not be equal. Therefore atoms in nominal terms also behave differently from 'bound variables' of systems such as HRSs, ERSs and CRSs.

  We call occurrences of a **abstracted** when they are in the scope of $[a]$-, and otherwise we may call them **unabstracted**.
- $(a\ b)\cdot X$ represents 'X with a and b swapped'. So $\pi\cdot[a]s \equiv [\pi\cdot a]\pi\cdot s$, and $(a\ b)\cdot[a][b]X \equiv [b][a](a\ b)\cdot X$. Therefore this permutation action is distinct from De Bruijn's transformers and other explicit substitutions, which avoid capture as they distribute under abstractions, and which do not satisfy the same simple algebraic laws [7].

We now come to some of the more technical machinery which gives nominal terms their power.

We call $a\#t$ a **freshness** constraint, and $s \approx_\alpha t$ an **equality** constraint. We will use letters P, Q to range over constraints. We introduce a notion of derivation as follows (below a, b denote two different atoms):

$$\frac{}{a\#b} \quad \frac{a\#s \quad a\#s_i \quad (1\leq i\leq n)}{a\#fs \quad a\#(s_1,\dots,s_n)} \quad \frac{}{a\#[a]s} \quad \frac{a\#s}{a\#[b]s} \quad \frac{\pi^{-1}(a)\#X}{a\#\pi\cdot X} \quad \frac{s\approx_\alpha t}{fs\approx_\alpha ft}$$

$$\frac{}{a\approx_\alpha a} \quad \frac{s\approx_\alpha t}{[a]s\approx_\alpha[a]t} \quad \frac{s\approx_\alpha(a\ b)\cdot t \quad a\#t}{[a]s\approx_\alpha[b]t} \quad \frac{ds(\pi,\pi')\#X}{\pi\cdot X\approx_\alpha\pi'\cdot X} \quad \frac{s_i\approx_\alpha t_i \quad (1\leq i\leq n)}{(s_1,\dots,s_n)\approx_\alpha(t_1,\dots,t_n)}$$

Here we write π^{-1} (the **inverse** of π) for the permutation obtained by reversing the order of the list of swappings; for example $((a\ b)\circ(c\ d))^{-1} = (c\ d)\circ(a\ b)$. Here $ds(\pi,\pi') \equiv \{n \mid \pi(n)\neq\pi'(n)\}$ is the **difference set**. For example $ds((a\ b),\mathbf{Id}) = \{a,b\}$ so $(a\ b)\cdot X \approx_\alpha X$ follows from $a\#X$ and $b\#X$. Also $[a]a \approx_\alpha [b]b$ and $[a]c \approx_\alpha [b]c$ but not $[a]c \approx_\alpha [a]a$; this is what we would expect.

Write Δ, ∇ for sets of freshness constraints of the form $a\#X$ and call these **freshness contexts**. We write Pr for an arbitrary set of freshness and equality constraints; we may call Pr a **problem**. Substitutions act on constraints and problems in the natural way. Write $\Delta \vdash P$ when a derivation of P exists using elements of Δ as assumptions, and extend this notation elementwise to $\Delta \vdash Pr$ for deducibility of all $P \in Pr$ (see [21,11] for algorithms to check constraints). For example, $\Delta \vdash \nabla\theta$ means that the constraints obtained by applying the substitution θ to each term in ∇ can be derived from Δ. We will use this notation in the definition of matching in Section 4.1.

The following result is one of the main technical correctness properties of nominal terms, and we should compare it with Theorem 8.

Theorem 1. *If $\Delta \vdash a\#s$ and $\Delta \vdash s\approx_\alpha t$ then $\Delta \vdash a\#t$.*

Proofs, and further properties of nominal terms, are in [21,12].

3   Typing

3.1   Types and Type Schemes

We consider denumerable sets of
- **base data types** (write a typical element δ), e.g. $\mathbb{N}$ is a base data type for numbers;
- **type variables** (write a typical variable α);
- **type-formers** (write a typical element C), e.g. *List* is a type-former.

Definition 1. *Types τ, type-schemes σ, and type-declarations (or **arities**) ρ are defined by:*

$$\tau ::= \delta \mid \alpha \mid \tau_1\times\dots\times\tau_n \mid C\,\tau \mid [\tau']\tau \qquad \sigma ::= \forall\overline{\alpha}.\tau \qquad \rho ::= (\tau')\tau$$

*$\overline{\alpha}$ denotes any finite set of type variables (if empty we omit the $\forall$ entirely); we call them **bound** in σ and call **free** any type variables mentioned in σ and not in $\overline{\alpha}$. We write $TV(\tau)$ for the set of type variables in τ, and $\equiv$ for **equality** modulo α-equivalence for bound variables[1].*

---

[1] We could express types too using nominal syntax. We use the standard informal treatment because we focus on the term language in this paper.

We call $\tau_1 \times \ldots \times \tau_n$ a **product** and $[\tau]\tau'$ an **abstraction type**. We say that $C\,\tau$ is a **constructed type**, and we associate a type declaration to each term-former. For example, we can have $succ : (\mathbb{N})\mathbb{N}$ and $0 : ()\mathbb{N}$ (we may write just $0 : \mathbb{N}$ in this case).

Basic type substitution. $\tau[\alpha \mapsto \tau']$ is the usual inductive replacement of τ' for every α in τ; base cases are $\alpha[\alpha \mapsto \tau] \equiv \tau$ and $\alpha[\beta \mapsto \tau] \equiv \alpha$. We extend the action elementwise to arities ρ, for example $((\tau)\tau')[\alpha \mapsto \tau''] \equiv (\tau[\alpha \mapsto \tau''])(\tau'[\alpha \mapsto \tau''])$, and to type-schemes σ in the usual capture-avoiding manner. For example:

$$([\alpha]\alpha)[\alpha \mapsto \tau] \equiv [\tau]\tau \quad (\forall \beta.[\beta]\alpha)[\alpha \mapsto \beta] \equiv \forall \beta'.[\beta']\beta \quad ((\alpha \times \beta)\alpha)[\alpha \mapsto \beta] \equiv (\beta \times \beta)\beta$$

Type substitutions. S, T, U are defined by $S ::= \mathbf{Id} \mid S[\alpha \mapsto \tau]$ where $\mathbf{Id}$ is the **identity** substitution: $\tau\mathbf{Id} \equiv \tau$ by definition (we also use $\mathbf{Id}$ for the identity substitution on terms, but the context will always indicate which one we need). S has an action on types τ, schemes σ, and arities ρ, given by the action of $\mathbf{Id}$ and by extending the basic action of the last paragraph. We write application on the right as τS, and write composition of substitutions just as SS', meaning apply S then apply S'. The **domain** of a substitution S, denoted $dom(S)$, is the set of type variables α such that $\alpha S \neq \alpha$.

Substitutions are partially ordered by instantiation. Write $\mathrm{mgu}(\tau, \tau')$ (**most general unifier**) for a least element S such that $\tau S \equiv \tau' S$ (if one exists). We refer the reader to [1] for detailed definitions and algorithms for calculating mgu.

Write $\sigma \succcurlyeq \tau$ when $\sigma \equiv \forall \overline{\alpha}.\tau'$ and $\tau'S \equiv \tau$ for some S which instantiates only type variables in $\overline{\alpha}$. τ may contain other type variables; only *bound* type variables in σ may be instantiated, for example $\forall \alpha.(\alpha \times \beta) \succcurlyeq (\beta \times \beta)$ but $(\alpha \times \beta) \not\succcurlyeq (\beta \times \beta)$.

Also write $\rho \succcurlyeq \rho'$ when $\rho S \equiv \rho'$ for some S. In effect all variables in arities are bound, but since they are *all* bound we do not write the $\forall$. For example $(\alpha \times \alpha)\alpha \succcurlyeq (\beta \times \beta)\beta \succcurlyeq (\mathbb{N} \times \mathbb{N})\mathbb{N}$.

The following useful technical result follows by an easy induction:

Lemma 2. *If $\sigma \succcurlyeq \tau$ then $\sigma[\alpha \mapsto \tau'] \succcurlyeq \tau[\alpha \mapsto \tau']$. Also if $\rho \succcurlyeq \rho'$ then $\rho \succcurlyeq \rho'[\alpha \mapsto \tau']$.*

3.2 Typing Judgements

A **typing context** Γ is a set of pairs $(a : \sigma)$ or $(X : \sigma)$ subject to the condition that if $(a : \sigma), (a : \sigma') \in \Gamma$ then $\sigma \equiv \sigma'$, similarly for X.

We write $\Gamma, a : \sigma$ for Γ **updated** with $(a : \sigma)$, where this means either $\Gamma \cup \{(a : \sigma)\}$ or $(\Gamma \setminus \{(a : \sigma')\}) \cup \{(a : \sigma)\}$ as well-formedness demands. Similarly we write $\Gamma, X : \sigma$. For example, if $\Gamma = a : \alpha$ then $\Gamma, a : \beta$ denotes the context $a : \beta$. We say that a (or rather its association with a type) is **overwritten** in Γ. We write ΓS for the typing context obtained by applying S to the types in Γ. Similarly, $\pi \cdot \Gamma$ denotes the context obtained by applying π to the atoms in Γ. $TV(\Gamma)$ denotes the set of type variables occurring free in Γ.

Definition 2. *A **typing judgement** is a tuple $\Gamma; \Delta \vdash s : \tau$ where Γ is a typing context, Δ a freshness context, s a term and τ a type (when Δ is empty we omit the separating ';').*

We inductively define **derivable typing judgements** *as follows:*

$$\frac{\sigma \succcurlyeq \tau}{\Gamma, a : \sigma; \Delta \vdash a : \tau} \qquad \frac{\sigma \succcurlyeq \tau \quad \Gamma; \Delta \vdash \pi \cdot X : \diamond}{\Gamma, X : \sigma; \Delta \vdash \pi \cdot X : \tau} \qquad \frac{\Gamma, a : \tau; \Delta \vdash t : \tau'}{\Gamma; \Delta \vdash [a]t : [\tau]\tau'}$$

$$\frac{\Gamma; \Delta \vdash t_i : \tau_i \quad (1 \le i \le n)}{\Gamma; \Delta \vdash (t_1, \dots, t_n) : \tau_1 \times \dots \times \tau_n} \qquad \frac{\Gamma; \Delta \vdash t : \tau' \quad f : \rho \succcurlyeq (\tau')\tau}{\Gamma; \Delta \vdash ft : \tau}$$

Here $\Gamma; \Delta \vdash \pi \cdot X : \diamond$ holds if, for any a such that $\pi \cdot a \ne a$, $\Delta \vdash a\#X$ or $a : \sigma$, $\pi \cdot a : \sigma \in \Gamma$ for some σ. The condition $f : \rho \succcurlyeq (\tau')\tau$ is shorthand for $f : \rho$ and $\rho \succcurlyeq (\tau')\tau$. The way we set things up, the arity of f is fixed and $\succcurlyeq$ is independent of Γ. In the rule for abstractions the type environment is updated.

We may write '$\Gamma; \Delta \vdash s : \tau$' for '$\Gamma; \Delta \vdash s : \tau$ is derivable'.

To understand the condition in the second rule, note that $\pi \cdot X$ represents an unknown term in which π permutes atoms. Unlike non-nominal α-equivalence, α-equivalence on nominal terms exists in the presence of unknowns X for which substitution does not avoid capture, as $([a]X)[X \mapsto s] = [a]s$ demonstrates and as the rules $\dfrac{s \approx_\alpha (a\ b) \cdot t \quad a\#t}{[a]s \approx_\alpha [b]t}$ and $\dfrac{ds(\pi, \pi')\#X}{\pi \cdot X \approx_\alpha \pi' \cdot X}$ show. Our typing system is designed to be compatible with this relation and so must be sophisticated in its treatment of permutations acting on unknowns. For concreteness take $\pi = (a\ b)$ and any term t. If $\Gamma \vdash t : \tau$ then $\Gamma \vdash (a\ b) \cdot t : \tau$ should hold when at least one of the following conditions is satisfied:

1. $(a\ b)\Gamma = \Gamma$ so that Γ cannot 'tell the difference between a and b in t'.
2. If a and b occur in t then they are abstracted, so that what Γ says about a and b gets overwritten.

Given Γ, Δ and s, if there exists τ such that $\Gamma; \Delta \vdash s : \tau$ is derivable, then we say $\Gamma; \Delta \vdash s$ is **typable**. Otherwise say $\Gamma; \Delta \vdash s$ is **untypable**.

The following are examples of derivable typing judgements:

$$a : \forall \alpha . \alpha, X : \beta \vdash (a, X) : \alpha \times \beta \qquad a : \forall \alpha . \alpha, X : \beta \vdash (a, X) : \beta \times \beta$$

$$\vdash [a]a : [\alpha]\alpha \qquad a : \beta \vdash [a]a : [\alpha]\alpha \qquad \vdash [a]a : [\zeta]\zeta$$

$$a : \alpha, b : \alpha, X : \tau \vdash (a\ b) \cdot X : \tau \qquad X : \tau; a\#X, b\#X \vdash (a\ b) \cdot X : \tau$$

$$X : \tau, a : \alpha, b : \alpha \vdash [a]((a\ b) \cdot X, b) : [\alpha](\tau \times \alpha)$$

$$a : \alpha, b : \beta \vdash (\quad [a](a, b), \quad [a][b](a, b), \quad a, \quad b, \quad [a][a](a, b)) :$$
$$[\alpha](\alpha \times \beta) \times [\alpha][\beta](\alpha \times \beta) \times \alpha \times \beta \times [\alpha][\alpha](\alpha \times \beta).$$

$$a : \alpha, b : \beta \vdash (\quad [a](a, b), \quad [a][b](a, b), \quad a, \quad b, \quad [a]([a](a, b), a)) :$$
$$[\alpha_1](\alpha_1 \times \beta) \times [\alpha_2][\beta_2](\alpha_2 \times \beta_2) \times \alpha \times \beta \times [\alpha_3]([\alpha_4](\alpha_4 \times \beta), \alpha_3).$$

The first line of examples just illustrates some basic typings, and the use of $\succcurlyeq$. The second line illustrates typing abstractions and how this overwrites in Γ. The third line illustrates how permutations are typed. The last three illustrate interactions between typing and (multiple) abstractions and free occurrences of atoms. Note that $a : \alpha, X : \tau \nvdash (a\ b) \cdot X : \tau$ and $a : \alpha, X : \tau; b\#X \nvdash (a\ b) \cdot X : \tau$.

Lemma 3. *If* $\Gamma; \Delta \vdash t : \tau$ *then* $\Gamma[\alpha \mapsto \tau']; \Delta \vdash t : \tau[\alpha \mapsto \tau']$.

Proof. By induction on the derivation, using Lemma 2 for the side-conditions.

Lemma 4. *If* $\Gamma;\Delta \vdash t{:}\tau$ *and* $a,b{:}\sigma \in \Gamma$ *for some* σ, *then* $(a\ b){\cdot}\Gamma;\Delta \vdash (a\ b){\cdot}t{:}\tau$.

Proof. By induction on the type derivation: The case for atoms is trivial. In the case of a variable, the side condition holds since a and b have the same type in Γ. The other cases follow directly by induction.

3.3   Principal Types

Definition 3. *A **typing problem** is a triple* (Γ, Δ, s), *written:* $\Gamma; \Delta \vdash s{:}?$. *A* **solution** *to* $\Gamma; \Delta \vdash s$:? *is a pair* (S, τ) *of a type-substitution* S *and a type* τ *such that* $\Gamma S; \Delta \vdash s{:}\tau$. *We write* $S|_\Gamma$ *for the restriction of* S *to* $TV(\Gamma)$.

For example, solutions to $X{:}\alpha, a{:}\beta, b{:}\beta \vdash (a\ b){\cdot}X{:}?$ are $(\mathbf{Id}, \alpha)$ and $([\alpha{\mapsto}\mathbb{N}], \mathbb{N})$. Note that a solution may instantiate type-variables in Γ.

Solutions have a natural ordering given by instantiation of substitutions:

$$(S, \tau) \leq (S', \tau') \quad \text{when} \quad \exists S''. S' \equiv SS'' \wedge \tau' \equiv \tau S'';$$

we call (S', τ') an *instance* of (S, τ) using S''. A minimal element in a set of solutions is called a **principal** solution. By our definitions there may be many principal solutions; $(\mathbf{Id}, \alpha)$ and $(\mathbf{Id}, \beta)$ are both principal for $X : \forall \alpha.\alpha \vdash X$:?. As in the case of most general unifiers, principal solutions for a typable $\Gamma; \Delta \vdash s$ are unique modulo renamings of type-variables. In a moment we show that the following algorithm calculates principal solutions:

Definition 4. *The partial function* $pt(\Gamma; \Delta \vdash s)$ *is defined inductively by:*

- $pt(\Gamma, a{:}\forall \overline{\alpha}.\tau; \Delta \vdash a) = (\mathbf{Id}, \tau)$, *where* $\alpha \in \overline{\alpha}$ *are assumed fresh (not in* Γ *) without loss of generality.*
- $pt(\Gamma, X{:}\forall \overline{\alpha}.\tau; \Delta \vdash \pi{\cdot}X) = (S, \tau S)$ *(again* $\alpha \in \overline{\alpha}$ *are assumed fresh) provided that for each* a *in* π *such that* $a \neq \pi{\cdot}a$, *we have* $\Delta \vdash a\#X$, *or otherwise* $a{:}\sigma, \pi{\cdot}a{:}\sigma' \in \Gamma$ *for some* σ, σ' *that are unifiable. The substitution* S *is the mgu of those pairs, or* $\mathbf{Id}$ *if all such* a *are fresh for* X.
- $pt(\Gamma; \Delta \vdash (t_1, \ldots, t_n)) = (S_1 \ldots S_n, \phi_1 S_2 \ldots S_n \times \ldots \times \phi_{n-1} S_n \times \phi_n)$ *where* $pt(\Gamma; \Delta \vdash t_1) = (S_1, \phi_1)$, $pt(\Gamma S_1; \Delta \vdash t_2) = (S_2, \phi_2)$, $\ldots$, $pt(\Gamma S_1 \ldots S_{n-1}; \Delta \vdash t_n) = (S_n, \phi_n)$.
- $pt(\Gamma; \Delta \vdash ft) = (SS', \phi S')$ *where* $pt(\Gamma; \Delta \vdash t) = (S, \tau)$, $f : \rho \equiv (\phi')\phi$ *where the type variables in* ρ *are chosen distinct from those in* Γ *and* τ, *and* $S' = mgu(\tau, \phi')$.
- $pt(\Gamma; \Delta \vdash [a]s) = (S|_\Gamma, [\tau']\tau)$ *where* $pt(\Gamma, a{:}\alpha; \Delta \vdash s) = (S, \tau)$, α *is chosen fresh,* $\alpha S = \tau'$.

    Here $\Gamma, a{:}\alpha$ *denotes* Γ *with any type information about* a *overwritten to* $a{:}\alpha$, *as discussed in Subsection 3.2.*

$pt(\Gamma; \Delta \vdash s)$ may be undefined; Theorem 5 proves that s is untypable in the environment $\Gamma; \Delta$.

The definition above generalises the Hindley-Milner system [9] for the λ-calculus with arbitrary function symbols f, as is standard in typing algorithms for rewrite systems [2], and with atoms and variables (representing 'unknown terms') with suspended atoms-permutations.

The treatment of typing for atoms, abstraction and moderated variables is new; for example if $\Gamma = X : \tau, a : \alpha, b : \alpha$, then $\Gamma \vdash [a](a\ b) \cdot X : [\alpha]\tau$ but not $\Gamma \vdash [a](a\ b) \cdot X : [\beta]\tau$. However, $\Gamma \vdash [a]X : [\beta]\tau$ as expected.

$pt(\Gamma; \Delta \vdash s)$ gives a static semantics in the form of a most general type to s, given a typing of atoms and variables mentioned in s, and information about what atoms may be forbidden from occurring in some variables.

Theorem 5.

1. *If $pt(\Gamma; \Delta \vdash s)$ is defined and equal to (S, τ) then $\Gamma S; \Delta \vdash s : \tau$ is derivable.*
2. *Let U be a substitution such that $dom(U) \subseteq TV(\Gamma)$. If $\Gamma U; \Delta \vdash s : \mu$ is derivable then $pt(\Gamma; \Delta \vdash s)$ is defined and (U, μ) is one of its instances.*

That is: (1) "$pt(\Gamma; \Delta \vdash s)$ solves $(\Gamma; \Delta \vdash s:?)$", and (2) "any solution to $(\Gamma; \Delta \vdash s:?)$ is an instance of $pt(\Gamma; \Delta \vdash s)$".

Proof. The first part is by a routine induction on the derivation of $pt(\Gamma; \Delta \vdash s) = (S, \tau)$ (using Lemma 3), which we omit. The second part is by induction on the syntax of s:

- Suppose $\Gamma U; \Delta \vdash \pi \cdot X : \mu$. Examining the typing rules, we see that $X : \forall \overline{\alpha}.\tau \in \Gamma$ and $X : \forall \overline{\alpha}.\tau U \in \Gamma U$, that $\Gamma U; \Delta \vdash \pi \cdot X : \diamond$, and that $\mu = \tau U S$ for some S acting only on $\overline{\alpha}$ (U acts trivially on $\overline{\alpha}$ because we assume $\overline{\alpha}$ was chosen fresh). Hence, for each a such that $a \neq \pi \cdot a$, we have $\Delta \vdash a \# X$, or, for some σ, $a : \sigma, \pi \cdot a : \sigma \in \Gamma U$, that is, $a : \sigma_1, \pi \cdot a : \sigma_2 \in \Gamma$. Let V be the mgu of all such pairs σ_1, σ_2, so $U = V S'$ (we take $V = \mathbf{Id}$ if there are no pairs to consider). Thus, $pt(\Gamma; \Delta \vdash \pi \cdot X) = (V, \tau V)$. Therefore, $(U, \mu) = (V S' S, \tau V S' S)$.
- Suppose $\Gamma U; \Delta \vdash a : \mu$. Clearly a is typable in the context $\Gamma; \Delta$ so (examining the typing rules) $a : \forall \overline{\alpha}.\tau$ must occur in Γ and $pt(\Gamma; \Delta \vdash a) = (\mathbf{Id}, \tau)$. It is now not hard to satisfy ourselves that $(\mathbf{Id}, \tau)$ is principal.
- Suppose $\Gamma U; \Delta \vdash [a]t : \mu$. This may happen if and only if $\mu \equiv [\mu']\mu''$ and $(U[\alpha \mapsto \mu'], \mu'')$ solves $\Gamma, a : \alpha; \Delta \vdash t:?$, where α is fresh for Γ. By inductive hypothesis $(U[\alpha \mapsto \mu'], \mu'')$ is an instance of $(S, \tau) = pt(\Gamma, a : \alpha; \Delta \vdash t)$, that is, there is a substitution U_a such that $U[\alpha \mapsto \mu'] = S U_a$, $\mu'' = \tau U_a$. By definition, $pt(\Gamma; \Delta \vdash [a]t) = (S|_\Gamma, [\tau']\tau)$ where $\tau' = \alpha S$. So $(S U_a)|_\Gamma = U$ and $\alpha S U_a = \mu'$. Therefore $(U, [\mu']\mu'')$ is an instance of $(S|_\Gamma, [\alpha S]\tau)$.
- The cases for $(t_1, \ldots, t_n)$ and ft are long, but routine.

Corollary 6. *$\Gamma; \Delta \vdash s$ is typable if and only if $pt(\Gamma; \Delta \vdash s) = (\mathbf{Id}, \tau)$ for some τ.*

3.4 α-Equivalence and Types

We now prove that α-equivalence respects types; so our static semantics correctly handles swappings and variables X whose substitution can move atoms into new abstracted (typing) contexts. We need some definitions: Given two type contexts

Γ and Γ', write Γ, Γ' for that context obtained by updating Γ with typings in Γ', overwriting the typings in Γ if necessary. For example if $\Gamma = a: \alpha$ and $\Gamma' = a: \beta, b: \beta$ then $\Gamma, \Gamma' = a: \beta, b: \beta$. If Γ and Γ' mention disjoint sets of atoms and variables (we say they are **disjoint**) then Γ, Γ' is just a set union.

Lemma 7

1. If $\Gamma; \Delta \vdash s : \tau$ then $\Gamma, \Gamma'; \Delta \vdash s : \tau$, provided that Γ' and Γ are disjoint. Call this **type weakening**.
2. If $\Gamma, a : \tau'; \Delta \vdash s : \tau$ then $\Gamma; \Delta \vdash s : \tau$ provided that $\Delta \vdash a \# s$. Call this **type strengthening** (for atoms).
3. If $\Gamma, X : \tau'; \Delta \vdash s : \tau$ then $\Gamma; \Delta \vdash s : \tau$ provided X does not occur in s. Call this **type strengthening** (for variables).

Proof. By induction on the derivation. For the second part, if a occurs in s under an abstraction, then $a : \tau'$ is overwritten whenever a is used.

Theorem 8. $\Delta \vdash s \approx_\alpha t$ and $\Gamma; \Delta \vdash s : \tau$ imply $\Gamma; \Delta \vdash t : \tau$.

Proof. By induction on the size of (s, t). The form of t is rather restricted by our assumption that $\Delta \vdash s \approx_\alpha t$ — for example if $s \equiv \pi \cdot X$ then $t \equiv \pi' \cdot X$ for some π'. We use this information without commment in the proof.

- Suppose $\Delta \vdash a \approx_\alpha a$ and $\Gamma; \Delta \vdash a : \tau$.   There is nothing to prove.
- Suppose $\Delta \vdash [a]s \approx_\alpha [a]t$ and $\Gamma; \Delta \vdash [a]s : \tau$.   Then $\Delta \vdash s \approx_\alpha t$, and $\tau \equiv [\tau']\tau''$, and $\Gamma, a : \tau'; \Delta \vdash s : \tau''$. We use the inductive hypothesis to deduce that $\Gamma, a : \tau'; \Delta \vdash t : \tau''$ and concluce that $\Gamma; \Delta \vdash [a]t : [\tau']\tau''$.
- Suppose $\Delta \vdash [a]s \approx_\alpha [b]t$ and $\Gamma; \Delta \vdash [a]s : \tau$.   Then by properties of $\approx_\alpha$ [21, Theorem 2.11] we know $\Delta \vdash s \approx_\alpha (a\ b) \cdot t, a \# t, (a\ b) \cdot s \approx_\alpha t, b \# s$. Also $\tau \equiv [\tau']\tau''$ and $\Gamma, a : \tau'; \Delta \vdash s : \tau''$. By Lemma 7 (Weakening) also $\Gamma, a : \tau', b : \tau'; \Delta \vdash s : \tau''$. By equivariance (Lemma 4) $(a\ b) \cdot \Gamma, b : \tau', a : \tau'; \Delta \vdash (a\ b) \cdot s : \tau''$. Since $\Delta \vdash (a\ b) \cdot s \approx_\alpha t$, by inductive hypothesis $(a\ b) \cdot \Gamma, b : \tau', a : \tau'; \Delta \vdash t : \tau''$.

  Now note that $(a\ b) \cdot \Gamma, b : \tau', a : \tau' = \Gamma, b : \tau', a : \tau'$ (because any data Γ has on a and b is overwritten). So $\Gamma, b : \tau', a : \tau'; \Delta \vdash t : \tau''$. We conclude that $\Gamma, b : \tau', a : \tau'; \Delta \vdash [b]t : [\tau']\tau''$. Since $\Delta \vdash a \# [b]t$ and $\Delta \vdash b \# [b]t$ by Lemma 7 (strengthening for atoms) we have $\Gamma; \Delta \vdash [b]t : [\tau']\tau''$.
- Suppose $\Delta \vdash \pi \cdot X \approx_\alpha \pi' \cdot X$ and suppose $\Gamma; \Delta \vdash \pi \cdot X : \tau$. Then $\Delta \vdash ds(\pi, \pi') \# X$, and $X : \sigma \in \Gamma$, and $\sigma \succcurlyeq \tau$, and $\Gamma; \Delta \vdash \pi \cdot X : \diamond$.

  $\Delta \vdash ds(\pi, \pi') \# X$ so $\Gamma; \Delta \vdash \pi' \cdot X : \diamond$ and $\Gamma; \Delta \vdash \pi' \cdot X : \tau$ follows.
- The cases of ft and $(t_1, \ldots, t_n)$ follow easily.

Corollary 9. $\Delta \vdash s \approx_\alpha t$ implies $pt(\Gamma; \Delta \vdash s) = pt(\Gamma; \Delta \vdash t)$ modulo renamings of type variables, for all Γ, Δ such that either side of the equality is defined.

4   Rewriting

We now consider how to make a notion of nominal rewriting (i.e., a theory of general directed equalities on nominal terms [12]) interact correctly with our type system. We start with some definitions taken from [12].

A **nominal rewrite rule** $R \equiv \nabla \vdash l \to r$ is a tuple of a freshness context ∇, and terms l and r such that $V(r, \nabla) \subseteq V(l)$. Write $R^{(a\,b)}$ for that rule obtained by swapping a and b in R throughout. For example, if $R \equiv b\#X \vdash [a]X \to (b\ a)\cdot X$ then $R^{(a\,b)} \equiv a\#X \vdash [b]X \to (a\ b)\cdot X$. Let $R^{\mathbf{Id}} \equiv R$ and $R^{(a\ b)\circ\pi} = (R^{(a\ b)})^\pi$. Let $\mathcal{R}$ range over (possibly infinite) sets of rewrite rules. Call $\mathcal{R}$ **equivariant** when if $R \in \mathcal{R}$ then $R^{(a\ b)} \in \mathcal{R}$ for all distinct atoms a, b. A **nominal rewrite system** is an equivariant set of nominal rewrite rules.

Say a term s has a **position at** X when it mentions X once, with the permutation $\mathbf{Id}$. We may call X a **hole**. Write $s[s']$ for $s[X \mapsto s']$.[2] We would usually call s a 'context' but we have already used this word so we will avoid it. For example, X and $[a]X$ have positions at X, but (X, X) and $(a\ b)\cdot X$ do not. We may make X nameless and write it just -.

Definition 5. *Suppose* $pt(\Phi; \nabla \vdash l) = (\mathbf{Id}, \tau)$ *and suppose that* Φ *mentions no type-schemes. All the recursive calls involved in calculating* $pt(\Phi; \nabla \vdash l)$ *have the form* $pt(\Phi, \Phi'; \nabla \vdash l')$ *where* l' *is a subterm of* l *and* Φ' *contains only type assumptions for atoms. We will call recursive calls of the form* $pt(\Phi, \Phi'; \nabla \vdash \pi\cdot X) = (S, \tau')$ ***variable typings of*** $\Phi; \nabla \vdash l : \tau$.

Note that there is one variable typing for each occurrence of a variable in l, and they are uniquely defined modulo renaming of type variables. Also, S may affect Φ' but not Φ since we assume that $pt(\Phi; \nabla \vdash l) = (\mathbf{Id}, \tau)$.

Definition 6. *Let* $pt(\Phi, \Phi'; \nabla \vdash \pi\cdot X) = (S, \tau')$ *be a variable typing of* $\Phi; \nabla \vdash l : \tau$, *and let* Φ'' *be the subset of* Φ' *such that* $\nabla \vdash A(\Phi'\backslash\Phi'')\#\pi\cdot X$. *We call* $\Phi, \Phi''S; \nabla \vdash \pi\cdot X : \tau'$ *an* ***essential typing*** *of* $\Phi; \nabla \vdash l : \tau$.

So the essential typings of $a : \alpha, b : \alpha, X : \tau \vdash ((a\ b)\cdot X, [a]X) : \tau \times [\alpha']\tau$ are:

$$a : \alpha, b : \alpha, X : \tau \vdash (a\ b)\cdot X : \tau \quad \text{and} \quad b : \alpha, a : \alpha', X : \tau \vdash X : \tau.$$

The essential typings of $a : \alpha, b : \alpha, X : \tau; a\#X \vdash ((a\ b)\cdot X, [a]X) : \tau \times [\alpha']\tau$ are:

$$b : \alpha, X : \tau \vdash (a\ b)\cdot X : \tau \quad \text{and} \quad b : \alpha, X : \tau \vdash X : \tau.$$

We will talk about the **typing at a position in a term**; for example the typing at Z in $(Z, [a]X)[Z \mapsto (a\ b)\cdot X]$ in the first example above (with hole named Z) is $a : \alpha, b : \alpha, X : \tau \vdash (a\ b)\cdot X : \tau$.

4.1   Matching Problems

Definition 7. *A **(typed) matching problem** $(\Phi; \nabla \vdash l) \ _?\approx (\Gamma; \Delta \vdash s)$ is a pair of tuples (Φ and Γ are typing contexts, ∇ and Δ are freshness contexts, l and s are terms) such that the variables and type-variables mentioned on the left-hand side are disjoint from those mentioned in Γ, Δ, s, and such that Φ mentions no atoms or type-schemes.*

---

[2] Here 'positions' are based on substitution and not paths in the abstract syntax tree as in [10]. The equivalence is immediate since substitution is grafting.

Below, l will be the left-hand side of a rule and s will be a term to reduce. The condition that Φ mentions no atoms or type-schemes may seem strong, but is all we need: we give applications in Section 4.3.

Intuitively, we want to make the term on the left-hand side of the matching problem α-equivalent to the term on the right-hand side. Formally, a **solution** to this matching problem, if it exists, is the least pair (S, θ) of a type- and term-substitution (the ordering on substitutions extends to pairs component-wise) such that:

1. $X\theta \equiv X$ for $X \notin V(\Phi,\nabla,l)$, $\alpha S \equiv \alpha$ for $\alpha \notin TV(\Phi)^3$, $\Delta \vdash l\theta \approx_\alpha s$ and $\Delta \vdash \nabla\theta$.
2. $pt(\Phi; \nabla \vdash l) = (\mathbf{Id}, \tau)$, $pt(\Gamma; \Delta \vdash s) = (\mathbf{Id}, \tau S)$, and for each $\Phi, \Phi'; \nabla \vdash \pi \cdot X : \phi'$ an essential typing in $\Phi; \nabla \vdash l : \tau$, we have $\Gamma, (\Phi'S); \Delta \vdash (\pi \cdot X)\theta : \phi'S$.

The first condition defines a matching solution for untyped nominal terms (see [21,12] for more details on untyped nominal matching algorithms). The last condition enforces type consistency: the terms should have compatible types, and the solution should instantiate the variables in a way that is compatible with the typing assumptions. When the latter holds, we say that (S, θ) **respects the essential typings** of $\Phi; \nabla \vdash l : \tau$ in the context $\Gamma; \Delta$.

For example, $(X{:}\alpha \vdash X) \approx (a{:}\mathbb{B} \vdash a)$ has solution $([\alpha \mapsto \mathbb{B}], [X \mapsto a])$, whereas $(X{:}\mathbb{B} \vdash X) \approx (a{:}\alpha \vdash a)$ has no solution — α on the right is too general.

To see why we need to check θ in the second condition, consider the term $g(f\ True)$ where $g : (\alpha)\mathbb{N}$ and $f : (\beta)\mathbb{N}$, that is both functions are polymorphic, and produce a result of type $\mathbb{N}$. Then the untyped matching problem $g(f\ X)) \approx g(f\ True)$ has a solution $(\mathbf{Id}, \{X \mapsto True\})$, but the typed matching problem $(X : \mathbb{N} \vdash g(f\ X)) \approx (\vdash g(f\ True))$ has none: $\{X \mapsto True\}$ fails the last condition since X is required to have type $\mathbb{N}$ but it is instantiated with a boolean.

4.2 Typed Rewriting

Definition 8. *A **(typed) rewrite rule** $R \equiv \Phi; \nabla \vdash l \rightarrow r : \tau$ is a tuple of a type context Φ which only types the variables in l and has no type-schemes (in particular, Φ mentions no atoms), a freshness context ∇, and terms l and r such that*

1. *$V(r, \nabla, \Phi) \subseteq V(l)$,*
2. *$pt(\Phi; \nabla \vdash l) = (\mathbf{Id}, \tau)$ and $\Phi; \nabla \vdash r : \tau$.*
3. *Essential typings of $\Phi; \nabla \vdash r : \tau$ are also essential typings of $\Phi; \nabla \vdash l : \tau$.*

The first condition is standard. The second condition says that l is typable using Φ and ∇, and r is typable with a type *at least* as general. The third condition ensures a consistency best explained by violating it: Let $f : ([\alpha]\mathbb{N})\mathbb{N}$, then $X : \mathbb{N} \vdash f([a]X) \rightarrow X : \mathbb{N}$ passes the first two conditions, but fails the third because in the right-hand side we have an essential typing $X : \mathbb{N} \vdash X : \mathbb{N}$ whereas in the left-hand side we have $X : \mathbb{N}, a : \alpha \vdash X : \mathbb{N}$. For comparison,

---

[3] So in particular, by the side-conditions on variables being disjoint between left and right of the problem, $X\theta \equiv X$ for $X \in V(\Gamma, \Delta, s)$ and $\alpha S \equiv \alpha$ for $\alpha \in TV(\Gamma)$.

$X : \mathbb{N} \vdash g([a]X) \to [a]X : [\alpha]\mathbb{N}$ with $g : ([\alpha]\mathbb{N})[\alpha]\mathbb{N}$ passes all three conditions and is a valid rewrite rule, as well as $X : \mathbb{N};\ a\#X \vdash f([a]X) \to X$.

The rewrite relation is defined on terms-in-context: Take $\Gamma; \Delta \vdash s$ and $\Gamma; \Delta \vdash t$, and a rule $R \equiv \Phi; \nabla \vdash l \to r : \tau$, such that $V(R) \cap V(\Gamma, \Delta, s, t) = \emptyset$, and $TV(R) \cap TV(\Gamma) = \emptyset$ (renaming variables in R if necessary). Assume $\Gamma; \Delta \vdash s$ is typable: $pt(\Gamma; \Delta \vdash s) = (\mathbf{Id}, \mu)$, $s \equiv s''[s']$ and $\Gamma'; \Delta \vdash s' : \mu'$ is the typing of s' at the corresponding position. We say that s **rewrites with** R **to** t **in the context** $\Gamma; \Delta$ and write $\Gamma; \Delta \vdash s \xrightarrow{R} t$ when:

1. $(\Phi; \nabla \vdash l) \ _? \approx (\Gamma'; \Delta \vdash s')$ has solution (S, θ).
2. $\Delta \vdash s''[r\theta] \approx_\alpha t$.

These conditions are inherited from nominal rewriting [10,12] and extended with types. Instantiation of X does not avoid capture, so an atom a introduced by a substitution may appear under different abstractions in different parts of a term. We must pay attention to the typing *at the position of the variable* in the rewrite; essential typings do just this. For example if $f : (\tau_1)\tau$, $g : (\tau)[\alpha]\tau$ and $R \equiv X : \tau \vdash gX \to [a]X : [\alpha]\tau$ then $pt(X : \tau \vdash gX) = (\mathbf{Id}, [\alpha]\tau)$ and $X : \tau \vdash [a]X : [\alpha]\tau$. R satisfies the first two conditions in the definition of typed rule but fails the third: the only essential typing in the left-hand side is $X : \tau \vdash X : \tau$, whereas in the right-hand side we have $X : \tau, a : \alpha \vdash X : \tau$. Notice that $a : \tau_1 \vdash g(fa) : [\alpha]\tau$ and the typed matching problem $(X : \tau \vdash gX) \ _? \approx (a : \tau_1 \vdash g(fa))$ has a solution $(\mathbf{Id}, \{X \mapsto fa\})$. So, if we ignore the third condition in the definition of typed rule, we have a rewriting step $a : \tau_1 \vdash g(fa) \to [a](fa)$ which does not preserve types: $a : \tau_1 \vdash g(fa) : [\alpha]\tau$ but $a : \tau_1 \nvdash [a](fa) : [\alpha]\tau$.

We need a lemma to prove Subject Reduction (Theorem 11):

Lemma 10. *Suppose that* $\Phi; \nabla \vdash r : \tau$, *where* Φ *types only variables in* r *(it mentions no atoms) and has no type schemes. Let* θ *be a substitution instantiating all variables in* r, *and such that* (S, θ) *respects the essential typings of* $\Phi; \nabla \vdash r : \tau$ *in the context* Γ, Δ, *that is, for each essential typing* $\Phi, \Phi'; \nabla \vdash \pi \cdot X : \tau'$ *of* $\Phi; \nabla \vdash r : \tau$, *it is the case that* $\Gamma, \Phi'S; \Delta \vdash (\pi \cdot X)\theta : \tau'S$. *Then* $\Gamma; \Delta \vdash r\theta : \tau S$.

Theorem 11 (Subject Reduction). *Let* $R \equiv \Phi; \nabla \vdash l \to r : \tau$. *If* $\Gamma; \Delta \vdash s : \mu$ *and* $\Gamma; \Delta \vdash s \xrightarrow{R} t$ *then* $\Gamma; \Delta \vdash t : \mu$.

Proof. It suffices to prove that if $pt(\Gamma; \Delta \vdash s) = (\mathbf{Id}, \nu)$ and $\Gamma; \Delta \vdash s \xrightarrow{R} t$ then $\Gamma; \Delta \vdash t : \nu$. Suppose $\Gamma; \Delta \vdash s \xrightarrow{R} t$. Then (using the notation in the definition of matching and rewriting above) we know that:

1. $s \equiv s''[s']$, $\Delta \vdash l\theta \approx_\alpha s'$, and $\Delta \vdash \nabla\theta$.
2. θ acts nontrivially only on the variables in R, not those in Γ, Δ, s.
3. Assuming $\Gamma'; \Delta \vdash s' : \nu'$ is the typing of s', then $\Gamma', \Phi'S; \Delta \vdash (\pi \cdot X)\theta : \phi'S$ for each essential typing $\Phi, \Phi'; \nabla \vdash \pi \cdot X : \phi'$ in $\Phi; \nabla \vdash l : \tau$ (therefore also for each essential typing in $\Phi; \nabla \vdash r : \tau$ since they are a subset).
4. $pt(\Phi; \nabla \vdash l) = (\mathbf{Id}, \tau)$ and $pt(\Gamma', \Delta \vdash s') = (\mathbf{Id}, \tau S)$ so there is some S' such that $\Gamma'S' = \Gamma'$ and $\tau SS' = \nu'$.
5. $\Delta \vdash s''[r\theta] \approx_\alpha t$.

By Theorem 8, from 3, 4 and 1 we deduce $\Gamma'; \Delta \vdash l\theta : \tau SS'$. Since $pt(\Phi; \nabla \vdash l) = (\mathbf{Id}, \tau)$, by our assumptions on rewrite rules also $\Phi; \nabla \vdash r : \tau$, and by Lemma 3 also $\Phi SS'; \nabla \vdash r : \tau SS'$. By Lemma 10, $\Gamma'; \Delta \vdash r\theta : \tau S$. Since $\Gamma'S' = \Gamma'$, by Lemma 3 also $\Gamma'; \Delta \vdash r\theta : \tau SS'$. Hence $\Gamma; \Delta \vdash s''[r\theta] : \nu$ as required.

4.3   Examples

Untyped λ-calculus. Suppose a type Λ and term-formers $lam : ([\Lambda]\Lambda)\Lambda$, $app : (\Lambda \times \Lambda)\Lambda$, and $sub : ([\Lambda]\Lambda \times \Lambda)\Lambda$, sugared to $\lambda[a]s$, $s\ t$, and $s[a \mapsto t]$. Rewrite rules satisfying the conditions in Definition 8 are:

$$X, Y{:}\Lambda \vdash (\lambda[a]X)Y \to X[a \mapsto Y]{:}\Lambda \qquad X, Y{:}\Lambda; a\#X \vdash X[a \mapsto Y] \to X{:}\Lambda$$
$$Y{:}\Lambda \vdash a[a \mapsto Y] \to Y{:}\Lambda \qquad X, Y{:}\Lambda; b\#Y \vdash (\lambda[b]X)[a \mapsto Y] \to \lambda[b](X[a \mapsto Y]){:}\Lambda$$
$$X, Y, Z{:}\Lambda \vdash (XY)[a \mapsto Z] \to X[a \mapsto Z]\,Y[a \mapsto Z]{:}\Lambda$$

The freshness conditions are exactly what is needed so that no atom moves across the scope of an abstraction (which might change its type). For instance, in rule $X, Y{:}\Lambda; a\#X \vdash X[a \mapsto Y] \to X{:}\Lambda$, X is under an abstraction for a in the left-hand side and not in the right-hand side, but we have $a\#X$.

The typed λ-calculus Suppose a type-former $\Rightarrow$ of arity 2 and term-formers $\lambda : ([\alpha]\beta)(\alpha \Rightarrow \beta)$, $\circ : (\alpha \Rightarrow \beta \times \alpha)\beta$, and $\sigma : ([\alpha]\beta \times \alpha)\beta$. Sugar as in the previous example, so, instead of $\sigma([a]s, t)$ we write $s[a \mapsto t]$. Then the following rewrite rules satisfy the conditions in Definition 8:

$$X : \alpha, Y : \beta;\ a\#X \vdash X[a \mapsto Y] \to X : \alpha \qquad Y : \gamma \vdash a[a \mapsto Y] \to Y : \gamma$$
$$X : \alpha \Rightarrow \beta,\ Y : \alpha,\ Z : \gamma \vdash (XY)[a \mapsto Z] \to (X[a \mapsto Z])(Y[a \mapsto Z]) : \beta$$
$$X : \beta,\ Y : \gamma;\ b\#Y \vdash (\lambda[b]X)[a \mapsto Y] \to \lambda[b](X[a \mapsto Y]) : \alpha \Rightarrow \beta$$

For the β-reduction rule we mention two variants; they give the same rewrites:

$$X{:}[\alpha]\beta, Y{:}\alpha \vdash (\lambda X)Y \to \sigma(X, Y) : \beta \qquad X{:}\beta, Y{:}\alpha \vdash (\lambda[a]X)Y \to \sigma([a]X, Y) : \beta$$

Assume types $\mathbb{B}$ and $\mathbb{N}$. Then $B : \mathbb{B},\ N : \mathbb{N} \vdash ((\lambda[a]a)B, (\lambda[a]a)N) : \mathbb{B} \times \mathbb{N}$ and

$$B : \mathbb{B},\ N : \mathbb{N} \vdash ((\lambda[a]a)B, (\lambda[a]a)N) \to (B, N) : \mathbb{B} \times \mathbb{N}.$$

$\lambda[a]a$ takes different types just like $\lambda x.x$ in the Hindley-Milner type system; $pt(\vdash \lambda[a]a) = (\mathbf{Id}, \alpha \Rightarrow \alpha)$. Our system will not type $B : \mathbb{B},\ N : \mathbb{N} \vdash BN$ or $\lambda[a]aa$ — the system for the *untyped* λ-calculus above, types the second term.

Surjective pairing. Rewrites for surjective pairing cannot be implemented by a compositional translation to λ-calculus terms [4]. Our system deals with rules defining surjective pairing directly; assume $fst : (\alpha \times \beta)\alpha$ and $snd : (\alpha \times \beta)\beta$:

$$X : \alpha,\ Y : \beta \vdash fst(X, Y) \to X : \alpha \qquad X : \alpha,\ Y : \beta \vdash snd(X, Y) \to Y : \beta$$
$$X : \alpha \times \beta \vdash (fst(X), snd(X)) \to X : \alpha \times \beta$$

Higher-order logic. Extend the typed λ-calculus above with a type *Prop* and term-formers $\top$: *Prop*, $\bot$: *Prop*, $=$: $(\alpha \times \alpha)Prop$, $\forall$: $([\alpha]Prop)Prop$, $\wedge$: $(Prop \times Prop)Prop$, and so on. Rewrite rules include:

$$X : \alpha \vdash X = X \rightarrow \top : Prop \qquad X : Prop; a\#X \vdash \forall[a]X \rightarrow X : Prop$$

$$X, Y : Prop \vdash \forall[a](X \wedge Y) \rightarrow \forall[a]X \wedge \forall[a]Y : Prop$$

Arithmetic. Extend further with a type $\mathbb{N}$, term-formers 0 : $\mathbb{N}$, $succ$: $(\mathbb{N})\mathbb{N}$, $+$: $(\mathbb{N} \times \mathbb{N})\mathbb{N}$ and $=:$ $(\alpha \times \alpha)Prop$. Observe that $\lambda[a]succ(a)$ has principal type $\mathbb{N} \Rightarrow \mathbb{N}$ whereas $\lambda[a]0$ has principal type $\alpha \Rightarrow \mathbb{N}$. Likewise, $\forall[a](a = 0)$ is typable (with type *Prop*) whereas $\forall[a](\lambda[a]succ(a) = 0)$ is not typable.

5   Conclusions and Future Work

We have defined a syntax-directed type inference algorithm for nominal terms. It smoothly resolves the tension between the denotational intuition of a type as a set, and the syntactic intuition of a variable in nominal terms as a term which may mention atoms. The algorithm delivers principal types (our function *pt*). The types produced resemble the Hindley-Milner polymorphic type system for the λ-calculus, but are acting on nominal terms which include variables X representing context holes as well as atoms a representing program variables, and such that the same atom may occur in many different abstraction contexts and thus may acquire different types in different parts of the term.

Theorem 8 proves our types compatible with the powerful notion of α-equivalence inherited from nominal terms [21]. Theorem 11 shows that a notion of typed nominal rewrite rule exists which guarantees preservation of types.

Our system is in Curry style; type annotations on terms are not required. We do rely on type declarations for function symbols (arities) and in future we may investigate inferring the type of a function from its rewrite rules.

Type inference is well-studied for the λ-calculus and Curry-style systems also exist for first-order rewriting systems [2] and algebraic λ-calculi (which combine term rewriting and λ-calculus) [3]. We know of no type assignment system for the standard higher-order rewriting formats (HRSs use a typed metalanguage, and restrict rewrite rules to base types).

Our type system has only rank 1 polymorphism (type-variables are quantified, if at all, only at the top level of the type). It should be possible to consider more powerful systems, for instance using rank 2 polymorphic types, or intersection types [6]. The latter have been successfully used to provide characterisations of normalisation properties of λ-terms. Normalisation of nominal rewriting using type systems is itself a subject for future work, and one of our long-term goals in this work is to come closer to applying logical semantics such as intersection types, to nominal rewriting.

References

1. Baader, F., Snyder, W.: Unification Theory. In: Robinson, A., Voronkov, A. (eds.) Handbook of Automated Reasoning, vol. I, ch. 8, pp. 445–532. Elsevier Science, Amsterdam (2001)
2. van Bakel, S., Fernández, M.: Normalization results for typable rewrite systems. Information and Computation 133(2), 73–116 (1997)
3. Barbanera, F., Fernández, M.: Intersection type assignment systems with higher-order algebraic rewriting. Theoretical Computer Science 170, 173–207 (1996)
4. Barendregt, H.P.: Pairing without conventional constraints. Zeitschrift für mathematischen Logik und Grundlagen der Mathematik 20, 289–306 (1974)
5. Barendregt, H.P.: Lambda Calculi With Types. In: Abramsky, S., Gabbay, D., Maibaum, T.S.E. (eds.) Handbook of Logic in Computer Science, Oxford University Press, Oxford (1992)
6. Barendregt, H.P., Coppo, M., Dezani-Ciancaglini, M.: A filter lambda model and the completeness of type assignment. Journal of Symbolic Logic 48(4), 931–940 (1983)
7. Berghofer, S., Urban, C.: A Head-to-Head Comparison of de Bruijn Indices and Names. In: LFMTP'06, pp. 46–59 (2006)
8. Curry, H.B., Feys, R.: Combinatory Logic, vol. 1. North-Holland, Amsterdam (1958)
9. Damas, L.M.M., Milner, R.: Principal Type Schemes for Functional programs. In: Conference Record of the Ninth Annual ACM Symposium on Principles of Programming Languages, ACM Press, New York (1982)
10. Fernández, M., Gabbay, M.J., Mackie, I.: Nominal Rewriting Systems. In: PPDP'04. ACM Symposium on Principles and Practice of Declarative Programming, ACM Press, New York (2004)
11. Fernández, M., Gabbay, M.J.: Nominal Rewriting with Name Generation: Abstraction vs. Locality. In: PPDP'05. ACM Symposium on Principles and Practice of Declarative Programming, ACM Press, New York (2005)
12. Fernández, M., Gabbay, M.J.: Nominal Rewriting. Information and Computation (to appear), available from http://dx.doi.org/10.1016/j.ic.2006.12.002
13. Gabbay, M.J., Pitts, A.M.: A New Approach to Abstract Syntax with Variable Binding. Formal Aspects of Computing 13, 341–363 (2002)
14. Gabbay, M.J.: A Theory of Inductive Definitions with Alpha-Equivalence. PhD Thesis, Cambridge University (2000)
15. Girard, J.-Y.: The System F of Variable Types, Fifteen Years Later, Theoretical Computer Science, 45 (1986)
16. Gosling, J., Joy, B., Steele, G.: The Java Language Specification. Addison-Wesley, Reading (1996)
17. Khasidashvili, Z.: Expression reduction systems. In: Tbisili. Proceedings of I.Vekua Institute of Applied Mathematics, vol. 36, pp. 200–220 (1990)
18. Klop, J.-W., van Oostrom, V., van Raamsdonk, F.: Combinatory reduction systems, introduction and survey. Theoretical Computer Science 121, 279–308 (1993)
19. Mayr, R., Nipkow, T.: Higher-order rewrite systems and their confluence. Theoretical Computer Science 192, 3–29 (1998)
20. Shinwell, M.R., Pitts, A.M., Gabbay, M.: FreshML: Programming with binders made simple. In: ICFP 2003, pp. 263–274 (2003)
21. Urban, C., Pitts, A.M., Gabbay, M.J.: Nominal unification. Theoretical Computer Science 323, 473–497 (2004)

(In)consistency of Extensions of Higher Order Logic and Type Theory

Herman Geuvers

Radboud University Nijmegen
herman@cs.ru.nl

Abstract. It is well-known, due to the work of Girard and Coquand, that adding polymorphic domains to higher order logic, HOL, or its type theoretic variant λHOL, renders the logic inconsistent. This is known as Girard's paradox, see [4]. But there is also another presentation of higher order logic, in its type theoretic variant called λPREDω, to which polymorphic domains can be added safely, Both λHOL and λPREDω are well-known type systems and in this paper we study why λHOL with polymorphic domains is inconsistent and why nd λPREDω with polymorphic domains remains consistent. We do this by describing a simple model for the latter and we show why this can not be a model of the first.

1 Introduction

We study extensions of higher order logic **HOL** in the context of typed lambda calculi. It is known that extensions of higher order logic with polymorphic domains are inconsistent. This was established by Girard [11] and later this result was refined by Coquand [7] who showed that quantification over the collection of all domains wasn't needed to obtain the inconsistency. On the other hand, there are systems like the Calculus of Constructions (**CC**, [6]), which are consistent extensions of higher order logic in which we have polymorphic types. It is not so easy to relate **CC** directly to **HOL**, because in **CC** there is no syntactic distinction between domains and propositions (and therefore between set objects and proof objects). In this paper we therefore study the addition of polymorphic types in the context of a system of higher order order logic, presented as a (isomorphic) type theory following the Curry Howard formulas-as-types isomorphism. See [14] for an overview of formulas-as-types.

We present two isomorphic type systems for higher order logic, λ**HOL** and λ**PRED**ω, and we show why in the first case, the addition of polymorphic sets leads to inconsistency, and in the second case it does not. This is done by describing a model for λ**HOL** (and therefore for λ**PRED**ω), and to see how that can be extended to a model for λ**PRED**ω$^+$: higher order logic with polymorphism.

The model construction that we use is a variation of models described in [16]. It uses sets of untyped terms as the interpretation of types and is closely related to a saturated sets model. However, we will not make this connection precise.

T. Altenkirch and C. McBride (Eds.): TYPES 2006, LNCS 4502, pp. 140–159, 2007.
© Springer-Verlag Berlin Heidelberg 2007

The main contribution of the paper is a clarification of the fact that $\lambda\mathbf{PRED}\omega$ with polymorphic sets is consistent. The clue is that the "standard" model for higher order logic, where arrow types are interpreted as set theoretic function spaces does not work anymore if one adds polymorphic types (Reynolds [15] result), but that one can shift the interpretation of types one level lower, interpreting the arrow type $\sigma \to \tau$ as the collection of terms (from a combinatory algebra) that map terms of σ to terms of τ. This is basically Tait's [17] construction for saturated sets.

2 Higher Order Logic as a Pure Type System

2.1 Higher Order Predicate Logic

Definition 1. *The language of* **HOL** *is defined as follows.*

1. *The set of* domains, *D is defined by*

$$D ::= \text{Base} \mid \Omega \mid D \to D,$$

   where Base *represents the set of* basic domains *(we assume that there are countably many basic domains) and Ω represents the* domain of propositions.
2. *For every $\sigma \in D$, the set of* terms of domain σ, *Term_σ is inductively defined as follows. (As usual we write $t : \sigma$ to denote that t is a term of domain σ.)*
   (a) the constants $c_1^\sigma, c_2^\sigma, \ldots$ are in Term_σ,
   (b) the variables $x_1^\sigma, x_2^\sigma, \ldots$ are in Term_σ,
   (c) if $\varphi : \Omega$ and x^σ is a variable, then $(\forall x^\sigma.\varphi) : \Omega$,
   (d) if $\varphi : \Omega$ and $\psi : \Omega$, then $(\varphi \Rightarrow \psi) : \Omega$,
   (e) if $M : \sigma \to \tau$ and $N : \sigma$, then $(MN) : \tau$,
   (f) if $M : \tau$ and x^σ is a variable, then $(\lambda x^\sigma.M) : \sigma \to \tau$.
3. *The set of terms of* **HOL**, Term, *is defined by* $\text{Term} := \cup_{\sigma \in D} \text{Term}_\sigma$.
4. *The set of formulas of* **HOL**, **form**, *is defined by* **form** $:= \text{Term}_\Omega$.

We adapt the well-known notions of *free* and *bound* variable, substitution, β-reduction and β-conversion to the terms of this system. The λ-abstraction is both used for defining functions of higher type, like $\lambda f^{(\sigma \to \sigma) \to \sigma}.f(\lambda x^\sigma.x)$: $(\sigma \to \sigma) \to \sigma \to \sigma$, and for *comprehension*. Comprehension is the *axiom scheme* $\exists X \forall \boldsymbol{x} (X\boldsymbol{x} \leftrightarrow \varphi)$, where $\boldsymbol{x}$ is the sequence of free variables of the formula φ. Comprehension holds, because we can always take $X := \lambda \boldsymbol{x}.\varphi$.

A predicate is represented as a function to Ω, following the idea (probably due to Church; it appears in [3]) that a predicate can be seen as a function that takes a value as input and returns a formula. So, a binary relation over σ is represented as a term in the domain $\sigma \to (\sigma \to \Omega)$. (If $R : \sigma \to (\sigma \to \Omega)$ and $t, q : \sigma$, then $((Rt)q) : \Omega$.) The logical connectives are just implication and universal quantification. Due to the fact that we have *higher order* universal quantification, we can constructively express all other quantifiers using just $\Rightarrow$ and $\forall$.

We fix the usual notational conventions that outside brackets are omitted and that in the domains we omit the brackets by letting them associate to the right, so $\sigma \to \sigma \to \Omega$ denotes $\sigma \to (\sigma \to \Omega)$. In terms we omit brackets by associating them to the left, so Rtq denotes $(Rt)q$.

The derivation rules of **HOL** are given in a natural deduction style.

Definition 2. *The notion of* provability, $\Gamma \vdash \varphi$, *for* Γ *a finite set of formulas (terms of domain* **form***) and* φ *a formula, is defined inductively as follows.*

(*axiom*)	$\overline{\Gamma \vdash \varphi}$	*if* $\varphi \in \Gamma$
($\Rightarrow$ *-introduction*)	$\dfrac{\Gamma \cup \varphi \vdash \psi}{\Gamma \vdash \varphi \Rightarrow \psi}$	
($\Rightarrow$ *-elimination*)	$\dfrac{\Gamma \vdash \varphi \quad \Gamma \vdash \varphi \Rightarrow \psi}{\Gamma \vdash \psi}$	
($\forall$*-introduction*)	$\dfrac{\Gamma \vdash \varphi}{\Gamma \vdash \forall x^\sigma . \varphi}$	*if* $x^\sigma \notin \mathrm{FV}(\Gamma)$
($\forall$*-elimination*)	$\dfrac{\Gamma \vdash \forall x^\sigma . \varphi}{\Gamma \vdash \varphi[t/x^\sigma]}$	*if* $t : \sigma$
(*conversion*)	$\dfrac{\Gamma \vdash \varphi}{\Gamma \vdash \psi}$	*if* $\varphi =_\beta \psi$

Remark 1. The rule (conversion) is an operationalization of the comprehension axiom. The rule says that we don't want to distinguish between β-equal propositions.

2.2   Extension with Polymorphic Domains

Extending higher order logic with polymorphic domains makes the system inconsistent. This extension amounts to the system U^-. Allowing also quantification over all domains yields the system U. Both systems were defined in [11] and it was shown there that U is inconsistent, which became known as *Girard's paradox*. Later it was shown by [7] and [13] that U^- is also inconsistent. We now define these systems.

Definition 3. *The set of* domains *of* U^- D_U *is defined by*

$$D ::= \mathrm{Base} \,|\, \mathrm{Var}^D \,|\, \Omega \,|\, D {\to} D \,|\, \Pi_A . D$$

where Var^D *is a set of variables ranging over* D_U *and* $A \in \mathrm{Var}^D$.

For $\sigma \in D_U$, *the set of* terms *of domain* σ *in* U^-, $\mathrm{Term}_\sigma^{U^-}$ *is inductively defined as follows.*

1. *if $t : \Pi_A.\tau$ and $\sigma \in D_U$, then $t\sigma : \tau[\sigma/A]$*
2. *if $t : \tau$, then $\lambda_A.t : \Pi_A.\tau$*

The derivation rules for U^- are the same as for **HOL**.

The system U is the extension of U^-, where the terms are extended as follows.

3. *if $\varphi : \Omega$, then $\forall_A.\varphi : \Omega$*

The additional derivation rules for U are:

$$
\boxed{
\begin{array}{ll}
(\forall_2\text{-introduction}) \quad \dfrac{\Gamma \vdash \varphi}{\Gamma \vdash \forall_A.\varphi} & \text{if } A \notin \mathrm{FV}(\Gamma) \\[2em]
(\forall_2\text{-elimination}) \quad \dfrac{\Gamma \vdash \forall_A.\varphi}{\Gamma \vdash \varphi[\sigma/A]} & \text{if } \sigma \in D_U
\end{array}
}
$$

The systems U and U^- are inconsistent, which can be phrased as "higher order logic with polymorphic domains is inconsistent". However, the Calculus of Constructions [6] contains both higher order logic *and* polymorphic domains but it is still consistent. How to understand this paradoxical situation will be explained in this paper, using a model of higher order logic, defined as a typed λ calculus.

The Calculus of Constructions is a type theory where no distinction between objects and proofs is made. The Curry-Howard formulas-as-types embedding gives an embedding of higher order logic into **CC**, but it is not conservative. However, **CC** is consistent and contains higher order logic, so it must be possible to extend **HOL** with polymorphic domains in a consistent way. To see that, we define **PREDω**, which is a variant of **HOL** and its extension **PREDω^+**, which is higher order logic with polymorphic domains.

Definition 4. *The set of* domains *of* **PREDω** D_ω *is defined by*

$$
\begin{aligned}
D_w &::= D_s \mid D_p \\
D_p &::= \Omega \mid D \to D_p \\
D_s &::= B \mid D \to D_s
\end{aligned}
$$

The rules for terms and the derivation rules are the same as for **HOL**.

The system **PREDω^+** *is the extension of* **PREDω**, *where the domains D_s are as follows.*

$$
D_s ::= \mathrm{Base} \mid \mathrm{Var}^D \mid D \to D_s \mid \Pi_A.D_s
$$

where Var^D is a set of variables ranging over D_s and $A \in \mathrm{Var}^D$.

For $\sigma \in D$, the set of terms *of type σ in* **PREDω^+**, $\mathrm{Term}_\sigma^{\mathbf{PRED}\omega^+}$ *is inductively defined as follows.*

1. *if $t : \Pi_A.\tau$ and $\sigma \in D_s$, then $t\sigma : \tau[\sigma/A]$*
2. *if $t : \tau$ and $\tau \in D_s$, then $\lambda_A.t : \Pi_A.\tau$*

The derivation rules for **PREDω^+** *are the same as for* **PREDω**.

It can be shown that **PREDω** is isomorphic to **HOL**. The system **PREDω⁺**
is in flavor very close to U^-, both being extensions of **HOL** with polymorphic
domains. However, **PREDω⁺** is consistent.

2.3   Pure Type Systems for Higher Order Logic

In type theory, one interprets formulas and proofs via the well-known 'formulas-
as-types' and 'proofs-as-terms' embedding, originally due to Curry, Howard and
de Bruijn. (See [12].) Under this interpretation, a formula is viewed as the type
of its proofs. It turns out that one can define a typed λ-calculus λ**HOL** that
represents **HOL** in a very precise way, see [1] or [8]. In this section we briefly
introduce the general framework of Pure Type Systems or PTSs. The reason for
defining the class of PTSs is that many known systems are (or better: can be
seen as) PTSs. Here we will focus on higher order logic seen as a PTS.

Definition 5. *For S a set (the set of sorts), $\mathcal{A} \subset S \times S$ (the set of axioms)and
$\mathcal{R} \subset S \times S \times S$ (the set of rules), the* Pure Type System $\lambda(S, \mathcal{A}, \mathcal{R})$ *is the typed
lambda calculus with the following deduction rules.*

$$(sort) \quad \vdash s_1 : s_2 \qquad\qquad\qquad if\ (s_1, s_2) \in \mathcal{A}$$

$$(var) \quad \frac{\Gamma \vdash T : s}{\Gamma, x{:}T \vdash x : T} \qquad\qquad if\ x \notin \Gamma$$

$$(weak) \quad \frac{\Gamma \vdash T : s \quad \Gamma \vdash M : U}{\Gamma, x{:}T \vdash M : U} \qquad if\ x \notin \Gamma$$

$$(\Pi) \quad \frac{\Gamma \vdash T : s_1 \quad \Gamma, x{:}T \vdash U : s_2}{\Gamma \vdash \Pi x{:}T.U : s_3} \qquad if\ (s_1, s_2, s_3) \in \mathcal{R}$$

$$(\lambda) \quad \frac{\Gamma, x{:}T \vdash M : U \quad \Gamma \vdash \Pi x{:}T.U : s}{\Gamma \vdash \lambda x{:}T.M : \Pi x{:}T.U}$$

$$(app) \quad \frac{\Gamma \vdash M : \Pi x{:}T.U \quad \Gamma \vdash N : T}{\Gamma \vdash MN : U[N/x]}$$

$$(conv_\beta) \quad \frac{\Gamma \vdash M : T \quad \Gamma \vdash U : s}{\Gamma \vdash M : U} \qquad T =_\beta U$$

*If $s_2 \equiv s_3$ in a triple $(s_1, s_2, s_3) \in \mathcal{R}$, we write $(s_1, s_2) \in \mathcal{R}$. In the derivation
rules, the expressions are taken from the set of* pseudo-terms $\mathcal{T}$ *defined by*

$$\mathcal{T} ::= S \mid \mathcal{V} \mid (\Pi \mathcal{V}{:}\mathcal{T}.\mathcal{T}) \mid (\lambda \mathcal{V}{:}\mathcal{T}.\mathcal{T}) \mid \mathcal{T}\mathcal{T}.$$

The pseudo-term T is legal *if there is a context Γ and a pseudo-term U such
that $\Gamma \vdash T : U$ or $\Gamma \vdash U : T$ is derivable. The set of legal terms of $\lambda(S, \mathcal{A}, \mathcal{R})$ is
denoted by* Term($\lambda(S, \mathcal{A}, \mathcal{R})$).

By convention we write $A{\rightarrow}B$ for $\varPi x{:}A.B$ if $x \notin \mathrm{FV}(B)$.

In the following, we describe a PTS by just listing the sorts, the axioms and the rules in a box. For higher order logic **HOL** this amounts to the following λ**HOL**.

λ**HOL**
$\mathcal{S} \star, \square, \varDelta$
$\mathcal{A} \star : \square, \square : \varDelta$
$\mathcal{R} (\star, \star), (\square, \square), (\square, \star)$

The formulas-as-types interpretation from higher order predicate logic **HOL** into λ**HOL** maps a formula to a type and a derivation (in natural deduction) of a formula φ to a typed λ-term (of the type associated with φ):

$$\boxed{\;\genfrac{}{}{0pt}{}{}{\varSigma}\;}\genfrac{}{}{0pt}{}{}{\psi} \mapsto [\![\varSigma]\!] : (\![\psi]\!)$$

where $(\![-]\!)$ denotes the interpretation of formulas as types and $[\![-]\!]$ denotes the interpretation of derivations as λ-terms. In a derivation, we use expressions from the logical language (e.g. to instantiate the $\forall$), which may contain free variables, constants and domains. In type theory, in order to make sure that all terms are well-typed, the basic items (like variables and domains) have to be declared explicitly in the context. Also, a derivation will in general contain non-discharged assumptions $(\varphi_1, \ldots, \varphi_n)$ that will appear as declarations of variables $(z_1 : \varphi_1, \ldots, z_n : \varphi_n)$ in the type theoretic context. So the general picture is this.

$$\genfrac{}{}{0pt}{}{\varphi_1 \ldots \varphi_n}{\boxed{\varSigma}}\genfrac{}{}{0pt}{}{}{\psi} \mapsto \varGamma_\varSigma, z_1 : \varphi_1, \ldots, z_n : \varphi_n \vdash [\![\varSigma]\!] : (\![\psi]\!),$$

where $\varGamma_\varSigma$ is the context that declares all domains, constants and free variables that occur in $\varSigma$.

The system λ**HOL** is stratified in the sense that one can alternatively define it "layer by layer", starting from the terms of type $\varDelta$, then the terms of type A where $A : \varDelta$ etc. We introduce some naming and notation conventions for λ**HOL**. Some of these depend on properties of λ**HOL** that we do not prove here.

Remark 2

1. There is only one term of type $\varDelta$, and that is $\square$.
2. A term of type $\square$ is called a *kind*. Typical kind names are $\sigma, \tau, \ldots$. So a kind is a term σ with $\varGamma \vdash \sigma : \square$ for some $\varGamma$. All kinds are of the shape $\sigma_1 {\rightarrow} \ldots {\rightarrow} \star$ or $\sigma_1 {\rightarrow} \ldots {\rightarrow} A$ with A a variable of type $\square$.
3. A term of type a kind is called a *constructor*. Typical constructor names are $P, Q, \ldots$. So a constructor is a term P with $\varGamma \vdash P : \sigma$, where σ is a kind (so $\varGamma \vdash \sigma : \square$) for some $\varGamma$. If $\sigma = \star$, then we call P a *type*. Typical type names are $\varphi, \psi, \ldots$. So a type is a term φ with $\varGamma \vdash \varphi : \star$ for some $\varGamma$.

4. An *object* is a term of a type, so a p with $\Gamma \vdash p : \varphi$ where φ is a type, for some Γ. Typical object names are $p, q, \ldots$.

Calling the kinds σ conforms with the use of these names in **HOL**, where the domains were called σ. A domain in **HOL** corresponds with a kind in λ**HOL**.

Remark 3. There are three "ways" of introducing a variable in λ**HOL**. For each of these cases we take the variables from a specific subset of Var and we use specific notations for these variables. So we assume Var to be the disjoint subset of Var^Δ, $\mathrm{Var}^\square$ and $\mathrm{Var}^\star$. The three cases are:

1. $A : \square$; we take these variables, the *kind variables*, from Var^Δ and we use A as a typical name for such a variable.
2. $\alpha : \sigma$ with $\sigma : \square$; we take these variables, the *constructor variables*, from $\mathrm{Var}^\square$ and we use α as a typical name for such a variable.
3. $x : \varphi$ with $\varphi : \star$; we take these variables, the *object variables*, from $\mathrm{Var}^\star$ and we use x as a typical name for such a variable.

Here is a visual way of thinking of these classes of terms

$$
\begin{array}{ccc}
 & \Delta \quad\quad \Delta & \\
 & \vdots \quad\quad\; \vdots & \\
 & \square \quad\quad\; \square & \\
 & \vdots \quad\quad\; \vdots & \\
\text{kinds} & \sigma \quad\quad\; \star & \text{a special kind} \\
 & \vdots \quad\quad\; \vdots & \\
\text{constructors} & P \quad\quad\; \varphi & \text{types (a special case of constructors)} \\
 & \quad\quad\; \vdots & \\
 & \quad\quad\; p & \text{objects}
\end{array}
$$

The systems U^- and U can easily be seen as a Pure Type System as follows.

Definition 6. *The systems λU^- and λU are defined by adding to λ**HOL** respectively the rule $(\Delta, \square)$ and the rules $(\Delta, \square)$ and $(\Delta, \star)$.*

Another way of looking at higher order logic is the system **PRED**ω. This can be turned into a PTS as follows (originally due to [2]).

λPREDω
$\mathcal{S}$ Set, Type^s, Prop, Type^p
$\mathcal{A}$ Set : Type^s, Prop : Type^p
$\mathcal{R}$ (Set, Set), (Set, Type^p), (Type^p, Type^p), (Prop, Prop), (Set, Prop), (Type^p, Prop)

It can be formally shown that λ**HOL** and λ**PRED**ω are isomorphic. We will come to that in Lemma 1. We can also view λ**PRED**ω in a stratified way and we could have introduced it in that way. This would be very close to starting off from the domains, then the terms and then the proofs, as we did for **PRED**ω.

$$
\begin{array}{c}
\mathsf{Type}^s \\
\vdots \\
\mathsf{Set}
\end{array}
\quad
\begin{array}{cc}
\mathsf{Type}^p & \mathsf{Type}^p \\
\vdots & \vdots \\
K & \mathsf{Prop} \\
\vdots & \vdots \\
P & \varphi \\
 & \vdots \\
 & p
\end{array}
$$

	Type^s			
	Set	Type^p	Type^p	
sets	σ	K	Prop	kinds (Prop is a special kind)
set objects	t	P	φ	constructors (a type is a special constr.)
			p	proof objects

We can also add polymorphic types to λ**PRED**ω, which amounts to the system λ**PRED**ω^+

Definition 7. *The system* λ**PRED**ω^+ *is defined by adding to* λ**PRED**ω *the rule* $(\mathsf{Type}^s, \mathsf{Set})$.

We introduce two other known type systems as PTSs: **F**ω of [11] and the Calculus of Constructions, **CC** of [6].

Fω	**CC**
$\mathcal{S}\ \star, \square$	$\mathcal{S}\ \star, \square$
$\mathcal{A}\ \star : \square$	$\mathcal{A}\ \star : \square$
$\mathcal{R}\ (\star, \star), (\square, \square), (\square, \star)$	$\mathcal{R}\ (\star, \star), (\star, \square), (\square, \star), (\square, \square)$

In view of higher order predicate logic, one can understand CC as the system obtained by smashing the sorts Prop and Set into one, $\star$. Hence, higher order predicate logic can be done inside the Calculus of Constructions. We describe the map from λPREDω to **CC** in detail in Definition 9.

The system **F**ω is known to be consistent. As a consequence, λ**HOL** is consistent: if we map all kind variables to $\star$, then the rules are preserved, so we have an embedding of λ**HOL** into **F**ω, where $\bot$ $(:= \Pi\alpha : \star.\alpha)$ is mapped to itself. As $\bot$ is not inhabited in **F**ω, it is not inhabited in λ**HOL**.

One can sometimes relate results of two different systems by defining an embedding between them. There is one very simple class of embeddings between PTSs.

Definition 8. *For* $T = \lambda(\mathcal{S}, \mathcal{A}, \mathcal{R})$ *and* $T' = \lambda(\mathcal{S}', \mathcal{A}', \mathcal{R}')$ *PTSs, a PTS-morphism from* T *to* T' *is a mapping* $f : \mathcal{S} \to \mathcal{S}'$ *that preserves the axioms and rules. That is, for all* $s_1, s_2 \in \mathcal{S}$, *if* $(s_1, s_2) \in \mathcal{A}$ *then* $(f(s_1), f(s_2)) \in \mathcal{A}'$ *and if* $(s_1, s_2, s_3) \in \mathcal{R}$ *then* $(f(s_1), f(s_2), f(s_3)) \in \mathcal{R}'$.

A PTS-morphism f from $\lambda(\mathcal{S}, \mathcal{A}, \mathcal{R})$ to $\lambda(\mathcal{S}', \mathcal{A}', \mathcal{R}')$ immediately extends to a mapping f on pseudo-terms and contexts. Moreover, this mapping preserves reduction in a faithful way: $M \to_\beta N$ iff $f(M) \to_\beta f(N)$. We have the following property.

Proposition 1. *For T and T' PTSs and f a PTS-morphism from T to T', if $\Gamma \vdash M : A$ in T, then $f(\Gamma) \vdash f(M) : f(A)$ in T'.*

Definition 9. *The following is a PTS-morphism h from $\lambda\mathbf{PRED}\omega$ (and also from $\lambda\mathbf{PRED}\omega^+$) into $\mathbf{CC}$.*

$$h(\mathsf{Prop}) = \star \qquad h(\mathsf{Set}) = \star,$$
$$h(\mathsf{Type}^p) = \square \quad h(\mathsf{Type}^s) = \square$$

Not all PTSs are strongly normalizing. We have the following well-known theorem.

Theorem 1. *The Calculus of Constructions, $\mathbf{CC}$, is strongly normalizing, and therefore the following PTSs are all strongly normalizing as well: all subsystems of $\mathbf{CC}$; $\lambda\mathrm{PRED}$; $\lambda\mathrm{PRED}\omega$; $\lambda\mathbf{PRED}\omega^+$.*

The proof that $\mathbf{CC}$ is strongly normalizing can be found in [10], [9], [2]. As a direct consequence we find that all sub-systems of $\mathbf{CC}$ are strongly normalizing and also all systems T for which there is a PTS-morphism from T to $\mathbf{CC}$, including $\mathbf{PRED}\omega^+$, see Definition 9 (Note that a PTS-morphism preserves infinite reduction paths.)

Well-known examples of PTSs that are not strongly normalizing are $\lambda\mathbf{U}$ and $\lambda\mathbf{U}^-$.

As a matter of fact, we now have two formalizations of higher order predicate logic as a PTS: $\lambda\mathbf{HOL}$ and $\lambda\mathrm{PRED}\omega$. We employ the notion of PTS-morphism to see that they are equivalent. From $\lambda\mathrm{PRED}\omega$ to $\lambda\mathbf{HOL}$, consider the PTS-morphism f given by

$$f(\mathsf{Prop}) = \star \qquad f(\mathsf{Set}) = \square,$$
$$f(\mathsf{Type}^p) = \square \quad f(\mathsf{Type}^s) = \Delta.$$

One immediately verifies that f preserves $\mathcal{A}$ and $\mathcal{R}$, hence we have

$$\Gamma \vdash_{\lambda\mathrm{PRED}\omega} M : A \implies f(\Gamma) \vdash_{\lambda\mathbf{HOL}} f(M) : f(A).$$

The inverse of f can almost be described as a PTS-morphism, but not quite. Define the PTS-morphism g from $\lambda\mathrm{PRED}\omega$ to $\lambda\mathbf{HOL}$ as follows.

$$g(\star) = \mathsf{Prop} \quad g(\square) = \mathsf{Set} \quad g(\Delta) = \mathsf{Type}^s$$

(In $\lambda\mathbf{HOL}$ the sort Δ can not appear in a context nor in a term on the left side of the ':'.) We extend g to derivable judgments of $\lambda\mathbf{HOL}$ in the following way.

$$g(\Gamma \vdash M : A) = g(\Gamma) \vdash g(M) : g(A), \text{ if } A \neq \mathsf{Type}^p,$$
$$g(\Gamma \vdash M : \mathsf{Type}^p) = g(\Gamma) \vdash g(M) : \mathsf{Set}, \text{ if } M \equiv \cdots \to \alpha, (\alpha \text{ a variable}),$$
$$g(\Gamma \vdash M : \mathsf{Type}^p) = g(\Gamma) \vdash g(M) : \mathsf{Type}^p, \text{ if } M \equiv \cdots \to \mathsf{Prop}.$$

By easy induction one proves that g preserves derivations. Furthermore, $f(g(\Gamma \vdash M : A)) = \Gamma \vdash M : A$ and $g(f(\Gamma \vdash M : A)) = \Gamma \vdash M : A$.

Lemma 1. *The systems* λPREDω *and* λ**HOL** *are equivalent.*

This equivalence implies that the system λ**HOL** is strongly normalizing as well. But we already knew that, because it is also a consequence of the embedding of λ**HOL** into **F**ω and the fact that **F**ω is strongly normalizing.

The connection between these typed λ-calculi for higher order logic is described in Figure 1.

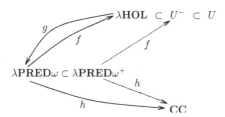

Fig. 1. Morphisms and inclusions between calculi for higer order logic

3 The Model Construction

3.1 Combinatory Algebras

To model the set of pseudo-terms of type theories we can use *combinatory algebras* (ca), or variants of combinatory algebras, like *partial combinatory algebras* (pca) or *conditionally partial combinatory algebras* (c-pca). We only list the important notions here.

Definition 10. *A combinatory algebra (ca) is a structure* $\langle \mathbf{A}, \cdot, \mathbf{k}, \mathbf{s}, =_A \rangle$ *satisfying*

$$(\mathbf{k} \cdot x) \cdot y =_A x , \qquad ((\mathbf{s} \cdot x) \cdot y) \cdot z =_A (x \cdot z) \cdot (y \cdot z)$$

The application $(\cdot)$ is usually not written. The set of terms over $\mathcal{A}$ (notation $\mathcal{T}(\mathcal{A})$) is defined by $\mathcal{T} ::= \mathrm{Var} \mid \mathbf{A} \mid \mathcal{T}\mathcal{T}$.

Every ca is *combinatory complete*, i.e., for every $T \in \mathcal{T}(\mathcal{A})$ with $\mathrm{FV}(T) \subset \{x\}$, there exists an $f \in \mathbf{A}$ such that

$$f \cdot a =_A T[a/x] \qquad \forall a \in \mathbf{A}.$$

Such an element f will be denoted by $\lambda x.T$ in the sequel. It is well-known that one can define λ as the standard abstraction λ^* with the help of the combinators $\mathbf{k}$ and $\mathbf{s}$ by induction on the structure of terms.

In the following we could restrict ourselves completely to the combinatory algebra $(\Lambda, \cdot, \lambda xy.x, \lambda xyz.xz(yz), =_\beta)$, where Λ is the set of (open, untyped) λ-terms and $\cdot$ denotes application, which we usually don't write. But the constructions apply more generally to other combinatory algebras, as long as they are *weakly extensional*, i.e. if the following holds

$$\mathcal{A} \models \forall x(T_1 = T_2) \rightarrow \lambda x.T_1 = \lambda x.T_2.$$

The combinatory algebra CL (combinatory logic) is not weakly extensional. The combinatory algebra $(\Lambda, \cdot, \lambda xy.x, \lambda xyz.xz(yz), =_\beta)$ is weakly extensional, if we take for $\lambda x.T$ just $\lambda x.T$ or the λ-abstraction definable by **k** and **s**. It is well-known that if we take for Λ the set of *closed* lambda terms, the ca is not weakly extensional. Another interesting example of a weakly extensional ca is $\Lambda(C)$, the set of open λ_C-terms (i.e. lambda-terms over some constant set C) modulo βc-equality, where the c-equality rule is defined by $cN =_c c$, $\lambda v.c =_c c$ (for all $c \in C$ and $N \in \Lambda_C$).

3.2   The Model for λHOL

The notion of λ**HOL**-*structure* and the interpretation of the typable terms of λ**HOL** are explained informally in the next paragraphs.

The typable terms of λ**HOL** are mapped into a (set-theoretical) hierarchical structure (called λ**HOL**-structure) according to their classification as objects, types, constructors, or kinds. The kinds of λ**HOL** are interpreted as sets from a *predicative structure* N, so Type^p is interpreted as N. Predicative structures are closed under set-theoretical function space construction. The impredicative universe Prop is interpreted as a collection P of subsets of the underlying ca. We call this collection *polystructure* and its elements *polysets*. P itself is an element of N and is closed under non-empty intersections and a function space construction (to be defined). Constructors are interpreted as elements of $\bigcup_{X \in N} X$ ($\bigcup N$ in short). Their interpretations are called *poly-functionals*. In particular, types are mapped to polysets.

Definition 11. *A* polyset structure *over the weakly extensional combinatory algebra $\mathcal{A}$ is a collection* $P \subseteq \wp(\mathbf{A})$ *such that*

1. $\mathbf{A} \in P$,
2. P *is closed under arbitrary non-empty intersection* $\bigcap$: *if* $I \neq \emptyset$ *and* $\forall i \in I(X_i \in P)$, *then* $\bigcap_{i \in I} X_i \in P$.
3. P *is closed under* function space, *i.e. if* $X, Y \in P$, *then* $X \to_0 Y \in P$, *where* $X \to_0 Y$ *is defined as*

$$\{a \in \mathbf{A} \mid \forall t \in X(a \cdot t \in Y)\}.$$

The elements of a polyset structure are called polysets.

Example 1

1. We obtain the *full polyset structure* over the weca $\mathcal{A}$ if we take $P = \wp(\mathbf{A})$.
2. The *simple polyset structure* over the weca $\mathcal{A}$ is obtained by taking $P = \{\emptyset, \mathbf{A}\}$. It is easily verified that this is a polyset structure.
3. Given the weca $\Lambda(C)$ as defined in the previous Section (so C is a set of constants), we define the *polyset structure generated from C* by

$$P := \{X \subseteq \Lambda(C) \mid X = \emptyset \vee C \subseteq X\}.$$

4. Given the weca $\mathcal{A}$ and a set $C \subseteq \mathbf{A}$ such that $\forall a, b \in \mathbf{A}(a \cdot b \in C \implies a \in C$, we define the *power polyset structure of C* by

$$\mathsf{P} := \{X \subseteq \mathbf{A} \mid X \subseteq C \lor X = \mathbf{A}\}.$$

5. The *degenerate polyset structure* is $\mathsf{P} := \{\mathbf{A}\}$, in which all types are interpreted as $\mathbf{A}$, so in this structure there are no empty types.

The function space of a polyset structure will be used to interpret types of the form $\varphi \rightarrow \psi$, where both φ and ψ are types. The intersection will be used to interpret types of the form $\Pi\alpha{:}\sigma.\varphi$, where σ is a kind and φ is a type. To interpret types we need a *predicative structure*.

Definition 12. *For P a polyset structure, the* predicative structure over P *is the collection of sets N defined inductively by*

1. $\mathsf{P} \in \mathsf{N}$,
2. *If $X, Y \in \mathsf{P}$, then $X \rightarrow_1 Y \in \mathsf{N}$, where $\rightarrow_1$ denotes the set-theoretic function space.*

Definition 13. *If $\mathcal{A}$ is a weakly extensional combinatory algebra, P a polyset structure over $\mathcal{A}$ and N the predicative structure over P, then we call the tuple $\langle \mathcal{A}, \mathsf{P}, \mathsf{N} \rangle$ a λ**HOL**-model.*

Remark 4. It is possible to vary on the notions of polystructure and predicative structure by requiring closure under dependent function spaces (in P and/or N). In that case we obtain models that can interpret dependent types. For details we refer to [16].

We now define the interpretation function $[\![-]\!]$, which maps kinds to elements of N, constructors to elements of $\bigcup \mathsf{N}$ (and types to elements of P, which is a subset of $\bigcup \mathsf{N}$) and objects to elements of the combinatory algebra $\mathcal{A}$. All these interpretations are parameterized by *valuations*, assigning values to the free variables (declared in the context).

Definition 14. *A* variable valuation *is a map from $\mathrm{Var}^\Delta \cup \mathrm{Var}^\square \cup \mathrm{Var}^\star$ to $\mathsf{N} \cup \bigcup \mathsf{N} \cup \mathbf{A}$ that consists of the union of an object variable valuation ρ_0 : $\mathrm{Var}^\star \rightarrow \mathbf{A}$, a constructor variable valuation $\rho_1 : \mathrm{Var}^\square \rightarrow \bigcup \mathsf{N}$ and a kind variable valuation $\rho_2 : \mathrm{Var}^\Delta \rightarrow \mathsf{N}$.*

Definition 15. *For ρ a valuation of variables, we define the map $([-])_\rho$ on the set of well-typed objects as follows. (We leave the model implicit.)*

$$([x])_\rho := \rho(x),$$
$$([tq])_\rho := ([t])_\rho \cdot ([q])_\rho, \text{ if } q \text{ is an object,}$$
$$([tQ])_\rho := ([t])_\rho, \text{ if } Q \text{ is a constructor,}$$
$$([\lambda x{:}\varphi.t])_\rho := \lambda v.([t])_{\rho(x:=v)}, \text{ if } \varphi \text{ is a type,}$$
$$([\lambda\alpha{:}\sigma.t])_\rho := ([t])_\rho, \text{ if } \sigma \text{ is a kind.}$$

Here, $\lambda v.([t])_{\rho(x:=v)}$ is the element of the $\mathcal{A}$ that is known to exist due to combinatory completeness, see the discussion after Definition 10.

Definition 16. *For ρ a valuation of variables, we define the maps $\mathcal{V}(-)_\rho$ and $[\![-]\!]_\rho$ respectively from kinds to N and from constructors to $\bigcup \mathsf{N}$ as follows. (We leave the model implicit.)*

$$\mathcal{V}(\star)_\rho := \mathsf{P},$$
$$\mathcal{V}(A)_\rho := \rho(A), \quad \text{if } A \text{ is a kind variable,}$$
$$\mathcal{V}(\sigma{\to}\tau)_\rho := \mathcal{V}(\sigma)_\rho {\to}_1 \mathcal{V}(\tau)_\rho,$$
$$[\![\alpha]\!]_\rho := \rho(\alpha),$$
$$[\![\Pi\alpha{:}\sigma.\varphi]\!]_\rho := \bigcap_{a \in \mathcal{V}(\sigma)_\rho} [\![\varphi]\!]_{\rho(\alpha:=a)}, \quad \text{if } \sigma \text{ is a kind,}$$
$$[\![\varphi{\to}\psi]\!]_\rho := [\![\varphi]\!]_\rho {\to}_0 [\![\psi]\!]_\rho, \quad \text{if } \varphi, \psi \text{ are a types,}$$
$$[\![PQ]\!]_\rho := [\![P]\!]_\rho([\![Q]\!]_\rho),$$
$$[\![\lambda\alpha{:}\sigma.P]\!]_\rho := \boldsymbol{\lambda} a \in \mathcal{V}(\sigma)_\rho.[\![P]\!]_{\rho(\alpha:=a)}.$$

Definition 17. *For Γ a $\lambda\mathbf{HOL}$-context, ρ a valuation of variables, we say that ρ fulfills Γ, notation $\rho \models \Gamma$, if for all $A \in \mathrm{Var}^\triangle$, $x \in \mathrm{Var}^\star$ and $\alpha \in \mathrm{Var}^\square$, $A \in \square \in \Gamma \Rightarrow \rho(A) \in \mathsf{N}$, $\alpha : \sigma \in \Gamma \Rightarrow \rho(\alpha) \in \mathcal{V}(\sigma)_\rho$ and $x : \varphi \in \Gamma \Rightarrow \rho(x) \in [\![\varphi]\!]_\rho$.*

It is (implicit) in the definition that $\rho \models \Gamma$ only if for all declarations $x{:}\varphi \in \Gamma$, $[\![\varphi]\!]_\rho$ is defined (and similarly for $\alpha{:}\sigma \in \Gamma$).

Definition 18. *The notion of truth in a $\lambda\mathbf{HOL}$-model, notation $\models^{\mathcal{S}}$ and of truth, notation $\models$ are defined as follows. For Γ a context, t an object, φ a type, P a constructor and σ a kind of $\lambda\mathbf{HOL}$,*

$$\Gamma \models^{\mathcal{S}} t : \varphi \quad if \forall \rho[\rho \models \Gamma \Rightarrow (\![t]\!)_\rho \in [\![\varphi]\!]_\rho],$$
$$\Gamma \models^{\mathcal{S}} P : \sigma \quad if \forall \rho[\rho \models \Gamma \Rightarrow [\![P]\!]_\rho \in \mathcal{V}(\sigma)_\rho].$$

Quantifying over the class of all $\lambda\mathbf{HOL}$-models, we define, for M an object or a constructor of $\lambda\mathbf{HOL}$,

$$\Gamma \models M : T \quad if \quad \Gamma \models^{\mathcal{S}} M : T \text{ for all } \lambda\mathbf{HOL}\text{-models } \mathcal{S}.$$

Soundness states that if a judgment $\Gamma \vdash M : T$ is derivable, then it is true in all models. It is proved by induction on the derivation in $\lambda\mathbf{HOL}$.

Theorem 2 (Soundness). *For Γ a context, M an object or a constructor and T a type or a kind of $\lambda\mathbf{HOL}$,*

$$\Gamma \vdash M : T \Rightarrow \Gamma \models M : T.$$

Example 2. Let $\mathcal{A}$ be a weca.

1. The *full $\lambda\mathbf{HOL}$-model* over $\mathcal{A}$ is $\mathcal{S} = \langle \mathcal{A}, \mathsf{P}, \mathsf{N} \rangle$, where P is the full polyset structure over $\mathcal{A}$ (as defined in Example 1).
2. The *simple $\lambda\mathbf{HOL}$-model* over $\mathcal{A}$ is $\mathcal{S} = \langle \mathcal{A}, \mathsf{P}, \mathsf{N} \rangle$, where P is the simple polyset structure over $\mathcal{A}$. (So $\mathsf{P} = \{\emptyset, \mathbf{A}\}$.)

3. The simple $\lambda\textbf{HOL}$-model over the degenerate $\mathcal{A}$ is also called the *proof-irrelevance model* or *PI-model* for $\lambda\textbf{HOL}$. Here $\textsf{P} = \{0, 1\}$, withe $0 = \emptyset$ and 1 is the one-element set.

4. For C a set of constants, the $\lambda\textbf{HOL}$-*model generated from* C is defined by $\mathcal{S} = \langle \Lambda(C), \textsf{P}, \textsf{N} \rangle$, where $\textsf{P}$ is the polyset structure generated from C.

4   Extending the Model Construction

4.1   Extensions of $\lambda\textbf{HOL}$

The model for $\lambda\textbf{HOL}$ can be extended to other type theories. First of all we remark that the rule $(\varDelta, \star)$ can easily be interpreted by putting

$$[\![\varPi A{:}\square.\varphi]\!]_\rho := \bigcap_{W \in \textsf{N}} [\![\varphi]\!]_{\rho(A:=W)}.$$

This can be interpreted in any model, so the extension of $\lambda\textbf{HOL}$ with the rule $(\varDelta, \star)$ is consistent.

The rule $(\varDelta, \square)$ makes $\lambda\textbf{HOL}$ inconsistent. This can be observed in the model, because the only possible interpretation in $\textsf{N}$ for $\varPi A{:}\square.\sigma$ would be

$$\mathcal{V}(\varPi A{:}\square.\sigma)_\rho := \bigcap_{W \in \textsf{N}} \mathcal{V}(\sigma)_{\rho(A:=W)},$$

which would only make sense if $\textsf{N}$ were also a polyset structure. (If $\textsf{N}$ were set theoretic, $\mathcal{V}(\varPi A{:}\square.\sigma)_\rho$ would just be empty.) But this can only be achieved by taking $\textsf{P} := \{\textbf{A}\}$, the degenerate polyset structure. (See Example 1.) then $\textsf{N} := \{\{\textbf{A}\}\}$, which can be seen as a predicative structure and is then closed under $\rightarrow_1$. In this model all types are interpreted as the non-empty set $\textbf{A}$, which conforms with the fact that $\lambda\textbf{U}^-$ is inconsistent.

4.2 $\lambda\textbf{PRED}\omega$ and Extensions

As $\lambda\textbf{HOL}$ is isomorphic to $\lambda\textbf{PRED}\omega$, we also have a model for $\lambda\textbf{PRED}\omega$. As we want to vary on the type theory $\lambda\textbf{PRED}\omega$, we make the model construction for $\lambda\textbf{PRED}\omega$ precise here. As a model we just take the definition of $\lambda\textbf{HOL}$-model as given in Definition 13.

Definition 19. *A valuation of variables is a map from* $\text{Var}^{\text{Type}^s} \cup \text{Var}^{\text{Type}^p} \cup \text{Var}^{\text{Set}} \cup \text{Var}^{\text{Prop}}$ *to* $\textsf{N} \cup \bigcup \textsf{N} \cup \textbf{A}$ *that consists of the union of an valuation of proof object variables* $\rho_0 : \text{Var}^{\text{Prop}} \rightarrow \textbf{A}$, *a valuation of constructor variables* $\rho_{1a} : \text{Var}^{\text{Type}^p} \rightarrow \bigcup \textsf{N}$, *a valuation of set object variables* $\rho_{1b} : \text{Var}^{\text{Set}} \rightarrow \bigcup \textsf{N}$ *and a valuation of set variables* $\rho_2 : \text{Var}^{\text{Type}^s} \rightarrow \textsf{N}$.

Definition 20. *For* ρ *a valuation of variables, we define the map* $(\![-]\!)_\rho$ *on the set of well-typed proof objects of* $\lambda\textbf{PRED}\omega$ *as follows. (We leave the model implicit.)*

$$\llbracket x \rrbracket_\rho := \rho(x),$$
$$\llbracket tq \rrbracket_\rho := \llbracket t \rrbracket_\rho \cdot \llbracket q \rrbracket_\rho, \ \textit{if } q \textit{ is a proof object,}$$
$$\llbracket tQ \rrbracket_\rho := \llbracket t \rrbracket_\rho, \ \textit{if } Q \textit{ is a constructor or a set object,}$$
$$\llbracket \lambda x{:}\varphi.t \rrbracket_\rho := \lambda v.\llbracket t \rrbracket_{\rho(x:=v)}, \ \textit{if } \varphi \textit{ is a type,}$$
$$\llbracket \lambda \alpha{:}U.t \rrbracket_\rho := \llbracket t \rrbracket_\rho, \ \textit{if } U \textit{ is a kind or a set.}$$

Definition 21. *For ρ a valuation of variables, we define the maps $\mathcal{V}(-)_\rho$ and $\llbracket - \rrbracket_\rho$ respectively from kinds of $\lambda\mathbf{PRED}\omega$ to N and from constructors and set objects of $\lambda\mathbf{PRED}\omega$ to $\bigcup \mathsf{N}$ as follows. (We leave the model implicit.)*

$$\mathcal{V}(\mathsf{Prop})_\rho := \mathsf{P},$$
$$\mathcal{V}(A)_\rho := \rho(A), \ \textit{if } A \in \mathrm{Var}^{\mathrm{Type}^s},$$
$$\mathcal{V}(\sigma{\to}\tau)_\rho := \mathcal{V}(\sigma)_\rho {\to}_1 \mathcal{V}(\tau)_\rho,$$
$$\llbracket \alpha \rrbracket_\rho := \rho(\alpha), \ \textit{if } \alpha \in \mathrm{Var}^{\mathrm{Type}^p} \cup \mathrm{Var}^{\mathrm{Set}},$$
$$\llbracket \Pi\alpha{:}U.\varphi \rrbracket_\rho := \bigcap_{a \in \mathcal{V}(U)_\rho} \llbracket \varphi \rrbracket_{\rho(\alpha:=a)}, \ \textit{if } U \textit{ is a kind or a set,}$$
$$\llbracket \varphi{\to}\psi \rrbracket_\rho := \llbracket \varphi \rrbracket_\rho {\to}_0 \llbracket \psi \rrbracket_\rho, \ \textit{if } \varphi, \psi \textit{ are a types,}$$
$$\llbracket PQ \rrbracket_\rho := \llbracket P \rrbracket_\rho(\llbracket Q \rrbracket_\rho), \ \textit{if } Q \textit{ is a constructor or a set object,}$$
$$\llbracket \lambda\alpha{:}U.P \rrbracket_\rho := \boldsymbol{\lambda} a \in \mathcal{V}(U)_\rho.\llbracket P \rrbracket_{\rho(\alpha:=a)} \ \textit{if } U \textit{ is a kind or a set.}$$

For Γ a $\lambda\mathbf{PRED}\omega$-context, ρ a valuation of variables, the notion of ρ *fulfills* Γ ($\rho \models \Gamma$), is the similar to the one for $\lambda\mathbf{HOL}$:

$A : \mathsf{Set} \in \Gamma \Rightarrow \rho(A) \in \mathsf{N}$, $\alpha : \sigma \in \Gamma \Rightarrow \rho(\alpha) \in \mathcal{V}(\sigma)_\rho$ (for σ a set), $\alpha : K \in \Gamma \Rightarrow \rho(\alpha) \in \mathcal{V}(K)_\rho$ (for K a kind) and $x : \varphi \in \Gamma \Rightarrow \rho(x) \in \llbracket \varphi \rrbracket_\rho$.

The notion of *truth* is the same as for $\lambda\mathbf{HOL}$ models (Definition 18) and we also have a soundness result, like Theorem 2.

To compare the situation fro $\lambda\mathbf{HOL}$ and $\lambda\mathbf{PRED}\omega$, we can take a look at the two figures 2 and 3. The first describes how the different "levels" of $\lambda\mathbf{HOL}$ and $\lambda\mathbf{PRED}\omega$ are interpreted in the model. (Forget about the part of dashed arrows for now.) The second describes how the function spaces are interpreted in the model. Again omit the dashed arrows and the part that is not between the two dashed lines on the right.

We now want to look at $\lambda\mathbf{PRED}\omega^+$, the extension of $\lambda\mathbf{PRED}\omega$ with polymorphic kinds (higher order logic with polymorphic domains). In this system we have types of the form $\Pi A{:}\mathsf{Set}.A{\to}A : \mathsf{Set}$ and the system is known to be consistent. According to the $\lambda\mathbf{PRED}\omega$ semantics, we would have to put

$$\mathcal{V}(\Pi A{:}\mathsf{Set}.A{\to}A)_\rho \in \mathsf{N},$$

but then we would have to interpret the Π either as a set-theoretic dependent function space (which is not possible for cardinality reasons) or as an intersection,

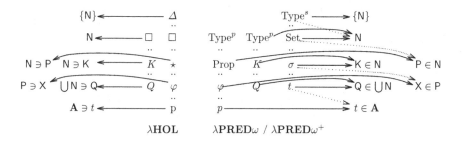

Fig. 2. Interpretation of the different object, types and sorts

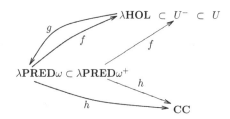

Fig. 3. Interpretation of the different (dependent) function spaces

and then N would have to be a polyset structure as well. The latter would amount to a model in which all types are interpreted as non-empty sets, which is not what we want.

The solution is to "shift" the interpretation of the sort Set one level down, interpreting Set as P and σ : Set as a polyset. Then the interpretation of ΠA:Set.$A{\rightarrow}A$ will be as follows.

$$[\![\Pi A\text{:Set}.A{\rightarrow}A]\!]_\rho = \bigcap_{X\in\mathsf{P}} X{\rightarrow}_0 X \in \mathsf{P}.$$

This amounts to the dashed arrows in the Figures 2 and 3. In order to define this interpretation we have to extend the notions of polyset structure and predicative structure a bit. As we now have *dependent types* at the polyset level, we need a *dependent function type* in P. Moreover, we have type dependent functions from polysets to predicative sets.

Remark 5. The model for $\lambda\mathbf{PRED}\omega^+$ that we describe below could also be seen as a model for **CC** – e.g. as described in [16] – that we turn into a $\lambda\mathbf{PRED}\omega^+$ model via the embedding h of Definition 9. (See also Figure 1.) However **CC** does not permit a definition in "levels", because kinds can depend on (proof)objects etc., and therefore **CC**-models are quite complicated. So we opt for a primitive notion of $\lambda\mathbf{PRED}\omega^+$-model, which is simpler.

However, the idea of using a **CC**-model does clarify the difference between $\lambda\mathbf{PRED}\omega$-models and $\lambda\mathbf{PRED}\omega^+$-models: a $\lambda\mathbf{PRED}\omega$-model is basically a $\lambda\mathbf{HOL}$-model via the embedding f, while a $\lambda\mathbf{PRED}\omega^+$-model is a **CC**-model via the embedding h.

Definition 22. *A polyset structure for $\lambda\mathbf{PRED}\omega^+$ is a polyset structure that is moreover closed under dependent function spaces, ($\rightarrow_p$ in Figure 3): if $F : X \rightarrow \mathsf{P}$ is a function such that $t =_\mathbf{A} q \Rightarrow F(t) = F(q)$, then P also contains*

$$\prod\nolimits_\mathbf{A}(X, F) := \{f \in \mathbf{A} \mid \forall n \in X(f \cdot n \in F(n))\}$$

For convenience $\prod_\mathbf{A}(X, F)$ will be denoted by $\prod_\mathbf{A} x \in X.F(x)$. Like in type theory, if F is a constant function on X, say $F(x) = Y$, then $\prod_\mathbf{A}(X, F)$ is just the function space $X \rightarrow_0 Y$.

Definition 23. *A predicative structure for $\lambda\mathbf{PRED}\omega^+$ over P is a predicative structure that is moreover closed under function spaces from P to N, ($\rightarrow_2$ in Figure 3): if $X \in \mathsf{P}$ and $\mathsf{K} \in \mathsf{N}$, then the following is also in N*

$$X \rightarrow_2 \mathsf{K} := \{h \mid \forall t, q \in X, t =_\mathbf{A} q \Rightarrow h(t) = h(q) \in \mathsf{K}\}.$$

Note that the elements of $X \rightarrow_2 \mathsf{K}$ are set-theoretic functions.

We now make the model construction for $\lambda\mathbf{PRED}\omega^+$ precise. As a model we just take the definition of $\lambda\mathbf{HOL}$-model, where the polyset structure and the predicative structure are as in Definitions 22, 23. The interpretations will be such that they conform with the dashed arrows in Figures 2 and 3.

Definition 24. *A valuation of variables for $\lambda\mathbf{PRED}\omega^+$ is a map from $\mathrm{Var}^{\mathsf{Type}^s} \cup \mathrm{Var}^{\mathsf{Type}^p} \cup \mathrm{Var}^{\mathsf{Set}} \cup \mathrm{Var}^{\mathsf{Prop}}$ to $\mathsf{N} \cup \bigcup \mathsf{N} \cup \mathbf{A}$ that consists of the union of an valuation of proof object variables $\rho_0 : \mathrm{Var}^{\mathsf{Prop}} \rightarrow \mathbf{A}$, a valuation of constructor variables $\rho_{1a} : \mathrm{Var}^{\mathsf{Type}^p} \rightarrow \bigcup \mathsf{N}$, a valuation of set object variables $\rho_{1b} : \mathrm{Var}^{\mathsf{Set}} \rightarrow \mathbf{A}$ and a valuation of set variables $\rho_2 : \mathrm{Var}^{\mathsf{Type}^s} \rightarrow \mathsf{P}$.*

Definition 25. *For ρ a valuation of variables for $\lambda\mathbf{PRED}\omega^+$, we define the map $(\![-]\!)_\rho$ on the set of well-typed proof objects and set objects of $\lambda\mathbf{PRED}\omega$ as follows.*

$$(\![x]\!)_\rho := \rho(x), \text{ if } x \in \mathrm{Var}^{\mathsf{Prop}} \cup \mathrm{Var}^{\mathsf{Set}},$$
$$(\![tq]\!)_\rho := (\![t]\!)_\rho \cdot (\![q]\!)_\rho, \text{ if } q \text{ is a proof object or set object,}$$
$$(\![tQ]\!)_\rho := (\![t]\!)_\rho, \text{ if } Q \text{ is a constructor or a set,}$$
$$(\![\lambda x{:}U.t]\!)_\rho := \lambda v.(\![t]\!)_{\rho(x:=v)}, \text{ if } U \text{ is a type or a set,}$$
$$(\![\lambda \alpha{:}K.t]\!)_\rho := (\![t]\!)_\rho, \text{ if } K \text{ is a kind or Set.}$$

Definition 26. *For ρ a valuation of variables, we define the maps $\mathcal{V}(-)_\rho$ and $[\![-]\!]_\rho$ respectively from kinds of $\lambda\mathbf{PRED}\omega^+$ and $\{\mathsf{Set}\}$ to N and from constructors and sets of $\lambda\mathbf{PRED}\omega^+$ to $\bigcup \mathsf{N}$ as follows.*

$$\mathcal{V}(\mathsf{Prop})_\rho := \mathsf{P},$$
$$\mathcal{V}(\mathsf{Set})_\rho := \mathsf{P},$$
$$\mathcal{V}(K_1 {\to} K_2)_\rho := \mathcal{V}(K_1)_\rho {\to}_1 \mathcal{V}(K_2)_\rho, \ \textit{if } K_1 \textit{ is a kind},$$
$$\mathcal{V}(\sigma {\to} K)_\rho := [\![\sigma]\!]_\rho {\to}_2 \mathcal{V}(K)_\rho, \ \textit{if } \sigma \textit{ is a set},$$
$$[\![A]\!]_\rho := \rho(A), \ \textit{if } A \in \mathrm{Var}^{\mathsf{Type}^s},$$
$$[\![\alpha]\!]_\rho := \rho(\alpha), \ \textit{if } \alpha \in \mathrm{Var}^{\mathsf{Type}^p},$$
$$[\![\Pi A{:}\mathsf{Set}.\sigma]\!]_\rho := \bigcap_{X \in \mathsf{P}} [\![\sigma]\!]_{\rho(A:=X)},$$
$$[\![K {\to} \sigma]\!]_\rho := [\![\sigma]\!]_\rho, \ \textit{if } K \textit{ is a kind},$$
$$[\![\Pi\alpha{:}K.\varphi]\!]_\rho := \bigcap_{a \in \mathcal{V}(K)_\rho} [\![\varphi]\!]_{\rho(\alpha:=a)}, \ \textit{if } K \textit{ is a kind},$$
$$[\![\Pi\alpha{:}\sigma.\varphi]\!]_\rho := \prod_{\mathbf{A}} t \in [\![\sigma]\!]_\rho . [\![\varphi]\!]_{\rho(\alpha:=t)}, \ \textit{if } \sigma \textit{ is a set},$$
$$[\![U {\to} T]\!]_\rho := [\![U]\!]_\rho {\to}_0 [\![T]\!]_\rho, \ \textit{if } T, U \textit{ are types or sets},$$
$$[\![PQ]\!]_\rho := [\![P]\!]_\rho ([\![Q]\!]_\rho), \ \textit{if } Q \textit{ is a constructor},$$
$$[\![Pt]\!]_\rho := [\![P]\!]_\rho (([\![t]\!])_\rho), \ \textit{if } t \textit{ is a set object},$$
$$[\![\lambda\alpha{:}K.P]\!]_\rho := \boldsymbol{\lambda} a \in \mathcal{V}(K)_\rho . [\![P]\!]_{\rho(\alpha:=a)} \ \textit{if } K \textit{ is a kind},$$
$$[\![\lambda\alpha{:}\sigma.P]\!]_\rho := \boldsymbol{\lambda} a \in [\![\sigma]\!]_\rho . [\![P]\!]_{\rho(\alpha:=a)} \ \textit{if } \sigma \textit{ is a set}.$$

Similar to $\lambda\mathbf{HOL}$ and $\lambda\mathbf{PRED}\omega$ there is a soundness result for $\lambda\mathbf{PRED}\omega^+$, saying that, if the valuation ρ *fulfills* the context Γ, then if $\Gamma \vdash P : K$ (K a kind or Set), then $[\![P]\!]_\rho \in \mathcal{V}(K)_\rho$ and if $\Gamma \vdash t : T$ (T a type or a set), then $([\![t]\!])_\rho \in [\![T]\!]_\rho$.

As a consequence of the model construction, $\lambda\mathbf{PRED}\omega^+$ is consistent, but we already knew that (because of the embedding into $\mathbf{CC}$). It is noteworthy that the model for $\lambda\mathbf{PRED}\omega^+$ is very different from the model for $\lambda\mathbf{PRED}\omega$. This is no surprise, because we know from [15] that polymorphism is not set-theoretic, so a $\lambda\mathbf{PRED}\omega$ model does not extend to a $\lambda\mathbf{PRED}\omega^+$ model in a direct way.

To illustrate this further we consider the following $\lambda\mathbf{PRED}\omega$ context

$$\Gamma := B : \mathsf{Set}, E : B {\to} \mathsf{Prop}, \varepsilon : \mathsf{Prop} {\to} B, h : \Pi\alpha{:}\mathsf{Prop}.E(\varepsilon\alpha) \leftrightarrow \alpha.$$

Here, $\leftrightarrow$ denotes bi-implication: $\varphi \leftrightarrow \psi := \varphi {\to} \psi \wedge \psi {\to} \varphi$. This context was considered by Coquand in [5] as a context of $\mathbf{CC}$, so $\Gamma := B : \star, E : B {\to} \star, \varepsilon : \star {\to} B, h : \Pi\alpha{:}\star E(\varepsilon\alpha) \leftrightarrow \alpha$. It was shown that Γ is inconsistent, because one can embed $\lambda\mathbf{U}^-$ into it. Here we use Γ to show the difference between $\lambda\mathbf{PRED}\omega$ and $\lambda\mathbf{PRED}\omega^+$.

Lemma 2. *Γ is consistent in $\lambda\mathbf{PRED}\omega$.*

Proof. Take a model in which $\emptyset \in P$ and take the valuation ρ as follows: $\rho(B) := P$, $\rho(E) = \rho(\varepsilon)$ is the identity, $\rho(h) := \langle \lambda x.x, \lambda x.x \rangle$, where $\langle -, - \rangle$ is the definable pairing constructor. Then $\rho \models \Gamma$ and $[\![\Pi\alpha{:}\mathsf{Prop}.\alpha]\!]_\rho = \emptyset$, so Γ is consistent[1].

Lemma 3. Γ *is inconsistent in* $\lambda\mathbf{PRED}\omega^+$.

It is instructive to first look at the interpretation of Γ in a $\lambda\mathbf{PRED}\omega^+$ model. Suppose $\rho(B) = \mathsf{B}$. Then $\mathcal{V}(B{\rightarrow}\mathsf{Prop})_\rho = \mathsf{B}{\rightarrow}_2 \mathsf{P}$ and $[\![\mathsf{Prop}{\rightarrow}B]\!]_\rho = \mathsf{B}$. So for a valuation ρ to fulfill Γ, we need that $\rho(\varepsilon) \in \mathsf{B}$ and $\rho(E) \in \mathsf{B}{\rightarrow}_2\mathsf{P}$ such that $\rho(E)\rho(\varepsilon) \leftrightarrow X$ for any $X \in \mathsf{P}$. This is only possible if $\mathsf{P} = \{\mathbf{A}\}$, the degenerate polyset structure in which all polysets are non-empty.

We now give the proof of the Lemma, which basically follows Coquand's proof in [5] (but Coquand's proof is for $\mathbf{CC}$).

Proof. We embed $\lambda\mathbf{U}^-$ into the context Γ of $\lambda\mathbf{PRED}\omega^+$ as follows.

$$\overline{\Delta} := \mathsf{Type}^s \quad \overline{\Pi A{:}\square.\sigma} := \Pi A{:}\mathsf{Type}^s.\overline{\sigma}$$
$$\overline{\square} := \mathsf{Set} \quad \overline{\sigma{\rightarrow}\tau} := \overline{\sigma}{\rightarrow}\overline{\tau}$$
$$\overline{\star} := B \quad \overline{\Pi\alpha{:}\sigma.\varphi} := \varepsilon(\Pi\alpha{:}\overline{\sigma}.E\overline{\varphi})$$
$$\overline{\varphi{\rightarrow}\psi} := \varepsilon(E\overline{\varphi}{\rightarrow}E\overline{\psi})$$

Now one can prove the following

$$\Gamma' \vdash_{\lambda\mathbf{U}^-} M : T \Rightarrow \Gamma, \overline{\Gamma'} \vdash_{\lambda\mathbf{PRED}\omega^+} \overline{M} : \overline{T} \text{ if } T : \square, \Delta$$
$$\Gamma' \vdash_{\lambda\mathbf{U}^-} M : T \Rightarrow \exists N[\Gamma, \overline{\Gamma'} \vdash_{\lambda\mathbf{PRED}\omega^+} N : E\overline{T}] \text{ if } T : \star$$

Therefore $\Gamma \vdash N : E(\overline{\Pi\alpha{:}\star.\alpha})$ for some N. But $E(\overline{\Pi\alpha{:}\star.\alpha}) = E(\varepsilon(\Pi\alpha{:}B.E\alpha))$ $\leftrightarrow \Pi\alpha{:}B.E\alpha$, so we have a term of $\Pi\alpha{:}B.E\alpha$. Taking $\varepsilon\bot$ for α, we have $E(\varepsilon\bot)$ and therefore $\bot$. $\square$

Acknowledgments. We thank the referees for their comments on the first version of this paper.

References

1. Barendregt, H.: Lambda calculi with types. In: Gabbai, D.M., Abramski, S., Maibaum, T.S.E. (eds.) Handbook of Logic in Computer Science, Oxford Univ. Press, Oxford (1992)
2. Berardi, S.: Type dependence and constructive mathematics. PhD thesis, Mathematical Institute, University of Torino, Italy (1990)
3. Church, A.: A formulation of the simple theory of types. JSL (1940)
4. Coquand, Th.: An analysis of Girard's paradox. In: Logic in Computer Science, pp. 227–236. IEEE, Los Alamitos (1986)

---

[1] Another way to prove the consistency is by using the morphism f to map Γ to $\lambda\mathbf{HOL}$ and to observe that the context $f(\Gamma)$ can be instantiated *inside* $\lambda\mathbf{HOL}$ by taking $B := \star$. So $f(\Gamma)$ is consistent in $\lambda\mathbf{HOL}$ and so Γ is consistent in $\lambda\mathbf{PRED}\omega$.

5. Coquand, Th.: Metamathematical investigations of a Calculus of Constructions. In: Odifreddi, P. (ed.) Logic and computer science, pp. 91–122. Academic Press, London (1990)

6. Coquand, T., Huet, G.: The Calculus of Constructions. Information and Computation 76, 96–120 (1988)

7. Coquand, Th.: A new paradox in type theory. In: Prawitz, D., Skyrms, B., Westerståhl, D. (eds.) Proc. 9th Int. Congress of Logic, Methodology and Phil. of Science, Uppsala, Sweden, August 1991, vol. 134, pp. 555–570. North-Holland, Amsterdam (1994)

8. Geuvers, H.: Logics and Type Systems. PhD thesis, University of Nijmegen, Netherlands (1993)

9. Geuvers, H.: A short and flexible proof of strong normalization for the Calculus of Constructions. In: Smith, J., Dybjer, P., Nordström, B. (eds.) TYPES 1994. LNCS, vol. 996, pp. 14–38. Springer, Heidelberg (1995)

10. Geuvers, H., Nederhof, M.-J.: A modular proof of strong normalization for the Calculus of Constructions. JoFP 1(2), 155–189 (1991)

11. Girard, J.-Y.: Interprétation fonctionelle et élimination des coupures dans l'arithmétique d'ordre supérieur. PhD thesis, Université Paris VII (1972)

12. Howard, W.A.: The formulae-as-types notion of construction. In: Hindley, J.R., Seldin, J.P. (eds.) Essays on Combinatory Logic, Lambda Calculus and Formalism, pp. 479–490. Academic Press, London (1980)

13. Hurkens, T.: A simplification of Girard's paradox. In: Dezani-Ciancaglini, M., Plotkin, G. (eds.) TLCA 1995. LNCS, vol. 902, pp. 266–278. Springer, Heidelberg (1995)

14. Urzyczyn, P., Sørensen, M.H.: Lectures on the Curry-Howard isomorphism. Studies in Logic and the Foundations of Mathematics, vol. 149. Elsevier, Amsterdam (2006)

15. Reynolds, J.: Polymorphism is not set-theoretic. In: Plotkin, G., MacQueen, D.B., Kahn, G. (eds.) Semantics of Data Types. LNCS, vol. 173, pp. 145–156. Springer, Heidelberg (1984)

16. Stefanova, M., Geuvers, H.: A simple model construction for the calculus of constructions. In: Berardi, S., Coppo, M. (eds.) TYPES 1995. LNCS, vol. 1158, pp. 249–264. Springer, Heidelberg (1996)

17. Tait, W.W.: Intensional interpretation of functionals of finite type I. JSL 32, 198–212 (1967)

Constructive Type Classes in Isabelle

Florian Haftmann* and Makarius Wenzel**

Technische Universität München
Institut für Informatik, Boltzmannstraße 3, 85748 Garching, Germany
http://www.in.tum.de/~haftmann/,
http://www.in.tum.de/~wenzelm/

Abstract. We reconsider the well-known concept of Haskell-style type classes within the logical framework of Isabelle. So far, axiomatic type classes in Isabelle merely account for the logical aspect as predicates over types, while the operational part is only a convention based on raw overloading. Our more elaborate approach to constructive type classes provides a seamless integration with Isabelle locales, which are able to manage both operations and logical properties uniformly. Thus we combine the convenience of type classes and the flexibility of locales. Furthermore, we construct dictionary terms derived from notions of the type system. This additional internal structure provides satisfactory foundations of type classes, and supports further applications, such as code generation and export of theories and theorems to environments without type classes.

1 Introduction

The well-known concept of type classes [18, 15, 6, 13, 10, 19] offers a useful structuring mechanism for programs and proofs, which is more light-weight than a fully featured module mechanism. Type classes are able to qualify types by associating operations and logical properties. For example, class *eq* could provide an equivalence relation = on type α, and class *ord* could extend *eq* by providing a strict order < etc.

Programming languages like Haskell merely handle the operational part [18, 15, 6]. In contrast, type classes in Isabelle [14, 12] directly represent the logical properties, but the associated operations are treated as a mere convention imposed on the user.

Recent Isabelle add-ons have demanded more careful support of type classes, most notably code generation from Isabelle/HOL to SML, or conversion of Isabelle/HOL theories and theorems to other versions of HOL. Here the target language lacks direct support for type classes, so the source representation in the Isabelle framework somehow needs to accommodate this, using similar techniques as those performed by the static analysis of Haskell.

How does this work exactly? Haskell is not a logical environment, and internal program transformations are taken on faith without explicit deductions. In traditional Isabelle type classes, the purely logical part could be directly embedded into the logic

---

* Supported by DFG project NI 491/10-1.
** Supported by BMBF project "Verisoft".

T. Altenkirch and C. McBride (Eds.): TYPES 2006, LNCS 4502, pp. 160–174, 2007.

[19], although some justifications were only done on paper. The operational aspect, which cannot be fully internalized into the logic, was explained by raw overloading.

Furthermore, the key disadvantage of raw type classes as "little theories" used to be the lack of flexibility in the signature part: operations being represented by polymorphic constants are fixed for any given type. On the other hand, in recent years the Isabelle infrastructure for structured specifications and proofs has been greatly improved, thanks to the Isar proof language [20, 21, 11] and locales [8, 1, 2]. We think it is time to reconsider the existing type class concepts, and see how they can benefit from these improvements without sacrificing their advantages.

The present work integrates Isabelle type classes and locales (by means of locale interpretation), and provides more detailed explanations of type classes with operations and logical propositions within the existing logical framework. Here we heavily re-use a careful selection of existing concepts, putting them into a greater perspective. We also reconstruct the essential relationship between the type system and its constructive interpretation by producing dictionary terms for class operations. The resulting concept of "constructive type classes" in Isabelle is both more convenient for the user, and more satisfactory from the foundational viewpoint.

2   Example

We demonstrate common elements of structured specifications and abstract reasoning with type classes by the algebraic hierarchy of semigroups and groups. Our background theory is that of Isabelle/HOL [12], which uses fairly standard notation from mathematics and functional programming. We also refer to basic vernacular commands for definitions and statements, e.g. **definition** and **lemma**; proofs will be recorded using structured elements of Isabelle/Isar [20, 21, 11], notably **proof/qed** and **fix/assume/show**.

Our main concern are the new **class** and **instance** elements used below — they will be explained in terms of existing Isabelle concepts later (§5). Here we merely present the look-and-feel for end users, which is quite similar to Haskell's class and instance [6], but augmented by logical specifications and proofs.

2.1   Class Definition

Depending on an arbitrary type α, class *semigroup* introduces a binary operation ∘ that is assumed to be associative:

> **class** *semigroup* =
> **fixes** *mult* :: $\alpha \Rightarrow \alpha \Rightarrow \alpha$ (**infix** ∘ 70)
> **assumes** *assoc*: $(x \circ y) \circ z = x \circ (y \circ z)$

This **class** specification consists of two parts: the *operational* part names the class operation (**fixes**), the *logical* part specifies properties on them (**assumes**). The local **fixes** and **assumes** are lifted to the theory toplevel, yielding the global operation *mult* :: $\alpha::semigroup \Rightarrow \alpha \Rightarrow \alpha$ and the global theorem *semigroup.assoc*: $\bigwedge x\,y\,z::\alpha::semigroup.$ $(x \circ y) \circ z = x \circ (y \circ z)$.

2.2   Class Instantiation

The concrete type *int* is made a *semigroup* instance by providing a suitable definition for the class operation *mult* and a proof for the specification of *assoc*.

> **instance** *int* :: *semigroup*
>   *mult-int-def*: $\bigwedge i\,j$ :: *int*. $i \circ j \equiv i + j$
> **proof**
>   **fix** $i\,j\,k$ :: *int* **have** $(i + j) + k = i + (j + k)$ **by** *simp*
>   **then show** $(i \circ j) \circ k = i \circ (j \circ k)$ **unfolding** *mult-int-def*.
> **qed**

From now on, the type-checker will consider *int* as a *semigroup* automatically, i.e. any general results are immediately available on concrete instances.

2.3   Subclasses

We define a subclass *group* by extending *semigroup* with additional operations *neutral* and *inverse*, together with the usual *left-neutral* and *left-inverse* properties.

> **class** *group* = *semigroup* +
>   **fixes** *neutral* :: α (**1**)
>   **and** *inverse* :: $\alpha \Rightarrow \alpha$ ((-^{-1}) [1000] 999)
>   **assumes** *left-neutral*: $1 \circ x = x$
>   **and** *left-inverse*: $x^{-1} \circ x = 1$

Again, type *int* is made an instance, by providing definitions for the operations and proofs for the axioms of the additional group specification.

> **instance** *int* :: *group*
>   *neutral-int-def*: $1 \equiv 0$
>   *inverse-int-def*: $i^{-1} \equiv -i$
> **proof**
>   **fix** i :: *int* **have** $0 + i = i$ **by** *simp*
>   **then show** $1 \circ i = i$ **unfolding** *mult-int-def* **and** *neutral-int-def*.
>   **have** $-i + i = 0$ **by** *simp*
>   **then show** $i^{-1} \circ i = 1$ **unfolding** *mult-int-def* **and** *neutral-int-def* **and** *inverse-int-def*.
> **qed**

2.4   Abstract Reasoning

Abstract theories enable reasoning at a general level, while results are implicitly transferred to all instances. For example, we can now establish the *left-cancel* lemma for groups, which states that the function $(x \circ)$ is injective:

> **lemma** (**in** *group*) *left-cancel*: $x \circ y = x \circ z \leftrightarrow y = z$
> **proof**
>   **assume** $x \circ y = x \circ z$
>   **then have** $x^{-1} \circ (x \circ y) = x^{-1} \circ (x \circ z)$ **by** *simp*
>   **then have** $(x^{-1} \circ x) \circ y = (x^{-1} \circ x) \circ z$ **using** *assoc* **by** *simp*

then show $y = z$ **using** *left-neutral* **and** *left-inverse* **by** *simp*
next
    assume $y = z$
    then show $x \circ y = x \circ z$ **by** *simp*
qed

Here the "**in** *group*" target specification indicates that the result is recorded within that context for later use. This local theorem is also lifted to the global one *group.left-cancel*: $\bigwedge x\, y\, z {::} \alpha {::} group.\ x \circ y = x \circ z \leftrightarrow y = z$. Since type *int* has been made an instance of *group* before, we may refer to that fact as well: $\bigwedge x\, y\, z {::} int.\ x \circ y = x \circ z \leftrightarrow y = z$.

3   Logical Foundations

We briefly review fundamental concepts of the Isabelle/Isar framework, from the Pure logic to Isar proof contexts (structured proofs) and locales (structured specifications).

3.1   The Isabelle/Pure Framework

The Pure logic [14] is an intuitionistic fragment of higher-order logic. In type-theoretic parlance, there are three levels of λ-calculus with corresponding arrows: $\Rightarrow$ for syntactic function space (terms depending on terms), $\bigwedge$ for universal quantification (proofs depending on terms), and $\Longrightarrow$ for implication (proofs depending on proofs).

Types are formed as simple first-order structures, consisting of type variables α or type constructor applications $\kappa\, \tau_1 \ldots \tau_k$ (where κ has always k arguments).

Term syntax provides explicit abstraction $\lambda x :: \alpha.\ b(x)$ and application $t\, u$, while types are usually implicit thanks to type-inference; terms of type *prop* are called propositions. Logical statements are composed via $\bigwedge x :: \alpha.\ B(x)$ and $A \Longrightarrow B$. Primitive reasoning operates on judgments of the form $\Gamma \vdash \varphi$, with standard introduction and elimination rules for $\bigwedge$ and $\Longrightarrow$ that refer to fixed parameters x and hypotheses A from the context Γ. The corresponding proof terms are left implicit, although they could be exploited separately [3].

The framework also provides definitional equality $\equiv\, :: \alpha \Rightarrow \alpha \Rightarrow prop$, with $\alpha\beta\eta$-conversion rules. The internal conjunction $\&\, :: prop \Rightarrow prop \Rightarrow prop$ allows to represent simultaneous statements with multiple conclusions.

Derivations are relative to a given theory Θ, which consists of declarations for type constructors κ (constructor name with number of arguments), term constants $c :: \sigma$ (constant name with most general type scheme), and axioms $\vdash \varphi$ (proposition being asserted). Theories are always closed by type-instantiation: arbitrary instances $c :: \tau$ of $c :: \sigma$ are well-formed; likewise for axiom schemes. Schematic polymorphism carries over to term formation and derivations, i.e. it is admissible to derive any type instance $\Gamma \vdash B(\tau)$ from $\Gamma \vdash B(\alpha)$, provided that α does not occur in the hypotheses of Γ.

3.2   Isar Proof Contexts

In judgments $\Gamma \vdash \varphi$ of the primitive framework, Γ essentially acts like a proof context. Isar elaborates this idea towards a higher-level notion, with separate information for type-inference, term abbreviations, local facts, and generic hypotheses (parameterized by specific discharge rules). For example, the context element **assumes** A introduces a hypothesis with $\Longrightarrow$ introduction as discharge rule; **notes** $a = b$ defines local facts; **defines** $x \equiv a$ and **fixes** $x :: \alpha$ introduce local terms.

Top-level theorem statements may refer directly to Isar context elements to establish a conclusion within an enriched environment; the final result will be in discharged form. For example, proofs of $\bigwedge x.\ B\ x$, and $A \Longrightarrow B$, and $B\ a$ can be written as follows:

lemma	**lemma**	**lemma**
fixes x	**assumes** A	**defines** $x \equiv a$
shows $B\ x\ \langle proof \rangle$	**shows** $B\ \langle proof \rangle$	**shows** $B\ x\ \langle proof \rangle$

There are separate Isar commands to build contexts within a proof body, notably **fix**, **assume** etc. These elements have essentially the same effect, only that the result lives still within a local proof body rather than the target theory context. For example:

{	{	{
fix x	**assume** A	**def** $x \equiv a$
have $B\ x\ \langle proof \rangle$	**have** $B\ \langle proof \rangle$	**have** $B\ x\ \langle proof \rangle$
}	}	}

Building on top of structured proof contexts, the Isar proof engine now merely imposes a certain policy for interpreting formal texts, in order to support structured proof composition [21, Chapter 3]. The very same notion of contexts may be re-used a second time for structured theory specifications, namely by Isabelle locales (see below).

3.3 Locales

Isabelle locales [8, 1] provide a powerful mechanism for managing local proof context elements, most notably **fixes** and **assumes**. For example:

> **locale** $l =$
> **fixes** x
> **assumes** $A\ x$

This defines both a predicate $l\ x \equiv A\ x$ (by abstracting the body of assumptions over the fixed parameters), and provides some internal infrastructure for structured reasoning. In particular, consequences of the locale specification may be proved at any time, e.g.:

> **lemma** (**in** l)
> **shows** b: $B\ x\ \langle proof \rangle$

The result b: $B\ x$ is available for further proofs within the same context. There is also a global version $l.b$: $\bigwedge x.\ l\ x \Longrightarrow B\ x$, with the context predicate being discharged.

Locale expressions provide means for high-level composition of complex proof contexts from basic principles (e.g. **locale** $ring = abelian\text{-}group\ R + monoid\ R + \ldots$). Expressions e are formed inductively as $e = l$ (named locale), or $e = e'\ x_1 \ldots x_n$ (renaming of parameters), or $e = e_1 + e_2$ (merge). Locale merges result in general acyclic graphs of sections of context elements — internally, the locale mechanism produces a canonical order with implicitly shared sub-graphs.

Locale interpretation is a separate mechanism for applying locale expressions in the current theory or proof context [2]. After providing terms for the **fixes** and proving the **assumes**, the corresponding instances of locale facts become available. For example:

> **interpretation** m: l $[a]$
> **proof** (*rule l.intro*)
>   **show** A a $\langle proof \rangle$
> **qed**

Here the previous locale fact $l.b$: $\bigwedge x.\, l\, x \Longrightarrow B\, x$ becomes $m.b$: $B\, a$. The link between the interpreted context and the original locale acts like a dynamic subscription: any new results emerging within l will be automatically propagated to the theory context by means of the same interpretation. For example:

> **lemma** (**in** l)
>   **shows** c: $C\, x$ $\langle proof \rangle$

This makes both $l.c$: $\bigwedge x.\, l\, x \Longrightarrow C\, x$ and $m.c$: $C\, a$ available to the current theory.

4   Type Classes and Disciplined Overloading

Starting from well-known concepts of order-sorted algebra, we recount the existing axiomatic type classes of Isabelle. Then we broaden the perspective towards explicit construction of dictionary terms, which explain disciplined overloading constructively.

4.1   An Order-Sorted Algebra of Types

The well-known concepts of order-sorted algebra (e.g. [17]) have been transferred early to the simply-typed framework of Isabelle [13, 10].

A type class c is an abstract entity that describes a collection of types. A sort s is a symbolic intersection of finitely many classes, written as expression $c_1 \cap \ldots \cap c_m$ (note that Isabelle uses the concrete syntax $\{c_1, \ldots, c_m\}$). We assume that type variables are decorated by explicit sort constraints α_s, while plain α refers to a vacuous constraint of the empty intersection of classes (the universal sort). An order-sorted algebra consists of a set C of classes, together with an acyclic subclass relation $<$, and a collection of type constructor arities $\kappa :: (s_1, \ldots, s_k)c$ (for constructor κ with k arguments). This induces an inductive relation $\tau : c$ on types τ and classes c by the rules given below (on sorts $\tau : c_1 \cap \ldots \cap c_m$ is defined as $\tau : c_i$ for all $i = 1, \ldots, m$ collectively).

$$\frac{\tau : c_1 \quad c_1 < c_2}{\tau : c_2} \ (classrel)$$

$$\frac{\tau_1 : s_1 \quad \ldots \quad \tau_k : s_k \quad \kappa :: (s_1, \ldots, s_k)c}{\kappa\, \tau_1 \ldots \tau_k : c} \ (constructor)$$

$$\frac{}{\alpha_{c_1 \cap \ldots \cap c_m} : c_i} \ (variable)$$

We also define canonical subclass and subsort relations on top of this: $c_1 \subseteq c_2$ iff $\forall \alpha.\ \alpha : c_1 \Longrightarrow \alpha : c_2$ for classes, and $s_1 \subseteq s_2$ iff $\forall \alpha.\ \alpha : s_1 \Longrightarrow \alpha : s_2$ for sorts.

Observe that class inclusion $c_1 \subseteq c_2$ is the reflexive-transitive closure of the original relation $c_1 < c_2$. Proof: consequences $\alpha : c_1 \Longrightarrow \alpha : c_2$ emerge exactly by zero or more application of the *classrel* rule.

Moreover, sort inclusion $s_1 \subseteq s_2$ for $s_1 = c_1 \cap \ldots \cap c_m$ and $s_2 = d_1 \cap \ldots \cap d_n$ can be characterized as $\forall j.\ \exists i.\ c_i \subseteq d_j$. Proof: $c_1 \cap \ldots \cap c_m \subseteq d_1 \cap \ldots \cap d_n$ is equivalent to $\alpha : c_1 \cap \ldots \cap c_m \Longrightarrow \alpha : d_1 \cap \ldots \cap d_n$, i.e. $(\forall i.\ \alpha : c_i) \Longrightarrow (\forall j.\ \alpha : d_j)$, which is equivalent to $\forall j.\ \exists i.\ (\alpha : c_i \Longrightarrow \alpha : d_j)$.

An order-sorted algebra is called *coregular* iff for all $c \subseteq c'$, any $\kappa :: (s_1, \ldots, s_k)c$ and $\kappa :: (s'_1, \ldots, s'_k)c'$ have related argument sorts $\forall i.\ s_i \subseteq s'_i$. Coregularity expresses the key correspondence of the global class hierarchy with individual type constructor arities. This achieves most general unification and principal type schemes [17, 13].

4.2   Axiomatic Type Classes in Isabelle

Axiomatic type classes [10, 19] are based on a purely logical interpretation of the order-sorted algebra of types as predicates. Any closed proposition $\varphi(\alpha)$ depending on exactly one type variable can be understood as a predicate on types. The trick is to represent predicate constants adequately in order to support type class definitions and abstract reasoning over type classes. Following [19], any type class $c \in C$ of the underlying algebra is turned into a logical constant *c-class* :: α *itself* $\Rightarrow$ *prop*, where α *itself* is an uninterpreted type with constant *TYPE* :: α *itself* as canonical representative.[1] Propositions of the form *c-class* (*TYPE* :: τ *itself*) shall be written as $(\!|\tau : c|\!)$.

The existing **axclass** mechanism defines type classes via $\vdash (\!|\alpha : c|\!) \equiv (\!|\alpha : d_1|\!)\ \&\ \ldots$ $\&\ (\!|\alpha : d_n|\!)\ \&\ A_1(\alpha)\ \&\ \ldots\ \&\ A_m(\alpha)$, where $d_1, \ldots, d_n$ are super-classes and $A_1, \ldots, A_m$ class axioms as intended by the user. From this the system derives an introduction rule $\vdash (\!|\alpha : d_1|\!) \Longrightarrow \ldots (\!|\alpha : d_n|\!) \Longrightarrow A_1(\alpha) \Longrightarrow \ldots A_m(\alpha) \Longrightarrow (\!|\alpha : c|\!)$ (for instantiation proofs), explicit class inclusions $\vdash (\!|\alpha : c|\!) \Longrightarrow (\!|\alpha : d_j|\!)$ (added to the order-sorted type algebra), and abstract lemmas $\vdash (\!|\alpha : c|\!) \Longrightarrow A_i(\alpha)$ (also called "class axioms"). The latter are represented compactly using sort constraints $\vdash A_i(\alpha_c)$. Isabelle inferences will use order-sorted type-unification in order to produce well-sorted instantiations $\vdash A_i(\tau)$ on the fly — this implicit reasoning is the main convenience of type classes.

It is easy to see that the interpretation of class membership $\tau : c$ as $(\!|\tau : c|\!)$ is correct in the sense that the notions of order-sorted type algebra approximate Pure derivations. In particular, the inference system for $\tau : c$ represents the following rules for $(\!|\tau : c|\!)$ (due to modus ponens and type instantiation):

$$\frac{(\!|\tau : c_1|\!) \quad \vdash (\!|\alpha : c_1|\!) \Longrightarrow (\!|\alpha : c_2|\!)}{(\!|\tau : c_2|\!)}$$

$$\frac{(\!|\tau_1 : s_1|\!) \quad \ldots \quad (\!|\tau_k : s_k|\!) \quad \vdash (\!|\alpha_1 : s_1|\!) \Longrightarrow \ldots (\!|\alpha_k : s_k|\!) \Longrightarrow (\!|\kappa\, \alpha_1 \ldots \alpha_k : c|\!)}{(\!|\kappa\, \tau_1 \ldots \tau_k|\!) : c}$$

$$\overline{(\!|\alpha : c_1|\!), \ldots, (\!|\alpha : c_m|\!) \vdash (\!|\alpha : c_i|\!)}$$

---

[1] This type could be defined explicitly as **datatype** α *itself* = *TYPE* in ML / Haskell / HOL, but Isabelle/Pure refrains from stating any specific properties.

Here the rule conditions $c_1 < c_2$ and $\kappa :: (\overline{s})c$ have been interpreted by schematic implications, and sort constraints of type variables have been turned into explicit hypotheses.

The general principle above is to interpret the inductive definition of $\tau : c$, by giving a constructive reading to its derivations. Thus inferences taking place during internal type-checking operations are turned into proofs of the Pure framework.

In conclusion, we observe that axiomatic type classes are able to model the logical part (**assumes**) of our **class** mechanism. The second half is proper management of class operations (**fixes**) which will be based on a disciplined version of overloaded definitions.

4.3   Disciplined Overloading for Isabelle

Simple Definitions essentially introduce abbreviations in terms of basic principles, by stating definitional equalities within the formal theory. A definitional theory extension $\Theta' = \Theta \cup c :: \sigma \cup \vdash c_\sigma \equiv t$ is well-formed iff c is a fresh constant name, t is a closed term that does not mention c, and all type variables occurring in t also occur in σ.

The latter condition ensures $\vartheta(t) = \vartheta'(t) \implies \vartheta(c_\sigma) = \vartheta'(c_\sigma)$ for arbitrary type instantiations ϑ and ϑ', i.e. there is a one-to-one relationship between the LHS and RHS. Due to substitution of $\equiv$, this means $\Gamma \vdash \varphi(\vartheta(c_\sigma))$ iff $\Gamma \vdash \varphi(\vartheta(t))$ in Θ'.

Moreover, $\Gamma \vdash \varphi(c)$ is derivable in Θ' iff $\Gamma \vdash \varphi(t)$ is derivable in Θ. Proof: (1) assume $\Gamma \vdash \varphi(c)$; hence $\Gamma \vdash \varphi(t)$ in Θ' by definition. Let $\psi = \varphi(t)$, which is a formula of Θ and theorem of Θ'. Show by induction over derivations that $\Gamma \cup \vartheta_1(c_\sigma \equiv t), \ldots,$ $\vartheta_n(c_\sigma \equiv t) \vdash \psi$ in Θ, for some collection of type instantiations $\vartheta_1, \ldots, \vartheta_n$ (stemming from instances of the definitional axiom occurring in the proof trees). Finally discharge these assumptions by reflexivity of $\equiv$. (2) the other direction is trivial.

A definitional theory may be presented in an incremental fashion, where later definitions refer to previously defined entities on the RHS. For example, define c_1, then c_2 in terms of c_1, then c_3 in terms of c_1, c_2 etc. Formally, we introduce a dependency relation between constant names: $c \to b$ iff constant b is mentioned on the RHS of the definition of c. Provided that $\to$ is well-founded, incremental definitions can be normalized such that the RHSes only mention basic principles. Thus simple definitions determine an immediate mapping from defined entities to basic principles.

Overloading (or "ad-hoc polymorphism") means to specify constants depending on the syntactic structure of their respective type instances. For example, $0 :: \alpha$ could be defined separately for 0_{nat}, 0_{bool}, $0_{\beta \times \gamma}$ (in terms of 0_β and 0_γ) etc. Unrestricted overloading sacrifices most of the key syntactic properties sketched above.

Subsequently, we borrow some notation from System F [16], notably type schemes $\forall \alpha. \sigma(\alpha)$ and type application $f[\tau]$. For example, the polymorphic identity function *id* $\equiv \lambda x.\ x$ can be given the most general type scheme $\forall \alpha.\ \alpha \Rightarrow \alpha$. System F also provides explicit type abstraction $\Lambda \alpha.\ t(\alpha)$, although this will not be required here, because naive polymorphism in the Pure framework is restricted to outermost constants (and axioms): instead of $id \equiv \Lambda \alpha.\ \lambda x{::}\alpha.\ x$ we write $id\ [\alpha] \equiv \lambda x{::}\alpha.\ x$ in applied form.

This quasi-polymorphic perspective allows an adequate view on constant declarations and type instances as required for overloading. Any declaration $c :: \sigma$ can be turned into an explicit type scheme $c :: \forall \overline{\alpha}.\ \sigma(\overline{\alpha})$ by presenting the type variables $\overline{\alpha}$ of the body σ in some canonical order. Type instances can now be written as $c\ [\overline{\tau}]$, where $\overline{\tau}$ emerges by matching against σ and putting the RHSes of the resulting substitution $[\alpha_1 \mapsto \tau_1, \ldots, \alpha_n \mapsto \tau_n]$ into the same canonical order.

Linear polymorphic declarations ($n = 1$) are an important special case of this. Here $c :: \forall \alpha.\ \sigma(\alpha)$ acts like a function that maps a type τ to a term $c\ [\tau]$ of type $\sigma(\tau)$.

Restricted Overloading is a theory extension $\Theta \cup c :: \forall \alpha.\ \sigma(\alpha) \cup \vdash c\ [\tau] \equiv t \cup \ldots$ that introduces a fresh constant declaration c, followed by a collection of specifications $\vdash c\ [\tau] \equiv t$ each, where t is a closed term, and all type variables of t also occur in τ. The defining equations for c are further restricted to

$$c\ [\kappa\ \overline{\alpha}] \equiv \ldots b\ [\overline{\tau}] \ldots d\ [\alpha_i] \ldots$$

such that the type argument of the LHS is a constructor κ applied to distinct variables $\overline{\alpha}$, and the RHS (after normalizations with respect to simple definitions) only mentions further constants as follows:

1. arbitrary instances of constants named b, provided that $c \to b$ holds according to a given well-founded dependency relation on constant names;
2. argument projections on overloaded constants $d\ [\alpha_i]$, selecting some α_i from $\overline{\alpha}$.

Moreover the following global conditions have to be observed:

- There is at most one specification $c\ [\kappa\ \overline{\alpha}] \equiv \ldots$ for each type constructor κ.
- Overloaded specifications are upwards-complete: for any $c_1 \to^+ c_2$, the presence of $c_1\ [\kappa\ \overline{\alpha}] \equiv \ldots$ implies the presence of $c_2\ [\kappa\ \overline{\alpha}] \equiv \ldots$.

Note that the restriction of $c \to b$ is independent of actual type instances and essentially decouples general interdependencies from overloading. For example, the specification of $c\ [nat] \equiv \ldots b\ [bool] \ldots$ and $b\ [nat] \equiv \ldots c\ [bool] \ldots$ is ruled out, due to the cycle $c \to b \to c$ on constant names.

The following example illustrates restricted overloading of constants eq and ord for types nat and $\times$:

$eq :: \forall \alpha.\ \alpha \Rightarrow \alpha \Rightarrow bool$
$eq\ [nat] \equiv \lambda m\ n.\ m = n$
$eq\ [\beta \times \gamma] \equiv \lambda p\ q.\ eq\ [\beta]\ (fst\ p)\ (fst\ q) \wedge eq\ [\gamma]\ (snd\ p)\ (snd\ q)$
$ord :: \forall \alpha.\ \alpha \Rightarrow \alpha \Rightarrow bool$
$ord\ [nat] \equiv \lambda m\ n.\ m < n$
$ord\ [\beta \times \gamma] \equiv \lambda p\ q.\ ord\ [\beta]\ (fst\ p)\ (fst\ q) \vee$
$\quad eq\ [\beta]\ (fst\ p)\ (fst\ q) \wedge ord\ [\gamma]\ (snd\ p)\ (snd\ q)$

In general, restricted overloading and simple definitions may be presented incrementally, with alternating dependencies of overloaded vs. non-overloaded constants. The resulting theory still describes a mapping from defined entities to basic principles — as sketched before for simple definitions alone. The key idea is to traverse the system along the lexicographic product of the global dependency relation $c \to b$ and the substructural order on types $\kappa\ \overline{\alpha} \to \alpha_i$, which is also well-founded.

Overloading as order-sorted type-algebra is a slightly more abstract view on the structure of interdependent overloaded specifications. After expanding all simple (non-overloaded) definitions, the resulting algebra of overloading is achieved as follows.

Classes: Each overloaded operation is turned into a type class of the same name.[2]

Class relation: The global dependency relation $\rightarrow$ is restricted to overloaded constants, i.e. $c_1 < c_2$ iff $c_1 \rightarrow^+ c_2$ on classes.

Constructor arities: The local dependencies of definitional equations are turned into constructor arities, i.e. $\kappa :: (s_1, \ldots, s_k)c$ for each constructor κ and class c, where $s_i = \bigcap d$ such that $d[\alpha_i]$ occurs on the RHS of some specification $c' [\kappa \, \alpha_1 \ldots \alpha_k] \equiv \ldots$ for some $c' \supseteq c$.

Observe that this algebra is coregular by construction, because the argument sorts of type arities account for the upwards-completion of definitions explicitly.

For example, the previous overloaded definitions of *eq* and *ord* result in the algebra consisting of classes $ord < eq$ with constructor arities $nat :: eq$, and $\times :: (eq, eq)eq$, and $nat :: ord$, and $\times :: (eq \cap ord, eq \cap ord)ord$.

We now employ the order-sorted algebra to expand disciplined overloading: for any $\vdash \varphi$ mentioning well-defined instances $c [\tau]$ of overloaded constants, we produce $\vdash \varphi'$ that refers only to basic principles. In the first stage, we normalize by all definitional equalities, which removes non-overloaded constants and reduces overloaded ones to occurrences c_α on type variables. In the second stage we construct dictionary terms.

A *dictionary* δ for class c is a collection of terms $[t_1, \ldots, t_n]$ that provide implementations for the class operations $[c_1, \ldots, c_n]$, for the collection of classes $c' \supseteq c$ presented in canonical order. The construction works by interpreting the derivation of $\tau : c$ for each $c [\tau]$ occurring in $\vdash \varphi$. The base case refers to locally fixed dictionary parameters $p^c :: \sigma(\alpha)$ for each $c [\alpha] :: \sigma(\alpha)$ in $\vdash \varphi$. The type constructor case refers to the collection $\overline{\varrho}$ of RHSes of all specifications $c' [\kappa \, \overline{\alpha} \equiv \ldots]$ for $c' \supseteq c$, as in the construction of type arities $\kappa :: (\overline{s})c$ above. The notation $\{\!|\delta : c|\!\}$ means that δ contains a dictionary term for c. We now get the following rules:

$$\frac{\{\!|\delta : c_1|\!\} \quad c_1 < c_2}{\{\!|\delta : c_2|\!\}} \; (classrel)$$

$$\frac{\{\!|\delta_1 : s_1|\!\} \quad \ldots \quad \{\!|\delta_k : s_k|\!\} \quad \kappa :: (s_1, \ldots, s_k)c}{\{\!|\overline{\varrho}(\delta_1, \ldots, \delta_k) : c|\!\}} \; (constructor)$$

$$\frac{}{\{\!|[p^{c_1}, \ldots, p^{c_m}] : c_i|\!\}} \; (variable)$$

For example, $\vdash P \ (ord \ [\beta \times \gamma])$ can be expanded to $\vdash P \ (\lambda p \, q. \ ord_1 \ (fst \ p) \ (fst \ q) \vee eq_1 \ (fst \ p) \ (fst \ q) \wedge ord_2 \ (snd \ p) \ (snd \ q))$, for new local variables eq_1, eq_2, ord_1, ord_2.

We see that disciplined overloading can be linked to the order-sorted type-algebra quite naturally. The key benefit is that well-definedness of $c [\tau]$ is reduced to well-sortedness $\tau : c$, while a constructive reading provides the dictionary expansion.

---

[2] We essentially assume that each type class corresponds to exactly one operation of the same name. Minor re-formulations will admit the more liberal scheme seen in practice (e.g. §2).

Thus we have managed to make "ad-hoc polymorphism less ad-hoc", although by quite different means than the original Haskell type class system [18]. In more general versions of type theory, the reconstruction of dictionary terms (for the operations) and proof terms (for the logical part) would have coincided anyway, but Pure has two distinctive categories of formal entities that appear to the user as **fixes** and **assumes**.

5   Integration

We are ready to integrate the concepts of §3 and §4 to explain our version of **class** and **instance**. Essentially, we shall introduce (I) a locale that manages both the **fixes** and **assumes** explicitly, (II) type class infrastructure that replaces the **fixes** by global operations according to disciplined overloading, and (III) a formal link between the locale and type class by locale interpretation. We illustrate this by the example from §2.

5.1   Class Definition

(I) The syntax for **class** specifications is the same as for **locale**, restricted to exactly one type variable α. Thus a **class** is literally made a **locale** of the same name. E.g.

> **locale** *semigroup* $=$
>   **fixes** *mult* $:: \alpha \Rightarrow \alpha \Rightarrow \alpha$ (**infix** $\circ$ 70)
>   **assumes** *assoc*: $(x \circ y) \circ z = x \circ (y \circ z)$

(II) The same specification is turned into type class infrastructure as follows.

1. For all class operations (**fixes**) introduce global operations (**consts**) with the same name and type. E.g.

   > **consts**
   >   *mult* $:: \alpha \Rightarrow \alpha \Rightarrow \alpha$ (**infix** $\circ$ 70)

2. Introduce an *axiomatic type class* whose axioms are the class premises (**assumes**), applied to the newly introduced **consts**. Since a locale definition already defines a predicate corresponding to the body, we can use a compact representation. E.g.

   > **axclass** *semigroup*
   >   *axiom*: *semigroup* $(mult :: \alpha \Rightarrow \alpha \Rightarrow \alpha)$

3. Restrict subsequent uses of the global operations to the new type class. E.g.

   > **constraints**
   >   *mult* $:: \alpha{::}semigroup \Rightarrow \alpha \Rightarrow \alpha$

   This is merely an extra-logical hint for type-inference, which ensures that occurrences of the operations will be well-defined.

(III) Finally link the locale and type class infrastructure by means of locale interpretation: the global operations (**consts**) are inserted for the local ones (**fixes**), and the (already derived) class axiom is inserted for the locale premises (**assumes**). E.g.

> **interpretation** *semigroup* [*mult* :: α::*semigroup* $\Rightarrow \alpha \Rightarrow \alpha$]
>   **by** (*rule semigroup-class.axiom*)

This reduces the generality of locale results by fixing the operations, but α remains free.

5.2   Class Instantiation

An instance provides term definitions and proofs on particular type patterns $\kappa \, \overline{\alpha}$. The class operations are introduced by the existing primitive for overloaded definitions, which is only used in the restricted sense of §4.3. E.g.

> **defs** (**overloaded**)
>   *mult-int-def*: (*i*::*int*) $\circ j \equiv i + j$

The actual instance proof uses the original **axclass** instantiation mechanism. E.g.

> **instance** *int* :: *semigroup*   — (existing version of axclass **instance**)
> **proof**
>   **fix** *i j k* :: *int* **have** $(i + j) + k = i + (j + k)$ **by** *simp*
>   **then show** $(i \circ j) \circ k = i \circ (j \circ k)$ **unfolding** *mult-int-def*.
> **qed**

5.3   Subclasses

(I) In order to derive a new class c from existing super-classes $b_1, \ldots, b_n$ we simply produce parallel hierarchies of locales and type classes. For locales this means to import the merge $b_1 + \ldots + b_n$ of the corresponding parent locales. E.g.

> **locale** *group* = *semigroup* +
>   **fixes** *neutral* :: α (**1**)
>   **and** *inverse* :: $\alpha \Rightarrow \alpha$ $((\text{-}^{-1})$ [1000] 999)
>   **assumes** *left-neutral*: $\mathbf{1} \circ x = x$
>   **and** *left-inverse*: $x^{-1} \circ x = \mathbf{1}$

(II) The type class setup is analogous; **axclass** treats super-classes as expected. E.g.

> **consts**
>   *neutral* :: α (**1**)
>   *inverse* :: $\alpha \Rightarrow \alpha$ $((\text{-}^{-1})$ [1000] 999)
>
> **axclass** *group* < *semigroup*
>   *axiom*: *group mult neutral inverse*
>
> **constraints**
>   *neutral* :: α::*group*
>   *inverse* :: α::*group* $\Rightarrow \alpha$

(III) The link between locale and class definition is again by interpretation. The implicit import of results established in parent locales [2] works without further ado. E.g.

> **interpretation**
> *group* [*mult* :: α::*group* ⇒ α ⇒ α *neutral* :: α::*group* *inverse* :: α::*group* ⇒ α]
> **by** (*rule group-class.axiom*)

5.4   Abstract Reasoning

Nothing special needs to be done here — we benefit directly from the existing mechanisms of locale lemmas. E.g. "**lemma** (*in group*) ..." refers to the target locale *group*, even if this happens to be related to a type class of the same name. Abstract reasoning is performed in full generality at the locale level relative to **fixes** and **assumes**.

6   Conclusion

Stocktaking. The present approach to constructive type classes in Isabelle integrates a fair amount of existing concepts into a coherent mechanism for the end-user, without having to extend the underlying logical foundations. Apart from collecting existing concepts, our main contribution is twofold: (1) explicit reconstruction of proofs and dictionary terms, guided by constructive interpretation of order-sorted type algebras, (2) relating **locale** and **class** concepts by means of interpretation.

The first aspect has foundational impact, the formal content of type classes is explained more thoroughly in terms of basic principles. Moreover, applications that build on the internal representations of theories and proofs may benefit from this additional structure (e.g. code generation for ML or proof export for other versions of HOL).

The second aspect is very important for user-level reasoning with type classes within the formal system. Our link to the locale mechanism [8, 1, 2] overcomes the former restriction of axiomatic type-classes to a fixed "signature" of overloaded constants. Our classes admit abstract reasoning in the general locale context, where operations are local parameters; results are implicitly passed down to the actual type class thanks to locale interpretation. Thus we essentially combine the best of both worlds.

Even more, several type classes can be linked to the same locale, using the additional **includes** element to refer to a renamed locale specification: e.g. **class** *abelian-group* = **includes** *group add* (**infix** + 60) **assumes** *commute*: ... etc. General lemmas established in *group* will then become available for both type classes *group* and *abelian-group*.

The present work has resulted in clarification of various Isabelle internals[3]. In particular, the constructive interpretation of order-sorted type-algebra is now explicit in the internal workings of **axclass**, so far some justifications have been only on paper [19]. There is now also a separation of constant declarations $c :: \forall \alpha.\ \sigma(\alpha)$, and extra-logical type-inference constraints $c :: \forall \alpha::c.\ \sigma(\alpha)$.

---

[3] See http://isabelle.in.tum.de/devel/ for a development snapshot.

Related work. Module systems (especially for theorem provers) provide a more general perspective on our work. Roughly speaking, the huge amount of existing approaches can be categorized as follows: (1) full / explicit module languages vs. (2) restricted / implicit structuring mechanisms. ML functors [16] and Coq modules [4, 5] represent the first kind, type classes in Haskell or Isabelle the second, more light-weight one. Our work helps to bridge the gap between these two extremes, by enhancing the basic type class concepts towards a more explicit notion of modules, thanks to the underlying locale infrastructure.

Compared to a full-grown module system, locales do have some limitations: no truly polymorphic parameters, no type-constructors as parameters. For example, a theory of monads would be hard to formalize. However, explaining locales (and classes) in terms of existing Isabelle/Pure concepts avoids tinkering with the logic itself.

Type classes have first appeared in Haskell [18, 15, 6], to make "ad-hoc polymorphism less ad-hoc". The underlying ideas have later been rephrased as a problem of Hindley-Milner type-checking within an order-sorted algebra of types [13], and integrated into the Isabelle/Pure type-checker [10]. Isabelle type classes acquired their first logical interpretation in [19]. Note that more recent extensions of the original Haskell type classes (including *constructor classes* and *multi-parameter classes* [7]) are not covered in this work, mostly due to fundamental limitations of the underlying logic.

Future Work. Our constructive combination of type classes and locales essentially organizes lemmas (proofs) that emerge in related contexts. This principle could be transferred to derived operations (terms). Recent experiments on "**definition** (*in l*)" for locales could be generalized to handle classes as well, by producing parallel definitions internally that refer either to locale parameters (**fixes**) or overloaded operations (**consts**).

Further considerations need to be spent on **instance** definitions. So far this is limited to simple definitions of Pure, but realistic applications demand more flexibility. The key question is how to combine derived definitional mechanisms with class instantiations in a modular fashion, without hardwiring one into the other. Then a package like [9] for general recursive functions could be used to specify class operations.

Acknowledgment. Alexander Krauss and Tobias Nipkow have commented on draft versions of this paper.

References

[1] Ballarin, C.: Locales and locale expressions in Isabelle/Isar. In: Berardi, S., Coppo, M., Damiani, F. (eds.) TYPES 2003. LNCS, vol. 3085, Springer, Heidelberg (2004)

[2] Ballarin, C.: Interpretation of locales in Isabelle: Theories and proof contexts. In: Borwein, J.M., Farmer, W.M. (eds.) MKM 2006. LNCS (LNAI), vol. 4108, Springer, Heidelberg (2006)

[3] Berghofer, S., Nipkow, T.: Proof terms for simply typed higher order logic. In: Aagaard, M.D., Harrison, J. (eds.) TPHOLs 2000. LNCS, vol. 1869, Springer, Heidelberg (2000)

[4] Chrzaszcz, J.: Modules in type theory with generative definitions. PhD thesis, Université Paris-Sud (2004)

[5] Courant, J.: $\mathcal{MC}_2$: A Module Calculus for Pure Type Systems. The Journal of Functional Programming (to appear, 2006)

[6] Hall, C., Hammond, K., Peyton Jones, S., Wadler, P.: Type classes in Haskell. ACM Transactions on Programming Languages and Systems 18(2) (1996)

[7] Jones, S., Jones, M., Meijer, E.: Type classes: an exploration of the design space (1997)

[8] Kammüller, F., Wenzel, M., Paulson, L.C.: Locales: A sectioning concept for Isabelle. In: Bertot, Y., Dowek, G., Hirschowitz, A., Paulin, C., Théry, L. (eds.) TPHOLs 1999. LNCS, vol. 1690, Springer, Heidelberg (1999)

[9] Krauss, A.: Partial recursive functions in higher-order logic. In: Furbach, U., Shankar, N. (eds.) IJCAR 2006. LNCS (LNAI), vol. 4130, Springer, Heidelberg (2006)

[10] Nipkow, T.: Order-sorted polymorphism in Isabelle. In: Huet, G., Plotkin, G. (eds.) Logical Environments, Cambridge University Press, Cambridge (1993)

[11] Nipkow, T.: Structured proofs in Isar/HOL. In: Geuvers, H., Wiedijk, F. (eds.) TYPES 2002. LNCS, vol. 2646, Springer, Heidelberg (2003)

[12] Nipkow, T., Paulson, L.C., Wenzel, M.: Isabelle/HOL – A Proof Assistant for Higher-Order Logic. In: Nipkow, T., Paulson, L.C., Wenzel, M. (eds.) Isabelle/HOL. LNCS, vol. 2283, Springer, Heidelberg (2002)

[13] Nipkow, T., Prehofer, C.: Type checking type classes. In: ACM Symp. Principles of Programming Languages, ACM Press, New York (1993)

[14] Paulson, L.C.: Isabelle: the next 700 theorem provers. In: Odifreddi, P. (ed.) Logic and Computer Science, Academic Press, London (1990)

[15] Peterson, J., Jones, M.P.: Implementing type classes. In: SIGPLAN. Conference on Programming Language Design and Implementation (1993)

[16] Pierce, B.: Types and Programming Languages. MIT Press, Cambridge (2002)

[17] Schmidt-Schauß, M.: Computational Aspects of an Order-Sorted Logic with Term Declarations. LNCS, vol. 395. Springer, Heidelberg (1989)

[18] Wadler, P., Blott, S.: How to make ad-hoc polymorphism less ad-hoc. In: ACM Symp. Principles of Programming Languages, ACM Press, New York (1989)

[19] Wenzel, M.: Type classes and overloading in higher-order logic. In: Gunter, E.L., Felty, A.P. (eds.) TPHOLs 1997. LNCS, vol. 1275, Springer, Heidelberg (1997)

[20] Wenzel, M.: Isar — a generic interpretative approach to readable formal proof documents. In: Bertot, Y., Dowek, G., Hirschowitz, A., Paulin, C., Théry, L. (eds.) TPHOLs 1999. LNCS, vol. 1690, Springer, Heidelberg (1999)

[21] Wenzel, M.: Isabelle/Isar — a versatile environment for human-readable formal proof documents. PhD thesis, Institut für Informatik, TU München (2002)

Zermelo's Well-Ordering Theorem in Type Theory

Danko Ilik

DCS Master Programme, Chalmers University of Technology

Abstract. Taking a 'set' to be a type together with an equivalence relation and adding an extensional choice axiom to the logical framework (a restricted version of constructive type theory) it is shown that any 'set' can be well-ordered. Zermelo's first proof from 1904 is followed, with a simplification to avoid using comparability of well-orderings. The proof has been formalised in the system AgdaLight.

1 Introduction

The well-ordering theorem is a proposition of set theory stating that any set can be well-ordered. A set M is well-ordered if there is a binary relation $<$ on M which is a linear order and for which every non-empty subset of M has a minimal element.

Georg Cantor, in the beginnings of his founding work on set theory, took this as a fact and called it the well-ordering *principle*. Later, he made attempts to prove it, thus suggesting that it should be regarded a theorem.

The first successful proof was displayed by Ernst Zermelo in a paper [1] published in 1904. In this proof he had used a principle which was later called the Axiom of Choice – he was the first to explicitly state this principle in a paper. A big debate over it arose between mathematicians in the following years.

The theorem itself was also controversial, since, for example, it implied that the set of real numbers could be well-ordered, though no such ordering was known[1]. Although that situation was controversial, it was not contradictory, since the principle of excluded middle allowed one to prove statements about the existence of an object, without requiring one to exhibit a sample of the object whose existence is proved.

The two mentioned principles, the one of excluded middle and the one of choice, were subject of much discussion in the field of foundation of mathematics during a large part of the 20th century, but in spite of that no one came to the idea that they could be simply correlated, and in this way: the principle of choice implies the principle of excluded middle. This was concluded for the first time by Diaconescu in 1975 for topos theory [4]; later there followed proofs in various other theories.

---

[1] Later [2] it was even shown that such an ordering can not be defined. More precisely, "there is no formula of ZF set theory which can be proved in ZFC to be a well-ordering of the reals" [3] p.423.

T. Altenkirch and C. McBride (Eds.): TYPES 2006, LNCS 4502, pp. 175–187, 2007.

1.1   Motivation for the Present Work

The goal of this work was to investigate whether it is possible to use type theory strengthened by an extensional choice axiom (ExtAC), instead of set theory, to prove the well-ordering theorem.

This addition allows us to derive the law of excluded middle (LEM). Thus, every proposition is decidable (in a constructive sense). As subsets in type theory are usually defined as propositional functions, we can equivalently define a subset to be a function into N_2, in which case the collection of subsets of a set, the *power-set*, is a set, hence it is possible to quantify over it.

Having ExtAC also allows us to define the function γ from Zermelo's proof, which takes a nonempty subset into one of its elements. This, and it being very clear and intuitive, makes Zermelo's first proof from 1904 a good candidate to follow. His second proof from 1908 [5], much like modern ones, use some set theoretic machinery which clutters the intuition behind it.

2   The Framework

We will work in constructive type theory (CTT) as explained in [6,7]. The theory we shall be concerned with will contain a base type Set and the constants $\Pi, \Sigma, N_0, N_1, N_2$: Set, together with their introduction and elimination rules. To these we will add the extensional axiom of choice and a constant T which lifts boolean values to propositions, defined by pattern-matching:

$$T : N_2 \to \text{Set}$$
$$T\ 0 \equiv N_0$$
$$T\ 1 \equiv N_1$$

We also need a small set universe containing codes for N_0 and N_1, for defining T and $=_2$ (below) by pattern-matching and for having $0 \neq 1$.

2.1   Axioms of Choice

The benefit of using an intensional theory like CTT is that we can make more distinctions, such as the one between *intensional* and *extensional* axioms of choice.

The first of these (IntAC) can be proved in the type theory. It reads:

$$\forall A, B^{\text{Set}}.\ \forall R^{A \to B \to \text{Set}}.\ \left(\forall x^A.\ \exists y^B.\ R\ x\ y\right)\ \to\ \exists f^{A \to B}.\ \forall x^A.\ R\ x\ fx \quad (1)$$

This is not surprising, because in type theory a proof of $\left(\forall x^A.\ \exists y^B.\ R\ x\ y\right)$ is exactly a function as the one required.

The second, the extensional axiom of choice (ExtAC), knows no justification in type theory. It reads:

$$\forall A, B^{\mathrm{Set}}. \ \forall R^{A \to B \to \mathrm{Set}}. \ \forall =_A^{A \to A \to \mathrm{Set}}. \ \forall =_B^{B \to B \to \mathrm{Set}}.$$

$$(=_A \text{ equivalence on } A) \to \ (=_B \text{ equivalence on } B) \to$$

$$\left(\forall x, y^A. \ x =_A y \to \forall z^B. \ R \ x \ z \to R \ y \ z\right) \to$$

$$\left(\forall x, y^B. \ x =_B y \to \forall z^A. \ R \ z \ x \to R \ z \ y\right) \to$$

$$\left(\forall x^A. \ \exists y^B. \ R \ x \ y\right) \to$$

$$\exists f^{A \to B}. \ \left(\forall x^A. \ R \ x \ fx\right) \wedge \left(\forall a, b^A. \ a =_A b \to fa =_B fb\right) \quad (2)$$

In words, what is required here in addition is that the function f must respect whatever equivalence relations may be defined on A and B, which the relation R preserves. It is this ability that allows one to 'smuggle in' non-constructive principles, by encoding them into an equivalence relation on which ExtAC is applied.

2.2   Derivation of the Law of Excluded Middle

The possibility of deriving the law of excluded middle (LEM) from ExtAC is well known. It has been carried out in various theories: topos theory [4], intuitionistic set theory [8] and intensional type theory [9].

The proof here is closest to the one from [10], the difference being that we use a (non-substitutive) equivalence relation, instead of the set Id. This relation, $=_2$, is defined only in terms of T and the elimination rules of N_2; it is defined to be N_1 for two elements of N_2 when they reduce to the same canonical element, and N_0 when they do not reduce to the same canonical element.

Now, let ExtAC be given and let P be a proposition (P : Set). Define a relation R (R : Rel N_2) as follows:

$$R \ a \ b \equiv a =_2 b \vee P$$

We show that there exists a function $f : N_2 \to N_2$ such that $P \leftrightarrow f0 =_2 f1$, meaning P is decidable.

We will use (2); let us satisfy the hypotheses: for A, B take N_2, for R take the R defined above, for the equivalence on A take R again and for the equivalence on B take $=_2$. Clearly, R is an equivalence relation on N_2.

By symmetry and transitivity of R, R is left-extensional for R itself. By transitivity of R and or-introduction, R is right-extensional for $=_2$. By reflexivity of R, for any $x : N_2$ there exists a $y : N_2$ such that $R \ x \ y$, namely x is such a y itself.

Thus, we get the following consequence of (2):

$$\exists f^{N_2 \to N_2}. \ \left(\forall x^{N_2}. \ R \ x \ fx\right) \wedge \left(\forall a, b^{N_2}. \ R \ a \ b \to fa =_2 fb\right) \quad (3)$$

From the definition of R we have $P \to R \ 0 \ 1$. From the right conjunct of (3) we have $R \ 0 \ 1 \to f0 =_2 f1$. Thus

$$P \to f0 =_2 f1 \quad (4)$$

To establish the other direction, first we prove $\forall a, b^{\mathrm{N}_2}.\ fa =_2 fb \to R\ a\ b$. Let a, b be given and let $fa =_2 fb$. From the fact that R is right-extensional for $=_2$ and the left conjunct of (3) we get $R\ a\ fb$. From the same conjunct and the symmetry of R we get $R\ fb\ b$. From these and the transitivity of R we get $R\ a\ b$.

From the definition of R and decidability of $=_2$ we get $R\ 0\ 1 \to P$. From this and the conclusion of the previous paragraph

$$f0 =_2 f1 \to P \tag{5}$$

(4) and (5) establish the decidability of P.

3   The Theorem

We present the definition of an extensional set and its subsets and define operations on them. After further definitions of special kinds of subsets, we state the well-ordering theorem in terms of those. The proof follows, divided into several propositions which are numbered in the same way as their parallels in Zermelo's proof from 1904 – the difference being that our proof is more detailed.

3.1   Representation of Sets and Subsets

We introduce the notion of *extensional set*, Xet[2]. An extensional set is an object of type Set, accompanied by a relation, accompanied by a proof that the relation is an equivalence one. Two such objects are equal if their Set-objects are equal and their relations are equal.

$$\text{Xet type}$$

All judgements $\Box$ we make will be hypothetical, i.e. of the form

$$\Box\ [X : \text{Set}][=_X : X \to X \to \text{Set}][=_X \text{ equivalence on } X]$$

but we will make this hypotheticalness implicit, in order to lighten the notation. So, let an object $(X, =_X, equivX) : \text{Xet}$ be given.

We define

$$\text{ext} : (X \to \text{Set}) \to \text{Set}$$
$$\text{ext } f \equiv \forall a, b^X.\ a =_X b \to fa \to fb$$

Subsets of a Xet object will be the boolean functions on X that are extensional:

$$\mathcal{P} : \text{Set}$$
$$\mathcal{P} \equiv \Sigma\, (X \to \mathrm{N}_2, \text{ext}\, ([U, x]\mathrm{T}\, (Ux)))$$

---

[2] As pointed out by one of the referees, this is just the known notion of *setoid*; see [11], for example.

Inhabited, or *non-empty*, subsets are subsets that contain an element:

$$\mathcal{P}' : \text{Set} \qquad\qquad\qquad \text{nonempty} : \mathcal{P} \to \text{Set}$$

$$\mathcal{P}' \equiv \Sigma\,(\mathcal{P}, \text{nonempty}) \qquad\qquad \text{nonempty}\ U \equiv \exists a^X.\ \text{T}\ (U.1\ a)$$

The suffix $.n$ of U is a selector, which picks the n-th component of an object of type Σ.

We also need some operations:

$$\in\,:\,X \to \mathcal{P} \to \text{Set} \qquad\qquad\qquad \in'\,:\,X \to \mathcal{P}' \to \text{Set}$$

$$a \in U \equiv \text{T}\ (U.1\ a) \qquad\qquad\qquad a \in' U \equiv a \in U.1$$

$$\subseteq\,:\,\mathcal{P} \to \mathcal{P} \to \text{Set} \qquad\qquad\qquad =\,:\,\mathcal{P} \to \mathcal{P} \to \text{Set}$$

$$U \subseteq V \equiv \forall a^X.\ a \in U \to a \in V \qquad\qquad U = V \equiv U \subseteq V \wedge V \subseteq U$$

$$\subseteq'\,:\,\mathcal{P}' \to \mathcal{P}' \to \text{Set} \qquad\qquad\qquad ='\,:\,\mathcal{P}' \to \mathcal{P}' \to \text{Set}$$

$$U \subseteq' V \equiv U.1 \subseteq V.1 \qquad\qquad\qquad U =' V \equiv U.1 = V.1$$

And some syntactic shortcuts:

$$\forall a \in U.\ \square \equiv \forall a^X.\ a \in U \to \square$$

$$\exists a \in U.\ \square \equiv \exists a^X.\ a \in U \wedge \square$$

LEM allows us to create subsets which consist of elements of X which satisfy a given extensional property; we will write this in a form of set comprehension:

$$\{|\}\,:\,\text{ExtPred} \to \mathcal{P}$$

$$\{x|P\} \equiv (\text{theSubset}\ P.1, \text{theExt}\ P.1)$$

where x is a placeholder for the free variable in P.1 (we want to mirror set theoretic notation), where $\text{ExtPred} \equiv \Sigma\,(X \to \text{Set}, \text{ext})$ and where

$$\text{theSubset} : (X \to \text{Set}) \to (X \to \text{N}_2)$$

$$\text{theSubset}\ P \equiv \text{IntAC}\ X\ \text{N}_2\ \left([x,b]\ \text{T}b \leftrightarrow \left(Px \wedge \forall y^X.\ y =_X x \to Py\right)\right)\ (\cdots)$$

$$\text{theExt} : (P : X \to \text{Set}) \to \text{ext}\,([x]T\,((\text{theSubset}\ P)\,x))$$

$$\text{theExt}\ P \equiv (\cdots)$$

To complete the proof of theSubset we need to prove $\forall x^X.\ \exists b^{\text{N}_2}.\ \text{T}b \leftrightarrow \left(Px \wedge \forall y^X.\ y =_X x \to Py\right)$, but this is immediate if we apply LEM to the right hand side of the equivalence – for b take 1 if it holds, 0 if it does not hold. The right hand side of the equivalence also gives us theExt immediately.

All predicates we shall apply comprehension on will be extensional.

Now, we have notation sufficient to mimic a set theoretic proof. We will just define a few more operators:

$$\complement : \mathcal{P} \to \mathcal{P} \qquad \backslash : \mathcal{P} \to \mathcal{P} \to \mathcal{P} \qquad \cap : \mathcal{P} \to \mathcal{P} \to \mathcal{P}$$
$$\complement U \equiv \{x \mid x \notin U\} \quad U \backslash V \equiv \{x \mid x \in U \wedge x \notin V\} \quad U \cap V \equiv \{x \mid x \in U \wedge x \in V\}$$

$$\{\} : X \to \mathcal{P} \qquad \cup : \mathcal{P} \to \mathcal{P} \to \mathcal{P} \qquad \emptyset : \mathcal{P}$$
$$\{a\} \equiv \{x \mid x =_X a\} \qquad U \cup V \equiv \{x \mid x \in U \vee x \in V\} \qquad \emptyset \equiv \{x \mid \mathrm{N}_0\}$$

3.2   Statement of the Theorem

We will need to be able to quantify over relations, thus we need *decidable* relations. We can lift a decidable relation into a normal one and vice-versa, since every proposition is decidable.

$$\text{DRel} : \text{Set} \qquad\qquad \text{Rel} : \mathit{Type}$$
$$\text{DRel} \equiv X \to X \to \mathrm{N}_2 \qquad \text{Rel} \equiv X \to X \to \text{Set}$$

$$\text{dRel} : \text{DRel} \to \text{Rel} \qquad\qquad \text{rRel} : \text{Rel} \to \text{DRel}$$
$$\text{dRel } D \equiv [a, b] \text{ T } (D\ a\ b) \qquad \text{rRel } R \equiv \cdots$$

For the definition of rRel we use IntAC; for details see the formalisation.

Now, some classes of relations *on subsets*. When a relation is trichotomous, transitive and linear:

trich : $\mathcal{P} \to \text{Rel} \to \text{Set}$
trich $U < \equiv \forall a, b \in U.\ (a < b \leftrightarrow b \not< a \wedge a \neq_X b) \wedge (a =_X b \leftrightarrow a \not< b \wedge b \not< a)$

trans : $\mathcal{P} \to \text{Rel} \to \text{Set}$
trans $U < \equiv \forall a, b, c \in U.\ a < b \to b < c \to a < c$

linear : $\mathcal{P} \to \text{Rel} \to \text{Set}$
linear $U < \equiv (\text{trich } U <) \wedge (\text{trans } U <)$

We also need a property expressing that a subset has a minimal element:

$$\text{hasLeast} : \mathcal{P} \to \text{Rel} \to \text{Set}$$
$$\text{hasLeast } U < \equiv \exists a \in U.\ \forall b \in U.\ b \not< a$$

And a property expressing that a subset is well-ordered:

$$\text{wellOrdered} : \mathcal{P} \to \text{Rel} \to \text{Set}$$
$$\text{wellOrdered } U < \equiv (\text{linear } U <) \wedge \left(\forall V^{\mathcal{P}'}.\ V.1 \subseteq U \to \text{hasLeast } V.1 < \right)$$

We are ready to state

Theorem 1 (Zermelo's Well-Ordering). *Any extensional set can be well-ordered.*

$$\left(\exists R^{\text{DRel}}.\ \forall U^{\mathcal{P}}.\ \text{wellOrdered } U\ (\text{dRel } R)\right)$$

3.3 Proof

We will now proceed with the proof following the one from 1904, enumerating the steps like it is done there. The key idea will be to use a choice function, γ, in defining a well-ordering relation $<$ in such a way that γ picks the $<$-least element of any subset.

(2) The Function γ. There is a function which takes a non-empty subset of X and gives an element of X which is contained in the subset. This function is extensional in respect to $=', =_X$. Formally:

$$\exists \gamma^{\mathcal{P}' \to X}.\ \left(\forall U^{\mathcal{P}'}.\ \gamma U \in' U\right) \wedge \left(\forall U, V^{\mathcal{P}'}.\ U =' V \to \gamma U =_X \gamma V\right)$$

Proof. We will use the extensional axiom of choice. Put $\mathcal{P}'$ for A, X for B, $(x \in' U)$ for R, $='$ for $=_A$ and $=_X$ for $=_B$. It is easy to see that $='$ is an equivalence relation, and $=_X$ is such by hypothesis. To get the desired function, we need only prove the following three things:

- $\forall U, V^{\mathcal{P}'}.\ U =' V \to \forall z^X.\ z \in' U \to z \in' V$. This we get immediately from the definition of $='$.
- $\forall x, y^X.\ x =_X y \to \forall W^{\mathcal{P}'}.\ x \in' W \to y \in' W$. This is immediate from the extensionality of subsets.
- $\forall U^{\mathcal{P}'}.\ \exists y^X.\ y \in' U$. This follows from the non-emptiness of U.

(3) γ-Sets. An *initial segment* of a subset, for a given element of X and a relation, is the subset of all those elements which are in relation with the given one. Formally:

$$\text{IS} : \mathcal{P} \to X \to \text{Rel} \to \mathcal{P}$$
$$\text{IS } U\ a < \equiv \{x \mid x \in U \wedge x < a\}$$

A subset is called a *γ-set*, for a given relation, if it is well-ordered by the relation and if for any element a therein, γ takes the complement of the initial segment for a, into a itself. Formally:

$$\text{GS} : \mathcal{P} \to \text{Rel} \to \text{Set}$$
$$\text{GS } U < \equiv (\text{wellOrdered } U <) \wedge$$
$$\left(\forall a \in U.\ \forall ne^{\text{nonempty} \mathsf{C}(\text{IS } U\ a<)}.\ a =_X \gamma\left(\left(\mathsf{C}\ (\text{IS } U\ a <)\right), ne\right)\right)$$

(4) Example Subsets of X Which Are γ-Sets. Suppose X is inhabited, $\mathcal{X} \equiv \{x \mid N_1\}$ is a subset containing all elements of X, ne : nonempty$\mathcal{X}$ and take the following subset:

$$M' : \mathcal{P} \qquad\qquad m_1 : X$$
$$M' \equiv \{x \mid x =_X m_1\} \qquad\qquad m_1 \equiv \gamma(\mathcal{X}, ne)$$

Define the ordering:

$$<_1 : \mathrm{Rel}$$
$$x <_1 y \equiv N_0$$

It is easy to check that $<_1$ makes M' a γ-set. Similarly, the subset $\{x \mid x =_X m_1 \vee x =_X m_2\}$, where $m_2 \equiv \gamma(\mathcal{X} \setminus \{m_1\}, ne')$ and ne' : nonempty$(\mathcal{X} \setminus \{m_1\})$, is a γ-set.

(5) If M_1, M_2 Are Different γ-Sets, Then One Is an Initial Segment of the Other. Formally:

$$\forall M_1, M_2^{\mathcal{P}}. \ \forall <_1, <_2^{\mathrm{Rel}}. \ \mathrm{GS}\ M_1 <_1 \wedge \mathrm{GS}\ M_2 <_2 \rightarrow$$
$$(M_1 = M_2) \vee (\exists x_1 \in M_1.\ S_1 x_1 = M_2) \vee (\exists x_2 \in M_2.\ S_2 x_2 = M_1)$$

where

$$S_\square \equiv [x]\mathrm{IS}\ M_\square\ x <_\square$$

In the original paper, as well as modern papers like [12], this step is proven by using the comparability of well-orderings, which grants that there is an order-preserving injection from one of M_1, M_2 into the other. Then one proceeds to prove that this injection must be the identity. As the comparability is not easier to prove than the well-ordering theorem itself, we provide a direct proof using well-founded induction.

Proof. Let $M_1, <_1, M_2, <_2$ be given and let them be γ-sets. We need the following lemma:

Lemma 1. *An initial segment of M_1 is an initial segment of M_2 or is M_2; or a smaller initial segment of M_1 is M_2.*

$$\forall x \in M_1.\ (x \in M_2 \wedge S_1 x = S_2 x) \vee (\exists x_1 \in S_1 x.\ S_1 x_1 = M_2) \vee (S_1 x = M_2)$$

Proof. M_1 is well-ordered, thus we can use well-founded induction on it[3]: for $P : X \rightarrow \mathrm{Set}$, we have that $(\forall x \in M_1.\ (\forall y \in M_1.\ y <_1 x \rightarrow Py) \rightarrow Px) \rightarrow \forall x \in$

---

[3] Suppose that a subset U is well-ordered and not well-founded, take the minimal element t of U for which P does not hold, and derive a contradiction. All the details are in the formalisation.

M_1. Px. For P we take the expression in the scope of the universal quantifier from the formulation of the lemma.

Let $x \in M_1$ be given. We use the classical tautology (for $R, A, B, C : X \to$ Set):

$$(\forall x^X.\ Rx \to Ax \lor Bx \lor Cx) \to$$
$$(\forall x^X.Rx \to Ax) \lor (\exists x^X.Rx \land Bx) \lor (\exists x^X.Rx \land Cx) \quad (6)$$

on the induction hypothesis, and get these 3 cases:

1. $\forall y \in M_1.\ y <_1 x \to (y \in M_2 \land S_1y = S_2y)$. Thus, $S_1x \subseteq M_2$. We look into the following two cases:
   (a) $S_1x = M_2$.
   (b) $S_1x \neq M_2$. $S_1x \subseteq M_2$. Let t be the least element of the subset $M_2 \setminus S_1x$ for $<_2$. We will show that $S_1x = S_2t$:
      - Let $a \in S_1x$. Then $a \in M_2$ and we need to show that $a <_2 t$. If $a \not<_2 t$, then $t <_2 a$ or $a =_X t$. $a =_X t$ is not possible as $t \notin S_1x$ by definition. If $t <_2 a$, $t \in S_2a$ and by the induction hypothesis $S_2a = S_1a \subseteq S_1x \ni t$, again a contradiction with the definition of t. So, $a \in S_2t$.
      - Let $b \in S_2t$. If $b \notin S_1x$, then $b \in M_2 \setminus S_1x$ and then, since t is minimal, it must be that $b \not<_2 t$, a contradiction with $b \in S_2t$. So, $b \in S_1x$.
      We have that $S_1x = S_2t$. From S_1, S_2 being initial segments of the γ-sets M_1, M_2, $x =_X \gamma(\complement(S_1x)) =_X \gamma(\complement(S_2t)) =_X t$. Thus, $S_1x = S_2x$.
2. $\exists y \in M_1.\ y <_1 x \land (\exists x_1 \in S_1y.\ S_1x_1 = M_2)$. Let such y, x_1 be given. Then $x_1 \in S_1x$, so we get the 2nd disjunct of Px.
3. $\exists y \in M_1.\ y <_1 x \land (S_1y = M_2)$. Taking y for x_1, we immediately get the 2nd disjunct of Px.

We use tautology (6) again, now on the lemma itself and thus get 3 cases:

1. $\forall x \in M_1.\ x \in M_2 \land S_1x = S_2x$. Define the following subsets:

$$S \equiv \{y \mid \exists z \in M_1.\ y \in S_1z\}$$
$$T_1 \equiv M_1 \setminus S$$
$$T_2 \equiv M_2 \setminus S$$

S is the subsets of all elements of M_1 which belong to some initial segment of M_1. T_1, T_2 contain the remaining elements. From the hypothesis we have $M_1 \subseteq M_2$ and $S \subseteq M_2$. We will distinguish on the emptiness of T_1, T_2:
   (a) $T_1 = \emptyset$. As $S \subseteq M_1$ and $\emptyset = M_1 \setminus S$, $M_1 = S$.
      i. $T_2 = \emptyset$. As $S \subseteq M_1 \subseteq M_2$ and $\emptyset = M_2 \setminus S$, $M_2 = S = M_1$.
      ii. $T_2 \neq \emptyset$. Let t be the least element of T_2 for $<_2$. We want to show that $S_2t = S = M_1$:
         - if $a \in S_2t$, $a <_2 t$, so $a \notin T_2$, because t is the minimal of T_2. As $a \in M_2$ and $a \notin T_2$, $a \in S$.

- let $a \in S$; then $a \notin T_2$. Does $a \in M_2$ and $a <_2 t$? $a \in M_1 \subseteq M_2$. Let $a \not<_2 t$:
  - if $t <_2 a$, then $t \in S$ as, from the hypothesis, $S_1 a = S_2 a$; but, $t \notin S$ by definition.
  - if $a =_X t$, $a \in T_2$, a contradiction.

  Thus, $a <_2 t$ and $a \in S_2 t$.

(b) $T_1 \neq \emptyset$. We show that T_1 can contain only one element: let $t_1, t_2 \in T_1$ and, without loss of generality, let $t_1 <_1 t_2$; then $t_1 \in S_1 t_2$, thus $t_1 \in S$, thus $t_1 \notin T_1$, which is a contradiction. So, T_1 has exactly one element; call it t.

From $M_1 \subseteq M_2$ and the definitions of T_1, T_2, $T_1 \subseteq T_2$ and $t \in T_2$.

  i. $T_2 = \{t\}$. Then $M_2 = S \cup T_2 = S \cup T_1 = M_1$.
  ii. $T_2 \neq \{t\}$. Let t' be the least element of the subset $T_2 \setminus \{t\}$. Then $t <_2 t'$: if $t' =_X t$, then $t \in T_2 \setminus \{t\}$, a contradiction; if $t' <_2 t$, by the main hypothesis $S_1 t = S_2 t$, so $t' \in S_1 t$ and $t \in S$, a contradiction.

  Thus, $t <_2 t'$ and we have that t is the minimal of T_2 for $<_2$. Using this, like in case (a.ii) we get $S = S_2 t$ and $M_1 = S \cup \{t\} = S_2 t \cup \{t\}$. We show that $M_1 = S_2 t'$:
  - if $x \in M_1$, then $x \in S_2 t$ or $x =_X t$. If $x \in S_2 t$, then $x <_2 t <_2 t'$, thus $x \in S_2 t'$. If $x =_X t$, then $x <_2 t'$, thus $x \in S_2 t'$.
  - if $x \in S_2 t'$, then $x <_2 t'$. We look at the 3 cases:
    - $x =_X t$. Then $x \in T_1 \subseteq M_1$.
    - $x <_2 t$. Then $x \in S_2 t = M_1$.
    - $t <_2 x$. Then $t <_2 x <_2 t'$, so t' is not the minimal element of $T_2 \setminus \{t\}$.

2. $\exists x \in M_1. \exists x_1 \in M_1. x_1 <_1 x \wedge S_1 x_1 = M_2$. Thus, M_2 is an initial segment of M_1. QED
3. $\exists x \in M_1. S_1 x = M_2$. Again, M_2 is an initial segment of M_1. QED

(6) A Consequence. If two γ-sets have an element a in common, then their initial segments for a are the same. Formally:

$$\forall M_1, M_2^{\mathcal{P}}. \forall <_1, <_2^{\text{Rel}}. \text{ GS } M_1 <_1 \wedge \text{ GS } M_2 <_2 \rightarrow$$
$$\forall a \in M_1 \cap M_2 \rightarrow (S_1 a = S_2 a)$$

Proof. We can use (5) to decide which of the γ-sets is an initial segment of the other. The required follows from the definition of initial segment.

(7) X Is Well-Ordered. Define the following relation on X:

$$a < b \equiv \exists M_a^{\mathcal{P}}. \exists D_a^{\text{DRel}}. \text{ GS } M_a \text{ (dRel } D_a) \wedge a \in M_a \wedge$$
$$\forall M_b^{\mathcal{P}}. \forall D_b^{\text{DRel}}. \text{ GS } M_b \text{ (dRel } D_b) \rightarrow b \in M_b \rightarrow$$
$$\exists \beta. \beta \in M_b \wedge \text{IS } M_b \ \beta \text{ (dRel } D_b) = M_a$$

This relates two elements of X, if they are γ-elements and a γ-set containing the first element is an initial segment of a γ-set containing the other.

Call $x : X$ a γ-*element* if there exists a γ-set, for the relation $<$, which contains it:

$$GE : X \to \text{Set}$$

$$GE\ x \equiv \exists M_\gamma^\mathcal{P}.\ x \in M_\gamma \wedge GS\ M_\gamma <$$

Let L_γ be the subset of all γ-elements:

$$L_\gamma : \mathcal{P}$$

$$L_\gamma \equiv \{x \mid GE\ x\}$$

To establish that X is well-ordered, in the following 5 lemmas, we show that L_γ is well-ordered and that $\mathcal{X} \subseteq L\gamma$. Recall that $\mathcal{X}$ is a subset containing all elements of X.

(7-I) $<$ **Is Trichotomous on L_γ.** First, we lighten the notation by writing $M_a \prec M_b$ when M_a is an initial segment of M_b, and by omitting the relations, which are always quantified together with their corresponding γ-sets.

- Let $a, b \in L_\gamma$ and $a < b$. Then there exist a γ-set M_a containing a, such that for any M_b containing γ-set of b, $M_a \prec M_b$.

    If $a =_X b$, then M_a is a containing γ-set of b as well, and we have $M_a \prec M_a$, which is not possible.

    If $b < a$, then there exists $L_b \ni b$ s.t. for every $L_a \ni a$, $L_b \prec L_a$. If we put L_b in place of M_b and M_a in place of L_a, we get both $L_b \prec M_a$ and $M_a \prec L_b$, which is not possible.
- Let $a, b \in L_\gamma$ and $b \not< a$ and $a \neq_X b$. From $b \not< a$ we have $\forall L_b.\ \exists L_a.\ L_b \not\prec L_a$. We will use the fact that b is a γ-element, to extract a containing W_b, which is a γ-set for $<$. We get that there exists a gamma set $L_a \ni a$ such that $W_b \not\prec L_a$. From step (5) we have that $W_b = L_a$ or $L_a \prec W_b$. In any case, $a \in W_b$ and we can use the hypotheses and the trichotomy of $<$ on W_b to complete the proof.

(7-II) $<$ **Is Linear.** We need to show that $<$ is transitive. Let $a, b, c \in L_\gamma$ and $a < b, b < c$. From $\exists M_a.\ \forall M_b.\ M_a \prec M_b$ and $\exists M_b.\ \forall M_c.\ M_b \prec M_c$, we have $\exists M_a.\ \forall M_c.\ M_a \prec M_c$, i.e. $a < c$.

(7-III) $<$ **Well-Orders L_γ.** Let $L' \subseteq L_\gamma, L' \neq \emptyset$. Pick $a \in L'$ and define $L'' \equiv \{x \mid (x \in \text{IS } L'\ a <) \vee (x =_X a)\}$. If M' is a witnessing γ-set of a, then (by step (6)) $L'' \subseteq M'$. Since L'' is not empty (it has at least a) and is a subset of a well-ordered M', L'' has a minimal element, which (because of the definition of L'') must be a minimal of L' as well.

(7-IV) L_γ Is a γ-Set. Let $a \in L_\gamma$, M_a be its witnessing γ-set for $<$; let $B \equiv$ IS M_a $a <$ and let $A \equiv$ IS L_γ $a <$. We will show that $A = B$; from this will follow that $\gamma(\complement A) =_X \gamma(\complement B) =_X a$.

- Let $x \in A$. Then $x < a$, so there is a containing γ-set M_x, such that $M_x \prec M_a$. So, $x \in M_a$ and $x < a$, thus $x \in B$.
- Let $x \in B$. Then it is a γ-element, since it belongs to M_a, so $x \in A$.

(7-V) $\mathcal{X} \subseteq L_\gamma$. If the Set X is not inhabited, then $\mathcal{X} = \emptyset$ and trivially $\mathcal{X} \subseteq L_\gamma$.

If the Set X is inhabited, then let $x \in \mathcal{X}$ and let $x \notin L_\gamma$. Then, $x \in \complement L_\gamma$, thus this complement is not empty, $ne : \text{nonempty}\complement L_\gamma$, and we can define

$$m : X$$
$$m \equiv \gamma(\complement L_\gamma, ne)$$

Now, the relation $<$ makes m larger than all elements in the subset:

$$L' : \mathcal{P}$$
$$L' \equiv \{x \mid (x \in L_\gamma) \vee (x =_X m)\}$$

It is not hard, but it takes some work to check that L' is a γ-set for $<$ (for details, see the formalisation). Since $m \in L'$, m is a γ-element, thus it must be that $m \in L_\gamma$, which is a contradiction. So, indeed, $\mathcal{X} \subseteq L_\gamma$, and $\mathcal{X}$ is well-ordered.

4   Formalisation

The presented proof served as a sketch for a formalisation [13] that was checked using AgdaLight [14], a version of the Agda [15] proof checker for constructive type theory.

In Agda a proof term is not constructed by using tactics, but is directly given. We use nested let-expressions and explicit type annotations to give structure to the proofs. This style of writing comes close to the requirements of Leslie Lamport's proof style[16]. We hope to have produced a readable formalised document.

Further work of the formalisation is possible, especially in respect to handling more systematically subsets created by set comprehension.

5   Related Work

In [17], Peter Aczel has shown how to interpret full Zermelo-Fraenkel set theory in constructive type theory + LEM. The type theory used is a standard one (with W), thus stronger than the one we use.

In [18], Per Martin-Löf shows that in type theory, the extensional axiom of choice is equivalent to Zermelo's axiom of choice. As a consequence of the work from Peter Aczel, full ZFC can be interpreted in constructive type theory + ExtAC.

Acknowledgements

The topic of this text came into being during the Types 2005 summer school in Göteborg, during a conversation with Thierry Coquand, my master's thesis' supervisor, and Per Martin-Löf, who on this school held a lecture in which he used constructive type theory to analyse Zermelo's axiom of choice. I thank Thierry for his good advises, the Swedish Institute for their financial support and the referees for the detailed comments.

References

1. Zermelo, E.: Beweis, daß jede menge wohlgeordnet werden kann. Mathematische Annalen 59, 514–516 (1904) English translation in van Heijenoort, 1967
2. Feferman, S.: Some applications of the notions of forcing and generic sets. Fundamenta Mathematicae 56, 325–345 (1964)
3. Beeson, M.J.: Foundations of Constructive Mathematics: Metamathematical Studies. Springer, Heidelberg (1985)
4. Diaconescu, R.: Axiom of choice and complementation. Proceedings of A.M.S. 51, 176–178 (1975)
5. Zermelo, E.: Neuer beweis für die möglichkeit einer wohlordnung. Mathematische Annalen 65, 107–128 (1908) English translation in van Heijenoort, 1967
6. Nordström, B., Petersson, K., Smith, J.M.: Martin-Löf's Type Theory. In: Handbook on Logic in Computer Science, vol. 5, Oxford University Press, Oxford (2000)
7. Martin-Löf, P.: On the meanings of the logical constants and the justifications of the logical laws. Nordic Journal of Philosophical Logic 1, 11–60 (1996) Text of lectures originally given in 1983 and distributed in 1985
8. Goodman, N.D., Myhill, J.: Choice implies excluded middle. Zeitschrift für Mathematische Logik und Grundlagen der Mathematik 24, 461 (1978)
9. Maietti, M.E., Valentini, S.: Can you add power-sets to martin-löf's intuitionistic set theory? Mathematical Logic Quarterly 45, 521–532 (1999)
10. Carlström, J.: EM + ext- + ACint is equivalent to ACext. Mathematical Logic Quarterly 50, 236–240 (2004)
11. Barthe, G., Capretta, V., Pons, O.: Setoids in type theory. Journal of Functional Programming 13, 261–293 (2003)
12. Kanamori, A.: Zermelo and set theory. The Bulletin of Symbolic Logic 10, 487–553 (2004)
13. Ilik, D.: Formalisation of zermelo's well-ordering theorem in type theory (2006), http://www.mdstud.chalmers.se/~danko/
14. Norell, U.: Agdalight (2006), http://www.cs.chalmers.se/~ulfn/agdaLight/
15. Coquand, C.: Agda (2000), http://agda.sourceforge.net/
16. Lamport, L.: How to write a proof (1993)
17. Aczel, P.: The type theoretic interpretation of constructive set theory. In: Macintyre, A., Pacholski, L., Paris, J. (eds.) Logic Colloquium '77, pp. 55–66. North-Holland Publishing Company, Amsterdam (1978)
18. Martin-Löf, P.: 100 years of zermelo's axiom of choice: what was the problem with it? (2004)

A Finite First-Order Theory of Classes

Florent Kirchner

LIX, École Polytechnique, 91128 Palaiseau, France
florent.kirchner@inria.fr

Abstract. We expose a formalism that allows the expression of any theory with one or more axiom schemes using a finite number of axioms. These axioms have the property of being easily orientable into rewrite rules. This allows us to give finite first-order axiomatizations of arithmetic and real fields theory, and a presentation of arithmetic in deduction modulo that has a finite number of rewrite rules. Overall, this formalization relies on a weak calculus of explicit substitutions to provide a simple and finite framework.

1 Introduction

In mathematics, some theories — such as arithmetic or set theory — are often expressed using an infinite number of axioms. This is achieved through the use of one or more *axiom schemes*, i.e. sets of axioms, often described within the meta-theory. For instance the induction scheme in arithmetic can be expressed as: for any proposition P,

$$P(0) \Rightarrow \forall y, (P(y) \Rightarrow P(S(y))) \Rightarrow \forall z, P(z).$$

This scheme, parametrized by the *schematic variable* P that takes values in the set of formulas of arithmetic, generates an infinite number of axioms.

The use of these axiom schemes can be avoided though, by introducing a new sort of objects, *classes* (which we will distinguish from other objects by using uppercase letters), and a membership symbol $\in$. Using classes as representatives for propositions, the induction scheme is re-written as a single axiom:

$$\forall E, (0 \in E \Rightarrow \forall y, (y \in E \Rightarrow S(y) \in E) \Rightarrow \forall z, z \in E).$$

However, in order for classes to soundly emulate propositions, one needs to guarantee that any proposition has an associated class. This is assured by the comprehension axiom scheme, which states: for any proposition P that is well formed in the original language,

$$\exists E, \forall x, (x \in E \Leftrightarrow P).$$

This extension of arithmetic is *conservative*, meaning that any formula that is provable in the theory of arithmetic plus classes, involving symbols of arithmetic only, was already provable in arithmetic. Also note that some axiom schemes might have propositions with *two* free variables or more, and that to deal with

T. Altenkirch and C. McBride (Eds.): TYPES 2006, LNCS 4502, pp. 188–202, 2007.
© Springer-Verlag Berlin Heidelberg 2007

this in the most general way, one needs not only classes of objects but classes of n-tuples of objects. Therefore, one could for instance introduce a sort of lists, list constructors and adapt the symbol $\in$ and the comprehension axiom scheme to this new structure.

In the end, however, the theory would still have an axiom scheme that generates an infinite number of axioms.

This is not a zero-sum game, though. Previous works [1–5] have shown that, in the case of set theory, it is possible to reduce the comprehension axiom scheme to a finite number of axioms. However we believe that the notion of class is independent of set theory and can be extended to express any theory containing axiom schemes in a finite first-order axiomatization. Moreover, unlike in the previous systems, we will see that the axioms in the theory of classes can easily be oriented as rewrite rules.

This work is part of a long term project investigating the possibility to base proof checkers on weaker frameworks, such as first-order logic. The maturity of these frameworks make them very secure centerpieces of formal tools in general, and proof checkers in particular; some designs (e.g. [6,7]) have already taken advantage of them. It is essential in this project to implement strong theories, such as arithmetic, real fields or set theory, with a finite number of axioms. The main contribution of this paper is a systematic way of conducting these implementations, using a theory of classes.

2   A Theory with the Comprehension Scheme

We consider a language $\mathscr{L}$ in first-order predicate logic with equality, and we call Σ its *finite* signature. Let $\mathcal{T}$ be an intuitionistic theory of this language that has one or more axiom schemes, i.e. axioms of the form:

$$s(P(t_1^1, \ldots, t_n^1), \ldots, P(t_1^p, \ldots, t_n^p)),$$

where n and p are natural numbers depending on each scheme, s is a p-ary first-order formula, the t_j^i are terms of $\mathscr{L}$ and P is a schematic formula variable. Remark that, although the axiom schemes in this paper only have one schematic variable P, the generalization of our results to a system with arbitrary axiom schemes is straightforward.

Definition 1 ($\mathscr{L}^{cs}$). *Let $\mathscr{L}^{cs}$ be a many-sorted language with three sort symbols: the sort of objects X, lists L, and classes C. In $\mathscr{L}^{cs}$, Σ is finitely extended into a signature Σ^{cs} with a predicate symbol $\in$ of rank (L, C) and two function symbols:*

$$nil : L$$

$$:: \ : (X, L)L.$$

Notation 1. *We use $\langle x_1, \ldots, x_n \rangle$ as syntactic sugar for the term $x_1 :: \ldots :: x_n :: nil$.*

Definition 2 ($\mathcal{T}^{cs}$). *Define $\mathcal{T}^{cs}$ as the theory of the language $\mathscr{L}^{cs}$ derived from $\mathcal{T}$ by adding the comprehension scheme:*

$$\exists E : C, \forall x_1, \ldots, \forall x_n : X, \qquad (\langle x_1, \ldots, x_n \rangle \in E \Leftrightarrow P) \qquad (\mathcal{T}_1^{cs})$$

where P is built with the symbols of $\mathscr{L}$, and may contain some of the x_i as free variables. Axiom schemes are replaced by axioms:

$$\forall E : C, \qquad s(\langle t_1^1, \ldots, t_n^1 \rangle \in E, \ldots, \langle t_1^p, \ldots, t_n^p \rangle \in E). \qquad (\mathcal{T}_2^{cs})$$

Example 1. Following the translation of Definition 2, the induction axiom scheme:

$$P(0) \Rightarrow \forall y, (P(y) \Rightarrow P(S(y))) \Rightarrow \forall z, P(z)$$

is replaced by the following axiom scheme and axiom:

$$\exists E : C, \forall x_1, \ldots, \forall x_n : X, (\langle x_1, \ldots, x_n \rangle \in E \Leftrightarrow P)$$
$$\forall E : C, \langle 0 \rangle \in E \Rightarrow \forall y, (\langle y \rangle \in E \Rightarrow \langle S(y) \rangle \in E) \Rightarrow \forall z, \langle z \rangle \in E.$$

Proposition 1. $\mathcal{T}^{cs}$ *is an extension of $\mathcal{T}$.*

Proof. We need to show that in $\mathcal{T}^{cs}$, for any P,

$$s(P(t_1^1, \ldots, t_n^1), \ldots, P(t_1^p, \ldots, t_n^p)).$$

is provable. This is immediate, by $(\mathcal{T}_1^{cs})$ and $(\mathcal{T}_2^{cs})$.   □

We have now all ingredients to move towards our goal, finding a finite axiomatization of the comprehension scheme.

3   Finite Class Theory

3.1   Setting and Notations

Definition 3 ($\mathscr{L}^{ws}$). *We extend $\mathscr{L}^{cs}$ with the function symbols:*

$$
\begin{array}{ll}
1 : X & \varnothing : C \\
S : (X)X & \cap, \cup, \supset : (C, C)C \\
\cdot[\cdot] : (X, L)X & \mathscr{P}, \mathscr{C} : (C)C.
\end{array}
$$

To each predicate symbol p is associated a function symbol $\dot{p}$ of similar arity, which constructs elements of sort C. We call $\mathscr{L}^{ws}$ this language, and Σ^{ws} its finite signature.

Definition 4 ($\mathcal{T}^{ws}$). *Let $\mathcal{T}^{ws}$ be the theory of the language $\mathscr{L}^{ws}$ formed with the following axioms for explicit substitutions:*

$$
\begin{array}{lll}
\forall x : X, & x[nil] = x & (\mathcal{T}_1^{ws}) \\
\forall \ell : L, \forall x : X, & 1[x :: \ell] = x & (\mathcal{T}_2^{ws}) \\
\forall \ell : L, \forall x, y : X, & S(y)[x :: \ell] = y[\ell] & (\mathcal{T}_3^{ws}) \\
\forall \ell : L, \forall x_1, \ldots, x_n : X, & f(x_1, \ldots, x_n)[\ell] = f(x_1[\ell], \ldots, x_n[\ell]) & (\mathcal{T}_4^{ws})
\end{array}
$$

and the axioms for proposition encoding:

$$\forall \ell : L, \forall x_1, \dots, x_n : X, \quad \ell \in \dot{p}(x_1, \dots, x_m) \Leftrightarrow p(x_1[\ell], \dots, x_m[\ell]) \qquad (\mathcal{T}_5^{ws})$$

$$\forall A, B : C, \forall \ell : L, \qquad \ell \in A \cap B \Leftrightarrow \ell \in A \wedge \ell \in B \qquad (\mathcal{T}_6^{ws})$$

$$\forall A, B : C, \forall \ell : L, \qquad \ell \in A \cup B \Leftrightarrow \ell \in A \vee \ell \in B \qquad (\mathcal{T}_7^{ws})$$

$$\forall A, B : C, \forall \ell : L, \qquad \ell \in A \supset B \Leftrightarrow \ell \in A \Rightarrow \ell \in B \qquad (\mathcal{T}_8^{ws})$$

$$\forall \ell : L, \qquad \ell \in \varnothing \Leftrightarrow \bot \qquad (\mathcal{T}_9^{ws})$$

$$\forall A : C, \forall \ell : L, \qquad \ell \in \mathscr{P}(A) \Leftrightarrow \exists x, x{::}\ell \in A \qquad (\mathcal{T}_{10}^{ws})$$

$$\forall A : C, \forall \ell : L, \qquad \ell \in \mathscr{C}(A) \Leftrightarrow \forall x, x{::}\ell \in A. \qquad (\mathcal{T}_{11}^{ws})$$

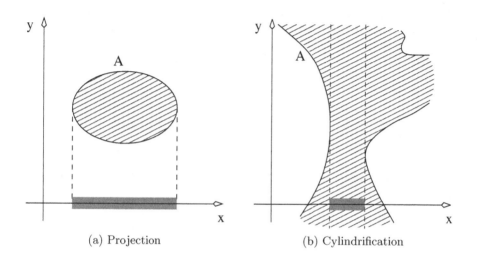

(a) Projection   (b) Cylindrification

Fig. 1. Two-dimensional operators

A couple of remarks on this formalization:

- The symbols 1 and S in $\mathscr{L}^{ws}$ are constructors of de Bruijn indices. It is important for them to be given the sort X, because the symbol S will need to be applied to non-substitutable variables of X to lift them out of the substitution's reach[1].
- This axiom system features a weak calculus of explicit substitutions [9]: the substitutions are propagated over the elements of the language via the symbols $\in$ and $\cdot[\cdot]$, and no lift is introduced by the $\mathscr{P}$ or $\mathscr{C}$ binders (axioms $(\mathcal{T}_{10}^{ws})$ and $(\mathcal{T}_{11}^{ws})$).
- Axiom scheme $(\mathcal{T}_4^{ws})$ (resp. $(\mathcal{T}_5^{ws})$) represents a finite number of first-order axioms, as there is one such axiom for each function (resp. predicate) symbol of arity n (resp. m) in the language $\mathscr{L}^{ws}$ (which as a finite extension of $\mathscr{L}^{cs}$ is finite).

---

[1] This operation is called *pre-cooking* in [8].

– The $\mathscr{P}$ and $\mathscr{C}$ operators are respectively called *projection* and *cylindrification* in Algebra. Figure 1 illustrates the semantics of these combinators in a two-dimensional space.

From now on in this paper, we will spare the type of variables in quantifiers when no ambiguities hold.

3.2   Expressiveness

We want to associate to each proposition a characteristic class, constructed with the symbols exposed in Definition 3.

Example 2. Assume given a predicate symbol $<$ and the associate characteristic set constructor $\dot{<}$. Using infix notations and the decimal representations 2 and 3 for the Peano numbers $S(1)$ and $S(S(1))$, the class of objects x such that $\exists y, \exists z, (x < z \wedge z < y)$ is written $\mathscr{P}\mathscr{P}(3\dot{<}1 \cap 1\dot{<}2)$. Indeed,

$$\langle x \rangle \in \mathscr{P}\mathscr{P}(3\dot{<}1 \cap 1\dot{<}2)$$
$$\Leftrightarrow \exists y, \exists z, (\langle z, y, x \rangle \in 3\dot{<}1 \cap 1\dot{<}2)$$
$$\Leftrightarrow \exists y, \exists z, (\langle z, y, x \rangle \in 3\dot{<}1 \wedge \langle z, y, x \rangle \in 1\dot{<}2)$$
$$\Leftrightarrow \exists y, \exists z, (x < z \wedge z < y).$$

We first need a little lemma to prove that the term substitution axioms are complete:

Lemma 1. *For all term u and variables $x_1, \ldots, x_n$ of $\mathscr{L}$, there exists a term t of $\mathscr{L}^{\mathsf{ws}}$ in which none of the x_i appear, such that $u = t[x_1 :: \ldots :: x_n :: nil]$ is provable in T^{ws}.*

Proof. We proceed inductively on the structure of u.

– If u is one of the x_i, then we take $S^{i-1}(1)$ for t and by using axioms (T_2^{ws}) and (T_3^{ws}), $x_i = S^{i-1}(1)[x_1 :: \ldots :: x_n :: nil]$ is provable.
– If u is a variable y different from the x_i, we take $t = S^n(y)$, and by axioms (T_1^{ws}), (T_2^{ws}) and (T_3^{ws}) we can prove $y = S^n(y)[x_1 :: \ldots :: x_n :: nil]$.
– Finally, if u is a term $f(u_1, \ldots, u_m)$, then by induction hypothesis there exist $t_1, \ldots, t_m$ such that for any $0 < i \leq m$, $u_i = t_i[x_1 :: \ldots :: x_n :: nil]$. Axiom (T_4^{ws}) allows to conclude $f(u_1, \ldots, u_m) = f(t_1, \ldots, t_m)[x_1 :: \ldots :: x_n :: nil]$. □

We now show that the comprehension axiom scheme holds in our formalism. Translated in our framework, the scheme states:

Proposition 2. *For any formula P built with the symbols of $\mathscr{L}$, the formula:*

$$\exists E, \forall x_1, \ldots, \forall x_n, (\langle x_1, \ldots, x_n \rangle \in E \Leftrightarrow P)$$

is provable in T^{ws}.

Proof. We show by induction on the structure of P that for any formula P there exists a term E of sort C such that, for all $x_1, \ldots, x_n$ of sort X, $\langle x_1, \ldots, x_n \rangle \in E \Leftrightarrow P$.

If P is an atomic proposition $p(u_1, \ldots, u_m)$, where p is a predicate symbol and $u_1, \ldots, u_n$ are terms, Lemma 1 allows us to equate this proposition to $p(t_1[l], \ldots, t_m[l])$ where $l = \langle x_1, \ldots, x_n \rangle$. Then by Axiom (T_5^{ws}), E is $\dot{p}(t_1, \ldots, t_m)$.

For the propositional connectors, using Axioms (T_6^{ws}) to (T_9^{ws}), we prove that:

- if $P = P' \wedge P''$ then we can take $E = E_{P'} \cap E_{P''}$,
- if $P = P' \vee P''$ then we can take $E = E_{P'} \cup E_{P''}$,
- if $P = \bot$ then we can take $E = \varnothing$,
- if $P = P' \Rightarrow P''$ then we can take $E = E_{P'} \supset E_{P''}$,

where, by induction hypothesis, $E_{P'}$ and $E_{P''}$ are the classes characterized by, respectively, the formulas P' and P''.

For the case $P = \exists y, A$, the bi-dimensional illustration of Fig. 1a provides the intuition: the class such that there exists a value y for which A holds — and E_A is characterized — is its projection $\mathscr{P}(E_A)$. Indeed, we can derive:

$$\langle x_1, \ldots, x_n \rangle \in E \Leftrightarrow \langle x_1, \ldots, x_n \rangle \in \mathscr{P}(E_A)$$
$$\Leftrightarrow \exists y, \langle y, x_1, \ldots, x_n \rangle \in E_A \qquad \text{by axiom } (T_{10}^{\text{ws}})$$
$$\Leftrightarrow \exists y, A \ .$$

Similarly for the universal quantifier: if $P = \forall y, A$ then $E = \mathscr{C}(E_A)$.   □

Remark that while our theory T^{ws} is intuitionistic, this proof unfolds just as well in a classical setting. In this case, we can additionally discard the $\cup$, $\cap$ and $\mathscr{C}$ symbols and related axioms, and to use the De Morgan equivalences to carry out the proof.

Because this entails that T^{ws} is an extension of T^{cs}, and that T^{cs} is an extension of T, we have:

Proposition 3 (Extension). T^{ws} *is an extension of* T.

This property assures that the formalism we proposed is expressive enough, however it should not be too strong: we should not be able to prove propositions that were not provable by using axiom schemes.

Proposition 4 (Conservativity). T^{ws} *is a conservative extension of* T.

Proof. We prove this proposition by showing that for each model $\mathcal{M}$ of T there is a model of T^{ws} validating the same T-built formulas.

Let $\mathcal{M}$ be a model of T, and consider the structure $\mathcal{M}'$ defined as follows:

- $[\![X]\!]^{\mathcal{M}'}$ is the set of functions from $\mathcal{M}^\omega$ to $\mathcal{M}$, where $\mathcal{M}^\omega$ is the set of infinite lists of elements of $\mathcal{M}$.

- For any n-ary function symbol f of $\mathscr{L}$, $[\![f]\!]^{\mathcal{M}'}$, is the function:

$$a_1 \mapsto \ldots \mapsto a_n \mapsto u \mapsto [\![f]\!]^{\mathcal{M}}(a_1(u), \ldots, a_n(u))$$

  where $\forall i, a_i \in [\![X]\!]^{\mathcal{M}'}$ and $u \in \mathcal{M}^\omega$.
- The denotation of a predicate symbol p of $\mathscr{L}$ in $\mathcal{M}'$ is the function mapping elements $a_1, \ldots, a_n$ of $[\![X]\!]^{\mathcal{M}'}$ to 1 if and only if for all infinite lists u, $[\![p]\!]^{\mathcal{M}}(a_1(u), \ldots, a_n(u)) = 1$.

Note that the denotation of a formula P of $\mathscr{L}$ in $\mathcal{M}'$ is entirely known when given $[\![X]\!]^{\mathcal{M}'}$ and the denotation of the symbols of $\mathscr{L}$.

Upon such a basis, we can define denotations for the symbols of $\mathscr{L}^{\mathsf{ws}}$:

- We say that a set e of sequences of elements of $[\![X]\!]^{\mathcal{M}'}$ is definable if and only if there exists a formula P in $\mathscr{L}$ such that the sequence $a_1, \ldots, a_n$ is a member of e if and only if $[\![P]\!]^{\mathcal{M}'}_{a_1/x_1, \ldots, a_n/x_n} = 1$.
- $[\![L]\!]^{\mathcal{M}'}$ is the set of finite sequences of elements of $[\![X]\!]^{\mathcal{M}'}$, and $[\![C]\!]^{\mathcal{M}'}$ is the set of definable subsets of $[\![L]\!]^{\mathcal{M}'}$.
- $[\![1]\!]^{\mathcal{M}'}$ is the function that to each infinite list associates its first element, and $[\![S]\!]^{\mathcal{M}'}$ maps a function f to a function g such that $g(u_1, u_2, \ldots) = f(u_2, u_3, \ldots)$.
- The denotation of $t[\ell]$ is the element of $[\![X]\!]^{\mathcal{M}'}$ that associates to any list $u = u_1, u_2, \ldots$ the term $[\![t]\!]^{\mathcal{M}'}(b_1(u), \ldots, b_n(u), u_1, u_2, \ldots)$ where the finite sequence $b_1, \ldots, b_n$ is the denotation of ℓ.
- $[\![\dot{p}]\!]^{\mathcal{M}'}$ is the set of sequences defined by the corresponding predicate symbol p.
- The denotation of the rest of the symbols of $\mathscr{L}^{\mathsf{ws}}$ is self-evident: $\cap$ is set intersection, $\in$ is set membership, etc.

Remark that this definition of definability and the denotation of the symbols of $\mathscr{L}^{\mathsf{ws}}$ make the following sentence a tautology: if e is the set of sequences defined by a proposition P then for all $a_1, \ldots, a_n$,

$$[\![\langle x_1, \ldots, x_n \rangle \in E]\!]^{\mathcal{M}'}_{a_1/x_1, \ldots, a_n/x_n, e/E} = [\![P]\!]^{\mathcal{M}'}_{a_1/x_1, \ldots, a_n/x_n}. \qquad (\star)$$

Indeed, both are interpreted as 1 when $a_1, \ldots, a_n$ is a member of e.

To prove that $\mathcal{M}'$ is a model of T^{ws}, we check that it validates axioms (T^{ws}_1) to (T^{ws}_{11}). We also check that the translation we did of the original axiom schemes (e.g. the induction scheme of Example 1) is also valid in $\mathcal{M}'$. Let

$$s(P(t^1_1, \ldots, t^1_{n_1}), \ldots, P(t^p_1, \ldots, t^p_{n_p}))$$

be an axiom scheme of T, and

$$\forall E, s(\langle t^1_1, \ldots, t^1_{n_1} \rangle \in E, \ldots, \langle t^p_1, \ldots, t^p_{n_p} \rangle \in E)$$

its translation in T^{ws}. We prove that for any definable set e and sequence $a_1, \ldots, a_n$,

$$[\![s(\langle t^1_1, \ldots, t^1_{n_1} \rangle \in E, \ldots, \langle t^p_1, \ldots, t^p_{n_p} \rangle \in E)]\!]^{\mathcal{M}'}_{a_1/x_1, \ldots, a_n/x_n, e/E} = 1.$$

Assume e is defined by the proposition Q of $\mathscr{L}$. Since the particular scheme instance $s(Q(t_1^1, \ldots, t_{n_1}^1), \ldots, Q(t_1^p, \ldots, t_{n_p}^p))$ is valid in $\mathcal{M}$,

$$[\![s(Q(t_1^1, \ldots, t_{n_1}^1), \ldots, Q(t_1^p, \ldots, t_{n_p}^p))]\!]_{a_1/x_1, \ldots, a_n/x_n}^{\mathcal{M}'}$$
$$= [\![s(Q(t_1^1, \ldots, t_{n_1}^1), \ldots, Q(t_1^p, \ldots, t_{n_p}^p))]\!]_{a_1/x_1, \ldots, a_n/x_n}^{\mathcal{M}}$$
$$= 1 \, ,$$

and all we need to do is to prove:

$$[\![s(\langle t_1^1, \ldots, t_{n_1}^1 \rangle \in E, \ldots, \langle t_1^p, \ldots, t_{n_p}^p \rangle \in E)]\!]_{a_1/x_1, \ldots, a_n/x_n, e/E}^{\mathcal{M}'}$$
$$= [\![s(Q(t_1^1, \ldots, t_{n_1}^1), \ldots, Q(t_1^p, \ldots, t_{n_p}^p))]\!]_{a_1/x_1, \ldots, a_n/x_n}^{\mathcal{M}'}.$$

But this is simply a consequence of $(\star)$. Hence $\mathcal{M}'$ is a model of $\mathcal{T}^{\mathsf{ws}}$.

Finally, a formula of the language of $\mathscr{L}$ has, obviously, the same denotation in $\mathcal{M}$ and in $\mathcal{M}'$. Thus we can conclude the conservativity of $\mathcal{T}^{\mathsf{ws}}$ over $\mathcal{T}$.   □

This last proof holds in classical logic. However it can be extended to an intuitionistic proof by using Heyting algebra based models instead of classical models.

4 Applications

The result of Propositions 3 and 4 can be applied to any theory that uses axiom schemes. For instance, Zermelo's set theory accepts a conservative extension, built by applying these propositions to the traditional formulation of the theory. The same holds for the binary replacement axiom scheme of Zermelo-Fraenkel's theory, or the three schemes that result from Dowek and Miquel's encoding of set theory in a theory of pointed graphs [10]. We detail two examples: arithmetic and real analysis.

4.1 A Finite Theory of Arithmetic

In the following, we will explore Heyting's arithmetic. While our formalism applies to the original formulation of the theory, HA, we consider here a slightly more elaborate presentation of the theory where the universe of discourse is not restricted to natural numbers. This theory, called HA_N, was presented in [11] by Dowek and Werner.

Definition 5 (HA_N). *The theory HA_N of arithmetic is defined in first-order logic using the symbols 0, $Succ$, $+$, $\times$, $Pred$, $=$, $Null$ and N. It consists of the axioms:*

$$N(0) \qquad\qquad \forall x, (N(x) \Rightarrow N(Succ(x)))$$
$$Pred(0) = 0 \qquad\qquad \forall x, (Pred(Succ(x)) = x)$$
$$Null(0) \qquad\qquad \forall x, (\neg Null(S(x)))$$
$$\forall y, (0 + y = y) \qquad\qquad \forall x, \forall y, (Succ(x) + y = Succ(x + y))$$
$$\forall y, (0 \times y = 0) \qquad\qquad \forall y, (Succ(x) \times y = x \times y + y)$$

and the axioms and schemes for equality and induction:

$$\forall x, (x = x)$$
$$\forall x, \forall y, (x = y \Rightarrow P(x) \Rightarrow P(y))$$
$$P(0/x) \Rightarrow \forall y, (P(y/x) \Rightarrow P(Succ(y)/x)) \Rightarrow \forall n, (N(n) \Rightarrow P(n/x)).$$

In [11], the authors define a translation $|\cdot|$ between the languages of HA and HA_N, and prove:

Proposition 5. HA_N *is a conservative extension of* HA *in the sense that if A is a closed proposition formed in the language of* HA *then A is provable in* HA *if and only if $|A|$ is provable in* HA_N.

Finitizing the presentation of these axiom schemes is achieved by introducing lists and classes and a set of axioms that allows one to express comprehension, as per Sect. 3.

Definition 6 ($\mathsf{HA}_N^{\mathsf{ws}}$). *Define* $\mathsf{HA}_N^{\mathsf{ws}}$ *as an extension of* HA_N, *composed of the ranked signature:*

$0, 1 : X$	$S, Succ, Pred : (X)X$
$+, \times : (X, X)X$	$= : (X, X)$
$N, Null : (X)$	$\cdot[\cdot] : (X, L)X$
$\varnothing : C$	$\mathscr{P}, \mathscr{C} : (C)C$
$\cap, \cup, \supset : (C, C)C$	$\in : (L, C).$

axioms $(\mathcal{T}_1^{\mathsf{ws}})$ *to* $(\mathcal{T}_{11}^{\mathsf{ws}})$, *the axioms of arithmetic:*

$N(0)$	$\forall x, (N(x) \Rightarrow N(Succ(x)))$
$Pred(0) = 0$	$\forall x, (Pred(Succ(x)) = x)$
$Null(0)$	$\forall x, (\neg Null(Succ(x)))$
$\forall y, (0 + y = y)$	$\forall x, \forall y, (Succ(x) + y = Succ(x + y))$
$\forall y, (0 \times y = 0)$	$\forall y, (Succ(x) \times y = x \times y + y)$

the equality and induction axioms:

$$\forall x, (x = x)$$
$$\forall x, \forall y, (x = y \Rightarrow \forall A, (\langle x \rangle \in A \Rightarrow \langle y \rangle \in A))$$
$$\forall n, \forall A, (\langle 0 \rangle \in A \Rightarrow \forall y, (\langle y \rangle \in A \Rightarrow \langle Succ(y) \rangle \in A) \Rightarrow \forall n, (N(n) \Rightarrow \langle n \rangle \in A)).$$

In particular, axiom $(\mathcal{T}_4^{\mathsf{ws}})$ has four instances ($Succ$, $Pred$, $+$ and $\times$) and axiom $(\mathcal{T}_5^{\mathsf{ws}})$ three ($=$, N and $Null$), for a total of 29 axioms.

Remark that there are two sets of integer constructors in $\mathsf{HA}_N^{\mathsf{ws}}$: the native arithmetic integers, build with 0 and $Succ$; and the de Bruijn indices formed by the symbols 1 and S.

Propositions 3 and 4 applied to HA_N, composed with Proposition 5, allow us to state:

Proposition 6. $\mathsf{HA}^{\mathsf{ws}}_N$ *is as a conservative extension of* HA.

We can define a slight variant of $\mathsf{HA}^{\mathsf{ws}}_N$ by replacing the class induction axiom and Leibniz's equality axiom by the equivalences:

$$\forall x, \forall y, (x = y \Leftrightarrow \forall A, (\langle x \rangle \in A \Rightarrow \langle y \rangle \in A))$$
$$\forall n, (N(n) \Leftrightarrow \forall A, (\langle 0 \rangle \in A \Rightarrow \forall y, (\langle y \rangle \in A \Rightarrow \langle Succ(y) \rangle \in A) \Rightarrow \forall n, \langle n \rangle \in A)).$$

And we can drop the three axioms $\forall x, (x = x)$, $N(0)$ and $\forall x, (N(x) \Rightarrow N(Succ(x)))$ that have become superfluous.

Definition 7 (HA^+). *Let* HA^+ *be this shortened theory.*

Lemma 2. HA^+ *is equivalent to* $\mathsf{HA}^{\mathsf{ws}}_N$, *and counts 26 axioms instead of 29.*

4.2 Arithmetic as a Theory Modulo

A theory modulo [12] is a theory in which formulas are identified modulo a congruence, defined as a rewriting system. In particular, the theory of arithmetic has been expressed in such a framework [11], but this formalization had an infinite number of rewrite rules. The goal of this section is to show how the result of Section 3 allows a finite formulation of arithmetic modulo.

Definition 8 ($\mathsf{HA}^{\mathsf{mod}}$). *The language of the theory* $\mathsf{HA}^{\mathsf{mod}}$ *is the same as the theory* HA^+. *The congruence* $\equiv_{\mathcal{R}}$ *associated with this theory is given by the rewrite system* $\mathcal{R}$ *of* Fig. 2.

The system is split between rules dealing with substitutions, rules for arithmetic operations and rules defining relations (equality, etc.). This formalism counts a total of 26 rules, which is reduced to 22 or 23 in classical logic using the fact that all the connectors and quantifiers can be defined from 2 or 3 primitive ones.

4.3 Application to Real Fields

Real numbers or their approximation are used in exact arithmetic, programming languages, computer algebra and formal systems. The following formalization is quite common [13], and is used *e.g.* in the proof assistant Coq to implement the theory of real numbers.

Definition 9 ($\mathbb{R}^{\mathsf{cs}}$). *The language of the theory of real numbers* $\mathbb{R}^{\mathsf{cs}}$ *is formed by the symbols* 0, 1, $+$, $\times$, *the opposite* $-$, *inverse* $1/\cdot$, *the symbol* $\lceil \cdot \rceil$ *that maps real numbers to natural numbers, and the predicates* $<$ *and* $=$. *We note* $\leq$ *the disjunction of the two aforementioned predicates. The axioms follow:*

Substitutions rules

$$t[nil] \rightarrow t \qquad (\text{HA}_1^{\text{mod}})$$

$$1[t :: \ell] \rightarrow t \qquad (\text{HA}_2^{\text{mod}})$$

$$S(n)[t :: \ell] \rightarrow n[\ell] \qquad (\text{HA}_3^{\text{mod}})$$

$$Succ(t)[\ell] \rightarrow Succ(t[\ell]) \qquad (\text{HA}_4^{\text{mod}})$$

$$Pred(t)[\ell] \rightarrow Pred(t[\ell]) \qquad (\text{HA}_5^{\text{mod}})$$

$$(t_1 + t_2)[\ell] \rightarrow t_1[\ell] + t_2[\ell] \qquad (\text{HA}_6^{\text{mod}})$$

$$(t_1 \times t_2)[\ell] \rightarrow t_1[\ell] \times t_2[\ell] \qquad (\text{HA}_7^{\text{mod}})$$

Arithmetic rules

$$Pred(0) \rightarrow 0 \qquad (\text{HA}_8^{\text{mod}})$$

$$Pred(Succ(x)) \rightarrow x \qquad (\text{HA}_9^{\text{mod}})$$

$$0 + y \rightarrow y \qquad (\text{HA}_{10}^{\text{mod}})$$

$$0 \times y \rightarrow 0 \qquad (\text{HA}_{11}^{\text{mod}})$$

$$Succ(x) + y \rightarrow Succ(x + y) \qquad (\text{HA}_{12}^{\text{mod}})$$

$$Succ(x) \times y \rightarrow x \times y + y \qquad (\text{HA}_{13}^{\text{mod}})$$

Proposition rules

$$\ell \in \dot{Null}(t) \rightarrow Null(t[\ell]) \qquad (\text{HA}_{14}^{\text{mod}})$$

$$\ell \in \dot{=}(t_1, t_2) \rightarrow t_1[\ell] = t_2[\ell] \qquad (\text{HA}_{15}^{\text{mod}})$$

$$\ell \in \dot{N}(t) \rightarrow N(t[\ell]) \qquad (\text{HA}_{16}^{\text{mod}})$$

$$x = y \rightarrow \forall A, (\langle x \rangle \in A \Rightarrow \langle y \rangle \in A)) \qquad (\text{HA}_{17}^{\text{mod}})$$

$$N(n) \rightarrow \forall A, (\langle 0 \rangle \in A \Rightarrow \forall y, (\langle y \rangle \in A \Rightarrow \langle Succ(y) \rangle \in A) \Rightarrow \langle n \rangle \in A) \qquad (\text{HA}_{18}^{\text{mod}})$$

$$Null(0) \rightarrow \top \qquad (\text{HA}_{19}^{\text{mod}})$$

$$Null(Succ(x)) \rightarrow \bot \qquad (\text{HA}_{20}^{\text{mod}})$$

$$\ell \in A \cap B \rightarrow \ell \in A \wedge \ell \in B \qquad (\text{HA}_{21}^{\text{mod}})$$

$$\ell \in A \cup B \rightarrow \ell \in A \vee \ell \in B \qquad (\text{HA}_{22}^{\text{mod}})$$

$$\ell \in A \supset B \rightarrow \ell \in A \Rightarrow \ell \in B \qquad (\text{HA}_{23}^{\text{mod}})$$

$$\ell \in \varnothing \rightarrow \bot \qquad (\text{HA}_{24}^{\text{mod}})$$

$$\ell \in \mathscr{P}(A) \rightarrow \exists n, n :: \ell \in A \qquad (\text{HA}_{25}^{\text{mod}})$$

$$\ell \in \mathscr{C}(A) \rightarrow \forall n, n :: \ell \in A \ . \qquad (\text{HA}_{26}^{\text{mod}})$$

Fig. 2. Rewrite system $\mathcal{R}$ for arithmetic

$$(1 = 0) \Rightarrow \bot \tag{$\mathbb{R}_1$}$$

$$\forall x, \forall y, \qquad x + y = y + x \tag{$\mathbb{R}_2$}$$

$$\forall x, \forall y, \forall z, \qquad (x + y) + z = x + (y + z) \tag{$\mathbb{R}_3$}$$

$$\forall x, \qquad x + (-x) = 0 \tag{$\mathbb{R}_4$}$$

$$\forall x, \qquad x + 0 = x \tag{$\mathbb{R}_5$}$$

$$\forall x, \forall y, \qquad x \times y = y \times x \tag{$\mathbb{R}_6$}$$

$$\forall x, \forall y, \forall z, \qquad (x \times y) \times z = x \times (y \times z) \tag{$\mathbb{R}_7$}$$

$$\forall x, \qquad ((x = 0) \Rightarrow \bot) \Rightarrow (1/x) \times x = 1 \tag{$\mathbb{R}_8$}$$

$$\forall x, \qquad 1 \times x = x \tag{$\mathbb{R}_9$}$$

$$\forall x, \forall y, \forall z, \qquad x \times (y + z) = x \times y + x \times z \tag{$\mathbb{R}_{10}$}$$

$$\forall x, \forall y, \qquad (x < y) \vee (x = y) \vee (y < x) \tag{$\mathbb{R}_{11}$}$$

$$\forall x, \forall y, \qquad x < y \Rightarrow y < x \Rightarrow \bot \tag{$\mathbb{R}_{12}$}$$

$$\forall x, \forall y, \forall z, \qquad x < y \Rightarrow y < z \Rightarrow x < z \tag{$\mathbb{R}_{13}$}$$

$$\forall x, \forall y, \forall z, \qquad y < z \Rightarrow x + y < x + z \tag{$\mathbb{R}_{14}$}$$

$$\forall x, \forall y, \forall z, \qquad 0 < x \Rightarrow y < z \Rightarrow x \times y < x \times z \tag{$\mathbb{R}_{15}$}$$

$$\forall x, \qquad x < \lceil x \rceil \wedge (\lceil x \rceil + (-x) \leq 1). \tag{$\mathbb{R}_{16}$}$$

One way of formulating the completeness theorem of real fields is to use classes and bounds. Thus we consider classes of reals, manipulated using the nil, :: and $\in$ symbols, and the comprehension axiom scheme:

$$\forall E, \forall x_1, \ldots, x_n, (\langle x_1, \ldots, x_n \rangle \in E \Leftrightarrow P). \tag{$\mathbb{R}_{17}$}$$

The last four axioms of this theory follow: the first three define the semantics of the predicate symbols $isUB(\cdot, \cdot)$, $bounded(\cdot)$ and $isLUB(\cdot, \cdot)$; the fourth is the completeness axiom.

$$\forall E, \forall m, (\forall x, \langle x \rangle \in E \Rightarrow x \leq m) \Leftrightarrow isUB(E, m) \tag{$\mathbb{R}_{18}$}$$

$$\forall E, (\exists m, isUB(E, m)) \Leftrightarrow bounded(E)) \tag{$\mathbb{R}_{19}$}$$

$$\forall E, \forall m, (isUB(E, m) \wedge (\forall b, isUB(E, b) \Rightarrow m \leq b)) \Leftrightarrow isLUB(E, m) \tag{$\mathbb{R}_{20}$}$$

$$\forall E, (bounded(E) \Rightarrow (\exists x, \langle x \rangle \in E) \Leftrightarrow (\exists m, isLUB(E, m))). \tag{$\mathbb{R}_{21}$}$$

Following the result of Sect. 3, we give a conservative finite first-order presentation of this theory.

Definition 10 ($\mathbb{R}^{\mathsf{ws}}$). *Define $\mathbb{R}^{\mathsf{ws}}$ as an extension of $\mathbb{R}^{\mathsf{cs}}$, formed with the ranked signature:*

$$0, 1 : X \qquad\qquad S, -, 1/\cdot : (X)X$$
$$+, \times : (X, X)X \qquad\qquad <, = : (X, X)$$
$$\lceil \cdot \rceil : (X)X \qquad\qquad \cdot[\cdot] : (X, L)X$$
$$\varnothing : C \qquad\qquad \mathscr{P}, \mathscr{C} : (C)C$$
$$\cap, \cup, \supset : (C, C)C \qquad\qquad \in : (L, C).$$

axioms (T_1^{ws}) to (T_{11}^{ws}), the axioms $(\mathbb{R}_1)$ to $(\mathbb{R}_{16})$ and $(\mathbb{R}_{18})$ to $(\mathbb{R}_{21})$.

In particular, axiom (T_4^{ws}) has six instances $(S, -, 1/\cdot, +, \times$ and $\lceil \cdot \rceil)$ and axiom (T_5^{ws}) two $(=$ and $<)$, for a total of 37 axioms. Propositions 3 and 4 applied to $\mathbb{R}$ allow us to state:

Proposition 7. $\mathbb{R}^{\text{ws}}$ is as a conservative extension of $\mathbb{R}^{\text{cs}}$.

Remark that in the ranked signature of $\mathbb{R}^{\text{ws}}$, $\lceil \cdot \rceil$ has the rank $(X)X$, which is too general. This is because there is no notion of natural numbers in the formalism of Definition 10. This can be rectified by introducing the appropriate sort X', and the language and theory of natural arithmetic as in the previous section; then writing $\lceil \cdot \rceil : (X)X'$. However a more lightweight way of solving this issue is to emulate natural numbers within the sort of real numbers X. Indeed, we can define N as the smallest class of real numbers that satisfy the conjunction of the formulas:

$$0 \in N \qquad\qquad \forall x : \mathbb{R}, x \in N \Rightarrow x + 1 \in N.$$

Now the signature of $\lceil \cdot \rceil$ would still read $(X)X$, however the semantics of the operator would restrict its values to elements of the class N.

5   Related Work

The work we present in the previous sections is related to von Neumann, Bernays and Gödel's formalism for set theory (NBG) [4] that rehabilitated the notion of class used by 19th century mathematicians [14]. However it improves on a couple of points:

- Classes and the NBG approach have largely been associated to set theory [14]. We generalize it to any theory that has axiom schemes.
- By clarifying the classes/set distinction, not only is the system simplified, but we also allow a more structured hierarchy of objects. In T^{ws} the sorts X, L and C are clearly separate entities, while in NBG the sorts of objects and classes are indistinctly embedded into one another.
- Using lists and explicit substitutions to instantiate predicate free variables also greatly clarifies the argument-passing process, and allows us to bypass a couple of permutation axioms. Also, because we use native lists we are spared the tedious process of re-encoding them using sets, as is done in NBG.

– There is no easy way to orient NBG's permutation axioms to generate a well-behaved rewrite system. On the contrary, and as illustrated in Section 4.2, the rules $(\mathcal{T}_1^{\mathsf{ws}})$ to $(\mathcal{T}_{11}^{\mathsf{ws}})$ are easily orientable.

What is more, our formalism applied to Zermelo's set theory would use 15 axioms (in the classical case) vs. 14 for NBG, thus we feel it allows for a more understandable presentation of set theory without being overly bloated.

Vaillant [5] gives a presentation of set theory using explicit substitutions (in the form of the $\lambda\sigma$-calculus) to manipulate classes. The axiomatization we propose differs from it in the following ways:

– While Vaillant's paper, following NBG, was focused on set theory, our method applies to any theory with one or more axiom schemes.
– We have shown that a weak substitution calculus is strong enough to allow the complete manipulation of substitutions in this type of framework. This allows us to greatly reduce both the language's signature (neither lift, shift nor compose operators) and the number of axioms in our presentation.

If a comparison of the size of the two formalizations could reinforce the reader's opinion that our system is lighter, consider that Vaillant's intuitionistic theory uses a total of 42 axioms, while ours would only require 18 and still express full-blown Zermelo set theory. These axiom numbers comparisons might seem pointless without any experimental data, thus irrelevant in the scope of automated reasoning. However one should note that this work is destined to be implemented in proof assitants, where the low number of axioms, in particular for variable substitution, will allow for faster and less tedious computational steps.

Finally, while Megill's work on a finite formal predicate calculus [15] also uses a form reification, its approach is more invasive than the ones presented above, as the whole logical system (including inference rules) is finitized. This would make this solution hard to implement in a existing, general purpose prover.

6   Conclusion

We have exposed a generic formalization of theories with axiom schemes, which has the property of being finite. This was achieved through the use of classes and the recourse to weak explicit substitutions to cope straightforwardly with variable instantiation. This operating protocol was applied to give a finite axiomatization of the theories of arithmetic and of real fields, and a finite formalization of the former in deduction modulo.

Comparing to other methods such as von Neumann, Bernays and Gödel's or Vaillant's, it appears this way of formalizing theories using axiom schemes has links with both works. The use of a weak calculus allows us to keep a reduced number of axioms, and provides an intuitive, direct mechanism for substitutions — all of which are highly desirable properties in a proof checking environment.

Such a result easily fits into the trend set by the previous works done to formalize arithmetic and set theory into computer proof assistants [16–18, 10].

Moreover, the fact that our axioms are easily orientable is a major asset when dealing with theories in deduction modulo, which is a rapidly growing research topic showing strong potential. An implementation of the theory of real numbers into the first-order proof manager Fellowship [6] using this technique is currently underway. We will then be able to tell if this attempt at a better understanding of class theory leads to simpler real-world implementations.

References

1. von Neumann, J.: Eine Axiomatisierung der Mengenlehre. Journal für die reine und angewandte Mathematik **154** (1925) 219–240
2. Gödel, K.: The Consistency of the Axiom of Choice and of the Generalized Continuum-Hypothesis with the Axioms of Set Theory. Volume 3 of Annals of Mathematics Studies. Princeton University Press (1940)
3. Bernays, P.: Axiomatic Set Theory. Dover Publications (1958)
4. Mendelson, E.: Introduction to mathematical logic (4th ed.). Chapman & Hall (1997)
5. Vaillant, S.: A finite first-order presentation of set theory. In: TYPES. (2002) 316–330
6. Kirchner, F.: Fellowship: who needs a manual anyway? (2005)
7. Ridge, T., Margetson, J.: A mechanically verified, sound and complete theorem prover for first order logic. In: Hurd, J., Melham, T. (eds.) TPHOLs 2005. LNCS, vol. 3603, pp. 294–309. Springer, Heidelberg (2005)
8. Dowek, G., Hardin, T., Kirchner, C.: HOL-$\lambda\sigma$: An intentional first-order expression of higher-order logic. Mathematical Structures in Computer Science **11**(1) (2001) 21–45
9. Hardin, T., Maranget, L., Pagano, B.: Functional back-ends within the lambda-sigma calculus. In: ICFP, ACM Press (1996) 25–33
10. Dowek, G., Miquel, A.: Cut elimination for Zermelo's set theory. Submitted to RTA 2006 (2006)
11. Dowek, G., Werner, B.: Arithmetic as a theory modulo. In: Giesl, J. (ed.) RTA 2005. LNCS, vol. 3467, pp. 423–437. Springer, Heidelberg (2005)
12. Dowek, G., Hardin, T., Kirchner, C.: Theorem proving modulo. Journal of Automated Reasoning **31**(1) (2003) 33–72
13. Lelong-Ferrand, J., Arnaudies, J.M.: Cours de Mathématiques. Tome 2 : Analyse. Dunod (1972)
14. Bourbaki, N.: Éléments de mathématique – Théorie des ensembles. Volume 1 à 4. Masson, Paris (1968)
15. Megill, N.: A finitely axiomatized formalization of predicate calculus with equality. Notre Dame Journal of Formal Logic **36**(3) (1995) 435–453
16. Belinfante, J.: Computer proofs in Gödel's class theory with equational definitions for composite and cross. Journal of Automated Reasoning **22**(2) (1999) 311–339
17. Boyer, R., Lusk, E., McCune, W., Overbeek, R., Stickel, M., Wos, L.: Set theory in first-order logic: Clauses for Gödel's axioms. Journal of Automated Reasoning **2**(3) (1986) 287–327
18. Quaife, A.: Automated deduction in von Neumann-Bernays-Gödel set theory. Journal of Automated Reasoning **8**(1) (1992) 91–147

Coinductive Correctness of Homographic and Quadratic Algorithms for Exact Real Numbers

Milad Niqui*

Institute for Computing and Information Sciences,
Radboud University Nijmegen, The Netherlands
milad@cs.ru.nl

Abstract. In this article we present a method for formally proving the correctness of the lazy algorithms for computing homographic and quadratic transformations — of which field operations are special cases— on a representation of real numbers by coinductive streams. The algorithms work on coinductive stream of Möbius maps and form the basis of Edalat–Potts exact real arithmetic. We build upon our earlier work of formalising the homographic and quadratic algorithms in constructive type theory via general corecursion. Based on the notion of *cofixed point* equations for general corecursive definitions we prove by coinduction the correctness of the algorithms. We use the machinery of the *Coq* proof assistant for coinductive types to present the formalisation. The material in this article is fully formalised in the *Coq* proof assistant.

1 Introduction

Exact real numbers constitute one of the prime examples of infinite objects in computer science. The ubiquity and theoretical importance of real numbers as well as recent safety-critical applications of exact arithmetic makes them an important candidate for applying various approaches to formal verification. Among such approaches, one that is tailor-made for infinite objects is the *coinductive reasoning*. A careful coinductive formalisation of real numbers has double advantage: (1) it provides a certified packet of exact arithmetic; (2) it gives valuable insight into various notions of coinductive proof principles that can contribute to the area of formal verification for infinite objects.

Coinductive reasoning is dual to the usual approach of using algebraic and inductive data types both for computation and reasoning and can be studied from set theoretical [1], category theoretical [20] or type theoretical [9] point of view. In all these settings the coinductive structure of real numbers is usually expressible as *streams* which have a simple and well-understood shape. Although there are other coinductive objects (e.g. *expression trees* [24]) modelling exact real numbers, the stream approach has proven to be expressible enough for most computational purposes. In this approach a real number r is represented by a

---

* Research supported by the Netherlands Organisation for Scientific Research (NWO).

T. Altenkirch and C. McBride (Eds.): TYPES 2006, LNCS 4502, pp. 203–220, 2007.

stream of nested intervals whose intersection is the singleton $\{r\}$. This approach has always been the basis of representing real numbers, as the usual decimal representation is an instance of this representation with digits denoting interval-contracting maps. Because of this, much work has been done in the study and implementation of various algorithms for specific stream based representations such as continued fraction arithmetic [29]. On top of these, Edalat and Potts [12, 27, 11] present the general framework of representations by *linear fractional transformations* that covers all representations of real numbers that are based on streams of nested intervals. In particular Edalat–Potts *normalisation algorithm* is a unified algorithm for computing all elementary functions on real numbers.

The present work is part of the ongoing project of the author for formalising and verifying Edalat-Potts normalisation algorithm and builds upon our earlier work [24, 22]. We use constructive type theory extended with coinductive types to implement and formalise the homographic and quadratic algorithms which form the basis of the Edalat–Potts algorithm. These two algorithms suffice for equipping the stream representations of real numbers with a field structure and thus are important in themselves both from a theoretical and a practical point of view.

We use the machinery of *Coq* proof assistant for coinductive types to present the formalisation. We start by presenting the notion of 'coinductive proof' in type theory of *Coq* in Section 2. In Section 3 we present the homographic and quadratic algorithms as *Haskell*-like specifications and in Sections 3.1–3.2 we sketch the way they were formalised in [22]. In Section 4 we introduce a coinductive predicate for stream representations that will be used in the coinductive proof of correctness. In Section 5 we present the coinductive proof of correctness of homographic and quadratic algorithms. In Section 6 we conclude the article by presenting some directions for further research. Throughout the article we use a syntax loosely based on *Coq* syntax, adapted for presenting in an article. In particular we use the uncurried version of the functions when they are presented in mathematical formulae. A complete *Coq* formalisation of the material in this paper can be found in [23].

Related Work. The stream representation of exact real numbers have been recently formalised in a coinductive setting by Ciaffaglione and Di Gianantonio [7], Bertot [5], Hou [19] and Gibbons [16]. Ciaffaglione and Di Gianantonio use *Coq* proof assistant to formalise a representation of real numbers in $[-1, 1]$ as ternary streams and to prove that they form a complete Archimedean ordered field. Bertot — using *Coq* as well— formalises a ternary representation of $[0, 1]$ using *affine maps* and formalises affine operations (multiplying by scalars), addition, multiplication and infinite sums. Hou studies two coinductive representations of signed ternary digits and Cauchy sequences considered as streams, proves their equivalence using set-theoretic coinduction and defines the addition via the average function. Gibbons, as an application of his notion of metamorphism, shows how one can transform various stream representations of real numbers and use the same algorithms for different representations.

Our work, while related, is different from all of the above, in that we formalise two powerful algorithms that give us all field operations on real numbers, including division which seems to be the most difficult one in the other approaches. Furthermore due to the expressiveness of Edalat–Potts framework, the algorithms that we formalise are in principal independent of any specific representation. For presentation purposes we use a specific representation, but our correctness proof can be adapted for other representations. This is because the correctness proofs have several layers and only one aspect of them is dependent upon the metric properties of the used representation. The coinductive aspect of our work is related to the above works. For example we follow Bertot's and Hou's idea of using a coinductive predicate to link real numbers and the streams representing them [3, 5, 19]. From a type theoretic point of view the notion of *cofixed point* equations has a central rôle in our development distinguishing it further from the above works.

In other related work, in [26] real numbers have been studied as final coalgebra leading to a classification of their order structure. In [13] the unit interval is constructed as an initial algebra and the Cauchy completeness is defined by uniqueness of a morphism from a coalgebra to an algebra. The big picture that we are working on, i.e., the formalisation of Edalat–Potts normalisation algorithm is related to the works in [25, 14] that reconcile the coalgebraic structure of real numbers with algebraic operations on them.

2   Type Theoretic Coinduction

The *Coq* proof assistant [8] is an implementation of the Calculus of Inductive Constructions (**CIC**) extended with coinductive types. This is an extension of Martin-Löf intensional type theory. Coinductive types were added to *Coq* by Giménez [17]. Their implementation follows the same philosophy as that of inductive types in **CIC**, namely there is a general scheme that allows for formation of coinductive types if their constructors are given, and if these constructors satisfy a strict positivity condition. For example, the type of streams of elements of a set A can be defined using[1] its constructor Cons as

```
CoInductive Streams (A : Set) : Set :=
| Cons :  A → Streams A → Streams A.
```

After a coinductive type is defined one can introduce its inhabitants and functions into it. Such definitions are given by a *cofixed point* operator cofix . This is an operator similar to the fixed point operator for structural recursion. This operator, when given a well-typed definition that satisfies a *guardedness condition*, will introduce an inhabitant of the coinductive type. Assuming that I is a coinductive type, when defining a function $f : T \longrightarrow I$ this condition requires

---

[1] Note that, as it is the case with algebraic and inductive data types, the type Stream and its constructor Cons are defined simultaneously.

each occurrence of f in the body of f after reduction, to be an argument of one of the constructors of I. This condition is due to Giménez [17] and is based on earlier work of Coquand [9]. Finally there is a reduction (in fact expansion) rule corresponding to the cofix operator that allows the expansion of a cofixed point only when a case analysis of the cofixed point is done.

Like other syntactic criteria, the guardedness condition of *Coq* is too restrictive a requirement to allow for formalisation of all productive functions[2], and thus one has to adhere to advanced type-theoretic methods to bypass this condition. This is similar to the application of dependent inductive types for formalising general recursion using structural recursion [10, 6]. For coinductive types this has led to the method of general corecursion for filter-like functions [4]. In Section 3.1 we sketch the general corecursion technique that we used in [22] for formalising homographic and quadratic algorithm in *Coq*.

The cofix operator and its expansion rule together with the guardedness condition constitute the machinery of *Coq* for coinductive types. This means that there is no separate tool for proofs by *coinduction*. This is in contrast to the set-theoretic greatest fixed point semantics for coinduction where for each coinductive object a coinduction proof principle is present which is inherent in the monotonicity of the set operator [1]. Instead in the type theoretic approach, where proofs are objects too, we use the cofix operator to directly *build* the coinduction proof as a proof object. This means that whenever we want to prove by coinduction, our goal should be a coinductive type. If necessary, specialised coinductive predicates should be created for formalising a coinduction proof. This additional predicates are in most cases straightforward reformulation of the corresponding set-theoretic proof principle (cf. the extensional equality $\cong$ below). However, sometimes special care has to be taken to overcome the restrictions put forward by guardedness condition (cf. rep in Section 4). As a result, *Coq*'s direct approach to coinduction makes the coinduction proofs easier than their set-theoretic counterparts as long the guardedness condition does not get in the way.

For proving equalities by coinduction, coalgebraic and set-theoretic settings rely on the notion of *bisimulation* [1, 20]. In the case of streams, a bisimulation is a binary relation R satisfying the property that

$$R(\alpha, \beta) \implies \mathsf{hd}(\alpha) = \mathsf{hd}(\beta) \land R(\mathsf{tl}(\alpha), \mathsf{tl}(\beta)).$$

Then one can prove that two streams are equal if they satisfy a bisimulation relation. The coinduction proof principle thus consists of finding a suitable bisimulation.

To translate this proof principle into type theoretic coinduction note that bisimulation relation leads to the extensional equality, which in the *intensional* type theories, such as **CIC**, is quite distinct from the built-in notion of equality. In fact each extensional equality should be defined and added to the type system. On the other hand, recall that we can only prove by coinduction in *Coq* if the

---

[2] Productive functions are those functions on infinite objects that produce provably infinite output.

goal of the proof has a coinductive type. This leads us to the following definition for a coinductive extensional equality on streams which we denote by $\cong$.

```
CoInductive ≅ :  A^ω → A^ω → Prop :=
|  ≅_c : ∀(α₁ α₂: A^ω), hd α₁ = hd α₂ → tl α₁ ≅ tl α₂ → α₁ ≅ α₂ .
```

(Here A^ω denotes the type of streams of elements of A). Note that the sole constructor of $\cong$ has the shape of a bisimulation relation. The proof that this is an equivalence relation can be found in the standard library of *Coq* [8]. Moreover, Giménez shows that this is a bisimulation equivalence relation and derives the usual principle of coinduction [17, § 4.2]. In the present work we use $\cong$ relation in our coinductive correctness proofs.

3 Homographic and Quadratic Algorithms

The homographic and quadratic algorithms are similar to Gosper's algorithm [18] for addition and multiplication on continued fractions and form the basis of Edalat–Potts approach to lazy exact real arithmetic [12, 27].

Here we use a representation which is much simpler than the continued fractions and is redundant enough to ensure the productivity. There is nothing special about this representation apart from the fact that the proof of its redundancy is easy to formalise. A treatment of the general case where we abstract away both the digit set and the compact subinterval of $[-\infty, +\infty]$ can be found in [21, § 5]. Thus, for presentational purposes, we consider a fixed representation for $[-1, 1]$ containing 3 digits, each of which a Möbius map. *Möbius maps* are maps of the form

$$x \longmapsto \frac{ax + b}{cx + d} \,,$$

where $a, b, c, d \in \mathbb{Z}$. A Möbius map is *refining* if it maps the closed interval $[-1, 1]$ into itself. Möbius maps are usually denoted by the matrix of their coefficients.

For our representation, we consider the set $\mathbf{DIG} = \{\, \mathbf{L}, \mathbf{R}, \mathbf{M} \,\}$ and denote the set of streams of elements of $\mathbf{DIG}$ by $\mathbf{DIG}^\omega$. We interpret each digit by a refining Möbius map as follows[3].

$$\mathbf{L} = \begin{bmatrix} 1 & -1 \\ 1 & 3 \end{bmatrix}, \quad \mathbf{R} = \begin{bmatrix} 1 & 1 \\ -1 & 3 \end{bmatrix}, \quad \mathbf{M} = \begin{bmatrix} 1 & 0 \\ 0 & 3 \end{bmatrix}.$$

The fact that $\mathbf{DIG}^\omega$ is a representation for $[-1, 1]$ is easily derivable form the properties of the Stern–Brocot representation [21, § 5.7] (see also Section .4).

The *homographic algorithm* is the algorithm that given a Möbius map μ and a stream $\alpha \in \mathbf{DIG}^\omega$ representing r_α, outputs a stream γ that represents r_γ such that $\mu(r_\alpha) = r_\gamma$. In order to present the homographic algorithm we need an *emission condition* $\mathbf{Incl}(\mu, d)$ for a digit d and μ which checks the inclusion of intervals $\mu([-1, 1]) \subseteq d([-1, 1])$. Note that since the endpoints of these intervals

---

[3] In fact these are the conjugates (under the conjugacy map $S(x) = \frac{x-1}{x+1}$) of the Stern–Brocot representation for $[0, +\infty]$ presented in [21, § 5.7].

are rational the emission condition is a decidable predicate. This enables us to state the homographic algorithm:

homographic μ $(x: xs) :=$

$$
\begin{cases}
\mathbf{L}: \text{homographic} \ (\mathbf{L}^{-1} \circ \mu) \ (x: xs) & \text{if } \mathbf{Incl}(\mu, \mathbf{L}), \\
\mathbf{R}: \text{homographic} \ (\mathbf{R}^{-1} \circ \mu) \ (x: xs) & \text{else if } \mathbf{Incl}(\mu, \mathbf{R}), \\
\mathbf{M}: \text{homographic} \ (\mathbf{M}^{-1} \circ \mu) \ (x: xs) & \text{else if } \mathbf{Incl}(\mu, \mathbf{M}), \\
\text{homographic} \ \mu \circ x \ \ xs & \text{otherwise.}
\end{cases}
$$

Here d^{-1} and $\circ$ denote the usual matrix inversion and matrix product. The first three branches (resp. the last branch) are called *absorption steps* (resp. *emission step*). Note that due to the redundancy of the representation, the case distinction need not be mutually exclusive, but this does not affect the outcome.

The intuition behind the algorithm is that we start by considering an infinite product of Möbius maps, of which all but the first one are digits. We start pushing μ towards infinity by absorbing digits (hence obtaining a new refining Möbius map) and emitting digits whenever the emission condition holds, i.e., whenever the range of Möbius map applied to the interval $[-1, 1]$ fits inside the range of a digit.

$$
\mu \circ d_0 \circ d_1 \circ \cdots \quad \leadsto \quad d \circ (d^{-1} \circ \mu) \circ d_0 \circ d_1 \circ \cdots \quad \text{if } \mathbf{Incl}(\mu, d).
$$

For a more formal semantics for the algorithm see the semantical proof of correctness that is given in [21, § 5.6].

To compute field operations we consider the *quadratic map* which is a map

$$
\xi(x, y) := \frac{axy + bx + cy + d}{exy + fx + gy + h},
$$

with $a, b, c, d, e, f, g \in Z$ and can be denoted by its $2 \times 2 \times 2$ tensor of coefficients. A *refining* quadratic map is a quadratic map ξ such that $\xi([-1, 1], [-1, 1]) \subseteq [-1, 1]$.

The *quadratic algorithm* is an algorithm that given a quadratic map ξ and two streams $\alpha, \beta \in \mathbf{DIG}^\omega$ representing r_α and r_β, outputs a stream γ that represents r_γ such that $\xi(r_\alpha, r_\beta) = r_\gamma$. Here too we need a decidable emission condition $\mathbf{Incl}(\xi, d)$ that checks the inclusion of intervals $\xi([-1, 1], [-1, 1]) \subseteq d([-1, 1])$ for each digit d. By $d \circ \xi$ we denote the composition of a Möbius map d and a quadratic map ξ (note that the outcome is again a quadratic map). Moreover we use $\xi \bullet_1 d$ and $\xi \bullet_2 d$ to denote the two different ways of composing a quadratic map and a Möbius map by considering the Möbius map as its first (resp. second) argument. With this notation we can present the quadratic algorithm:

quadratic ξ $(x: xs)$ $(y: ys) :=$

$$
\begin{cases}
\mathbf{L}: \text{quadratic} \ (\mathbf{L}^{-1} \circ \xi) \ (x: xs) \ (y: ys) & \text{if } \mathbf{Incl}(\xi, \mathbf{L}), \\
\mathbf{R}: \text{quadratic} \ (\mathbf{R}^{-1} \circ \xi) \ (x: xs) \ (y: ys) & \text{else if } \mathbf{Incl}(\xi, \mathbf{R}), \\
\mathbf{M}: \text{quadratic} \ (\mathbf{M}^{-1} \circ \xi) \ (x: xs) \ (y: ys) & \text{else if } \mathbf{Incl}(\xi, \mathbf{M}), \\
\text{quadratic} \ (\xi \bullet_1 x \bullet_2 y) \ xs \ ys & \text{otherwise.}
\end{cases}
$$

The intuition behind this algorithm is similar to the homographic algorithm. The homographic algorithm can be used to compute the unary field operations of opposite and inverse, while the quadratic algorithm can be used for binary field operations of addition and multiplication (e.g. taking $\xi := \left[\begin{smallmatrix} 1 & 0 & 0 & 0 \\ 0 & 0 & 0 & 1 \end{smallmatrix}\right]$ it gives the multiplication).

Note that here we are concerned with the real numbers in $[-1, 1]$. Transferring the computation to the whole real line is quite straightforward (e.g. by first moving to $[0, +\infty]$ via the inverse of the conjugacy map and then adding a redundant sign bit as done in [27]).

3.1 General Corecursive Version

The algorithms of previous section do not satisfy guardedness condition. In earlier work [22] we showed how to formalise homographic and quadratic using a general corecursion technique to bypass the guardedness condition. Some of the elements of that technique are needed for the correctness proofs, therefore we briefly sketch the technique.

Let $\mathbb{M}$ (resp. $\mathbb{T}$) be objects denoting the set of Möbius maps (resp. quadratic maps)[4]. For homographic algorithm we are seeking to define a function $h: \mathbb{M} \times \mathbf{DIG}^\omega \longrightarrow \mathbf{DIG}^\omega$. But h is a partial function and might not be productive at every point. So instead of defining h we shall define a map

$$\bar{h}: \Pi(\mu: \mathbb{M})(\alpha: \mathbf{DIG}^\omega).\ P_h\ \mu\ \alpha \longrightarrow \mathbf{DIG}^\omega \tag{1}$$

where $(P_h\ \mu\ \alpha)$ is a predicate (i.e., a term of type Prop) with the intended meaning that the specification of homographic algorithm is productive when applied to μ and α. In other words it specifies the domain of the partial function h.

The definition of P_h is based on the modulus of productivity. This modulus is a recursive function $m_h: \mathbb{M} \times \mathbf{DIG}^\omega \longrightarrow \mathbf{DIG} \times \mathbb{M} \times \mathbf{DIG}^\omega$ with the intended meaning that $m_h(\mu, \alpha) = \langle d, \langle \mu', \alpha' \rangle \rangle$ if and only if

$$\mathsf{homographic}\ \mu\ \alpha \quad \leadsto \quad d: \mathsf{homographic}\ \mu'\ \alpha',$$

where '$\leadsto$' denotes multiple reduction steps after which d is output (so after output of d there are no more digits absorbed in μ'). We would like this to be a function with recursive calls on α, but this is not possible. The reason is that α has a coinductive type while in the structural recursion one needs an element of an inductive type. Hence we use the inductive domain predicate method for general recursion [10, 6] to define an inductively defined predicate $E_h(\mu, \alpha)$ with the intended meaning that μ and α are in the domain of m_h which in turn means that the homographic algorithm should emit at least one digit when applied to μ and α.

The definition of above terms are given in Appendix and the *Coq* version can be found in [23]. Finally we can formalise the homographic algorithm as the

---

[4] They can be considered as $\mathbb{Z}^4$ and $\mathbb{Z}^8$.

function $\bar{h}$ in (1) that accommodates the proof of its own productivity as one of its arguments. Here CoFixpoint denotes that we are using the cofix operator to build a term (π_{ij} denotes the projections of j-tuples, for $P_h\_E_h$ and $\overline{m}_h\_P_h$ see Appendix).

CoFixpoint $\bar{h}$ $(\mu\colon \mathbb{M})$ $(\alpha\colon \text{DIG}^\omega)$ $(p\colon P_h\ \mu\ \alpha)$ $\colon \text{DIG}^\omega$ $:=$
    Cons $\pi_{13}(\overline{m}_h\ \mu\ \alpha\ (P_h\_E_h\ \mu\ \alpha\ p))$
        $(\bar{h}\ \pi_{23}(\overline{m}_h\ \mu\ \alpha\ (P_h\_E_h\ \mu\ \alpha\ p))$
           $\pi_{33}(\overline{m}_h\ \mu\ \alpha\ (P_h\_E_h\ \mu\ \alpha\ p))$
           $(\overline{m}_h\_P_h\ \mu\ \alpha\ (P_h\_E_h\ \mu\ \alpha\ p)\ p))$.

For the quadratic algorithm a similar procedure is followed and results in the following formalised version. These definition are explained in [22].

CoFixpoint $\bar{q}$ $(\xi\colon \mathbb{T})$ $(\alpha\ \beta\colon \text{DIG}^\omega)$ $(p\colon P_q\ \xi\ \alpha\ \beta)$ $\colon \text{DIG}^\omega$ $:=$
    Cons $\pi_{14}(\overline{m}_q\ \xi\ \alpha\ \beta\ (P_q\_E_q\ \xi\ \alpha\ \beta\ p))$
        $(\bar{q}\ \pi_{24}(\overline{m}_q\ \xi\ \alpha\ \beta\ (P_q\_E_q\ \xi\ \alpha\ \beta\ p))$
           $\pi_{34}(\overline{m}_q\ \xi\ \alpha\ \beta\ (P_q\_E_q\ \xi\ \alpha\ \beta\ p))$
           $\pi_{44}(\overline{m}_q\ \xi\ \alpha\ \beta\ (P_q\_E_q\ \xi\ \alpha\ \beta\ p))$
           $(\overline{m}_q\_P_q\ \xi\ \alpha\ \beta\ (P_q\_E_q\ \xi\ \alpha\ \beta\ p)\ p))$.

3.2 Cofixed Point Equations

In the inductive domain predicate approach to general recursion [6], to formalise a general recursive function f first a function $\bar{f}$ is defined and then it is proven that it satisfies the fixed point equation of f. This fixed point equations are stated using the inductive (*Leibniz*) equality of *Coq*.

In our general corecursive technique, we follow a similar path but we use the coinductive equality $\cong$ of Section 2. I.e., we prove that $\bar{h}$ satisfies the *Haskell*-like specification of previous section for homographic [5] modulo $\cong$. Here is the lemma stating the equation corresponding to the first branch of homographic:

Lemma $\bar{h}_L\colon \forall(\mu\colon \mathbb{M})\,(\alpha\colon \text{DIG}^\omega)\,(p\colon P_h\ \mu\ \alpha),\ \textbf{Incl}(\mu, \text{L}) \to$
    $\forall(p'\colon P_h\ (\text{L}^{-1}\circ\mu)\ \alpha),\ \bar{h}\ \mu\ \alpha\ p \cong \text{Cons L}\ (\bar{h}\ (\text{L}^{-1}\circ\mu)\ \alpha\ p')$.

We call this a *cofixed point equation* of the homographic algorithm, a notion that is applicable to any function on streams[6]. The lemma can be proven using type theoretic coinduction and its proof is based on a result (also provable by coinduction) which we call the *extensional proof irrelevance* of $\bar{h}$ (See [22]). For the proof of the above lemma as well as remaining equations see [23, 22].

---

[5] Of course the additional proof obligation $p\colon (P_h\ \mu\ \alpha)$ prevents us from obtaining the exact specification; but since this proof obligation is a term living in the Prop universe of *Coq*, it has no computational content.

[6] It is possible to define this notion for any coinductive type that corresponds to polynomial functors. One only needs to define a coinductive equivalence relation for each such type. But this is out of the scope of the present work.

4   Representation

To prove that the algorithms are correct, first we should prove that every stream in $\mathbf{DIG}^\omega$ represents a real number in $[-1, 1]$. This means that there exists a total[7] map ρ from $\mathbf{DIG}^\omega$ to $[-1, 1]$ such that for all $f_0 f_1 \cdots \in \mathbf{DIG}^\omega$ we have

$$\{ \rho(f_0 f_1 \ldots) \} = \bigcap_{i=1}^{\infty} f_0 \circ \ldots f_i([-1, 1]) \ .$$

This can be proven by coinduction, but one needs to define a coinductive predicate that captures the existence of ρ. This leads to the following definition for a binary predicate $\mathsf{rep} \colon \mathbf{DIG}^\omega \times [-1, 1] \longrightarrow \mathsf{Prop}$ with the intended meaning that $\mathsf{rep}(\alpha, r)$ holds if $\rho(\alpha) = \{r\}$.

```
CoInductive rep  :  DIGᵂ → ℝ → Prop  :=
 | rep_L  :  ∀ (α β : DIGᵂ) (r : ℝ),  −1≤r≤1→
                    rep α r  →  β ≅ Cons L α  →  rep β (r − 1)/(r + 3)
 | rep_R  :  ∀ (α β : DIGᵂ) (r : ℝ),  −1≤r≤1→
                    rep α r  →  β ≅ Cons R α  →  rep β (r + 1)/(−r + 3)
 | rep_M  :  ∀ (α β : DIGᵂ) (r : ℝ),  −1≤r≤1→
                    rep α r  →  β ≅ Cons M α  →  rep β r/3.
```

The constructors of this coinductive predicate spell out the effect of each digit and as such depend on the choice of the digits. However, they can be easily adapted or generalised for working with other digit sets. The predicate is similar to the predicate $\mathsf{represents}$ of Bertot [3, 5] and (to a lesser extent to the predicate $\sim'$ of Hou [19]) but has a notable difference: the clause $\beta \cong \mathsf{Cons}\ d\ \alpha$ that is added to each constructor. The purpose of this clause is to facilitate the use of cofixed point equations. Without this clause rep would still have the intended metric semantics in terms of ρ, but it would not be usable in the coinductive proof of correctness of next section. The reason is due to the guardedness condition of *Coq*: even without the $\cong$ clause in the constructors of rep we could prove by coinduction that

$$X : \forall \alpha \beta r, \ \mathsf{rep}(\alpha, r) \to \alpha \cong \beta \to \mathsf{rep}(\beta, r), \tag{2}$$

which is the basic property of rep that *should have been* enough for the correctness proof. But upon rewriting (2) in the course of coinductive proof Δ of correctness we would violate the guardedness condition. This would happen because we would have supplied a recursive occurrence of the coinductive proof Δ which is guarded as X (rep_d Δ) (where rep_d is a constructor of rep). This is not allowed because X is itself a cofixed point whose expansion takes the coinductive proof Δ as an argument in its recursive occurrence in a way that the

---

[7] In fact $\mathbf{DIG}^\omega$ is a *representation* which means ρ is also surjective. This is easily provable [21, § 5] but is not needed in the correctness proofs for our algorithms.

guardedness condition is rejected. Using cofixed point equations instead of (2) we will not land in this situation. Thus we have decided to add the $\cong$ clause which will eliminate the need to (2) and instead use the cofixed point equations in the correctness proofs.

Note that (2) is still a correct statement and can be used in other situations. In fact we can use it to prove that the inverse of constructors of rep hold, e.g.:

$$\forall \alpha r, \quad \mathsf{rep}(\mathsf{Cons}\ \mathbf{L}\ \alpha, r) \rightarrow \mathsf{rep}(\alpha, 3r + 1).$$

The inversion lemma in turn are used in proving the link between a stream and its future tails. Let α_n (resp. $\alpha|_n$) denote the $n + 1$-st digit of α (resp. the stream obtained by dropping the first n digits of α). Then we can prove by induction on n and using the inversion lemmas that

$$\forall \alpha r, \quad \mathsf{rep}(\alpha, r) \rightarrow \mathsf{rep}(\alpha|_n, \alpha_{n-1}^{-1} \circ \ldots \alpha_0^{-1}(r)). \tag{3}$$

To show that rep satisfies its metric property we have to define a function $[\![\_]\!]$ that evaluates a stream and obtains the real number which is represented by it (cf. `real_value` in [5]). In fact this function calculates the limit of converging sequence of shrinking intervals that is obtained by successive application of the digits starting from the base interval. To be able to define $[\![\_]\!]$ we should show this converging property. This proof is directly dependent on the metric properties of the specific digit set that we have chosen. Setting $\mathsf{diam}([a, b]) = b - a$ we have to show that

$$\max\{\,\mathsf{diam}\,(d_0 \circ d_1 \circ \ldots d_{k-1}([-1, 1]))\,|\,d_i \in \mathbf{DIG}\,\} \leq \frac{2}{k+1}.$$

This is provable by induction on k [21, Corollary 5.7.9] and it entails that the diameters of the intervals form a Cauchy sequence, and so do their endpoints. Hence if we define $l_k(\alpha)$ (resp. $u_k(\alpha)$) to be the lower bound (resp. upper bound) of the interval $\alpha_0 \circ \alpha_1 \circ \ldots \alpha_{k-1}([-1, 1])$ we can define[8]

$$[\![\alpha]\!] = \lim_{i \to \infty} l_i(\alpha).$$

We can prove (by induction on k) that

$$\forall \alpha k r, \mathsf{rep}(\alpha, r) \rightarrow r \in [l_k(\alpha), u_k(\alpha)];$$

and hence

$$\forall \alpha k r, \mathsf{rep}(\alpha, r) \rightarrow r \in [-1, 1]. \tag{4}$$

Furthermore using the properties of limit we can prove

$$[\![\mathsf{Cons}\ d\ \alpha]\!] = d([\![\alpha]\!]), \tag{5}$$

$$[\![\alpha]\!] \in [-1, 1]. \tag{6}$$

---

[8] Note that we could have equivalently used the upper bounds.

Thus we can prove the following by an easy coinduction on the structure of rep.

$$\forall \alpha, \mathsf{rep}(\alpha, [\![\alpha]\!]). \tag{7}$$

Finally we can prove the main properties of rep

$$\forall \alpha r, \ [\![\alpha]\!] = r \to \mathsf{rep}(\alpha, r); \tag{8}$$

$$\forall \alpha r, \ \mathsf{rep}(\alpha, r) \to [\![\alpha]\!] = r. \tag{9}$$

The proof of (8) follows from (7) and (6) while (9) needs in addition some properties of the limit.

Hence we have shown that rep satisfies its intended metric property with respect to the map ρ defined in the beginning of this section. We conclude the section by pointing out what rep does *not* entail. The most important aspect is that our representation $\mathbf{DIG}^\omega$ is an admissible representation, i.e., it contains enough redundancy so that the usual computable functions are computable with respect to this representation [21, Corollary 5.7.10]. However, the $\cong$ equality does not know anything about this redundancy and it distinguishes the two streams representing the same real number. Therefore for two different representations α_1, α_2 of a real number r, there are two different proofs $\mathsf{rep}(\alpha_1, r)$ and $\mathsf{rep}(\alpha_2, r)$ that do not have any syntactic relation with each other. This, of course, is not an issue for our application of rep in the correctness proofs of the next section.

5   Coinductive Correctness

As it is the case with all algorithms, 'to prove the correctness' of the homographic and quadratic algorithms can point to different concepts:

(i) To prove that the algorithms satisfy their *Haskell*-like specification.
(ii) To prove that the algorithms turn the set $\mathbf{DIG}^\omega$ to a field and behave as Möbius and quadratic maps on this field.
(iii) To prove that the algorithms correspond to Möbius and quadratic maps on $[-1, 1]$.

Concept (i) tantamounts to proving the cofixed point equations and was carried out in [22]. Concept (ii) requires that we focus on the field operations (via specific tensors for $+, \times$) and prove that they satisfy the field axioms. Concept (iii) requires the use of a model of real numbers and indicates that we will project the algorithm to functions on this standard model. It is clear that (iii) is much less work, as we only have to prove the correspondence of the algorithms once and can reduce every question on $\mathbf{DIG}^\omega$ to a question on the standard model of $\mathbb{R}$. This way we do not have to prove one-by-one all the field axioms for $\mathbf{DIG}^\omega$.

We already used the standard model $\mathbf{R}$ in the definition of rep, and we are going to prove the correctness in the sense of (iii). We base our correctness proofs on the coinductive predicate rep and we prove that for functions $\bar{h}$ and $\bar{q}$ of Section 3.1 we have

$$\forall \mu \alpha p r, \ \mathsf{rep}(\alpha, r) \to \mathsf{rep}(\bar{h}(\mu, \alpha, p), \mu(r)); \tag{10}$$

$$\forall \xi \alpha \beta p r_1 r_2, \ \mathsf{rep}(\alpha, r_1) \to \mathsf{rep}(\alpha, r_2) \to \mathsf{rep}(\bar{q}(\xi, \alpha, \beta, p), \xi(r_1, r_2)). \tag{11}$$

It is clear that once we have proven these, applying the Properties (8)–(9) of rep, we can derive

$$\forall \mu \alpha p r, \quad [\![\alpha]\!] = r \rightarrow [\![\bar{h}(\mu, \alpha, p)]\!] = \mu(r);$$
$$\forall \xi \alpha \beta p r_1 r_2, \quad [\![\alpha]\!] = r_1 \rightarrow [\![\beta]\!] = r_2 \rightarrow [\![\bar{q}(\xi, \alpha, \beta, p)]\!] = \xi(r_1, r_2).$$

In the remainder of this paper we show how to prove (10) and (11).

5.1 Homographic Algorithm

We want to prove (10), that means in addition to μ, α and r we are also given a proof obligation $p \colon P_h \mu \alpha$ that ensures the productivity of $\bar{h}(\mu)$ at α. We use p to obtain some auxilliary tools that we will need in the proof of (10). We will also use the terms that were used in the general corecursion technique (see Section 3.1 and Appendix). First we need a function $\bar{\delta}_h \colon \Pi(\mu \colon \mathbf{M})(\alpha \colon \mathbf{DIG}^\omega).E_h(\mu, \alpha) \longrightarrow \mathbb{N}$ that counts the number of absorption steps before the first (eventually) coming emission step. Note the resemebelence with the definition of the modulus of productivity $\overline{m}_h$.

```
Fixpoint δ̄ₕ (μ: M) (α: DIG^ω) (t: Eₕ μ α){struct t}: N:=
match Incl_dec(μ,L) with
| left _⇒ 0
| right tₗ⇒
  match Incl_dec(μ,R) with
  | left _⇒ 0
  | right tᵣ⇒
    match Incl_dec(μ,M) with
    | left _⇒ 0
    | right tₘ⇒1+δ̄ₕ (μ∘(hd α)) (tl α) (Eₕₐᵦ_inv μ α tₗ tᵣ tₘ t)
    end
  end
end.
```

Here `Fixpoint` (resp. `struct`) are *Coq* keywords to denote a recursive definition (resp. recursive argument of structural recursive calls). First of all note that we use $\oplus \colon \mathsf{Prop} \times \mathsf{Prop} \longrightarrow \mathsf{Set}$ to transfer the proposition **Incl** to a boolean sum — with `left` and `right` its coprojections— on which we can pattern match[9]. Second, note that the function is defined using general recursion, otherwise its recursive argument would have a coinductive type which is not allowed in structural recursion. We will also need to prove the proof irrelevance of $\bar{\delta}_h$ (i.e., its value is independent of t), its fixed point equation and its relationship with $\overline{m}_h$. We state the latter:

---

[9] This is a consequence of the characteristic property of **CIC** that does not allow one to obtain computationally informative elements (e.g. natural numbers) by case analysis on propositions (e.g. **Incl**).

Lemma $\overline{\delta}_h\_\overline{m}_h \colon \forall(\mu\colon \mathbb{M})\,(\alpha\colon \mathrm{DIG}^\omega)\,(t\colon E_h\ \mu\ \alpha)\ (n\colon \mathbb{N})$,
$\overline{\delta}_h\ \mu\ \alpha\ t = n\ \rightarrow\ \exists d\colon \mathrm{DIG},\ \overline{m}_h\ \mu\ \alpha\ t = \langle d, \langle d^{-1}{\circ}\mu{\circ}\alpha_0{\circ}\ldots{\circ}\alpha_{n-1}, \alpha|_n\rangle\rangle$.

Then we need to prove that if the value of $\overline{\delta}$ is n then after n steps emission will occur, i.e., the emission condition will be satisfied:

Lemma $\overline{\delta}_h\_\mathbf{Incl}_{\mathrm{dec}}\colon \forall(\mu\colon \mathbb{M})\,(\alpha\colon \mathrm{DIG}^\omega)\,(t\colon E_h\ \mu\ \alpha)\ (n\colon \mathbb{N})$,
$\overline{\delta}_h\ \mu\ \alpha\ t = n\ \rightarrow\ \mathbf{Incl}(\mu{\circ}\alpha_0{\circ}\ldots{\circ}\alpha_{n-1}, \mathtt{L}) \wedge\ \pi_{13}(\overline{m}_h\ \mu\ \alpha\ t) = \mathtt{L}\ \oplus$
$\mathbf{Incl}(\mu{\circ}\alpha_0{\circ}\ldots{\circ}\alpha_{n-1}, \mathtt{R}) \wedge\ \pi_{13}(\overline{m}_h\ \mu\ \alpha\ t) = \mathtt{R}\ \oplus$
$\mathbf{Incl}(\mu{\circ}\alpha_0{\circ}\ldots{\circ}\alpha_{n-1}, \mathtt{M}) \wedge\ \pi_{13}(\overline{m}_h\ \mu\ \alpha\ t) = \mathtt{M}.$

Both lemmas above are proven by induction on n. All this machinery is used in proving the following lemma which describes the observable (hence the use of $\cong$) situation of the homographic algorithm at the moment of emission. It explicitly mentions the new input Möbius map passed to the homographic algorithm, the emission condition and the necessary proof obligation.

Lemma $\overline{h}_{em}\colon \forall(\mu\colon \mathbb{M})\,(\alpha\colon \mathrm{DIG}^\omega)\,(p\colon P_h\ \mu\ \alpha),\ \exists n\colon \mathbb{N}\ \exists d\colon \mathrm{DIG},$
$P_h\ (d^{-1}{\circ}\mu{\circ}\alpha_0{\circ}\ldots{\circ}\alpha_{n-1})\ \alpha|_n\ \wedge$
$\mathbf{Incl}(\mu{\circ}\alpha_0{\circ}\ldots{\circ}\alpha_{n-1}, d)\ \wedge$
$\forall\ p',\ \overline{h}\ \mu\ \alpha\ p \cong \mathtt{Cons}\ d\ (\overline{h}\ (d^{-1}{\circ}\mu{\circ}\alpha_0{\circ}\ldots{\circ}\alpha_{n-1})\ \alpha|_n\ p').$

Finally we need a property of refining Möbius maps whose proof is immediate [11], but we state it explicitly to highlight its use.

Lemma 1. *If μ is refining and $\mathbf{Incl}(\mu, d)$ then $d^{-1} \circ \mu$ is refining.*

Now we have the necessary tools for proving the correctness of the homographic algorithm:

Theorem 1. *Let $\mu \in \mathbb{M}$, $\alpha \in \boldsymbol{DIG}^\omega$, $r \in \mathbb{R}$ and $p\colon P_h\ \mu\ \alpha$. If $rep(\alpha, r)$ then $rep(\overline{h}(\mu, \alpha, p), \mu(r))$.*

Proof. By Lemma $\overline{h}_{em}$ above there exist n, d and p' such that

$$\mathbf{Incl}(\mu{\circ}\alpha_0{\circ}\ldots{\circ}\alpha_{n-1}, d), \tag{12}$$

$$\overline{h}(\mu, \alpha, p) \cong \mathtt{Cons}\ d\ \overline{h}(d^{-1}{\circ}\mu{\circ}\alpha_0{\circ}\ldots{\circ}\alpha_{n-1}, \alpha|_n, p'). \tag{13}$$

By Property (3) of rep (Section 4) we have

$$rep(\alpha|_n, \alpha_{n-1}^{-1} \circ \ldots \circ \alpha_0^{-1}(r)).$$

Whence by coinduction applied to

$$\mu_0 := d^{-1}{\circ}\mu{\circ}\alpha_0{\circ}\ldots{\circ}\alpha_{n-1},$$
$$\alpha_0 := \alpha|_n \qquad p_0 := p',$$
$$r_0 := \alpha_{n-1}^{-1}{\circ}\ldots{\circ}\alpha_0^{-1}(r);$$

we obtain

$$\mathsf{rep}\big(\overline{h}(d^{-1}\!\circ\!\mu\circ\alpha_0\circ\ldots\circ\alpha_{n-1},\alpha|_n,p'),d^{-1}\!\circ\!\mu\circ\alpha_0\circ\ldots\circ\alpha_{n-1}\circ\alpha_{n-1}^{-1}\circ\ldots\alpha_0^{-1}(r)\big) \tag{14}$$

Let $r_1 := \mu_0\circ\alpha_{n-1}^{-1}\circ\ldots\alpha_0^{-1}(r)$. According to Lemma 1, from (12) it follows that μ_0 is refining. Note that by Properties (4) and (3) of rep we have

$$\alpha_{n-1}^{-1}\circ\ldots\alpha_0^{-1}(r) \in [-1,1];$$

and thus according to the refining property $r_1 \in [-1,1]$.

From here and (14), according to the statement of the constructor rep_d of rep applied to r_1 and

$$\alpha_1 := \overline{h}(d^{-1}\!\circ\!\mu\circ\alpha_0\circ\ldots\circ\alpha_{n-1},\alpha|_n,p'),$$
$$\beta_1 := \overline{h}(\mu,\alpha,p);$$

we obtain

$$\mathsf{rep}\big(\overline{h}(\mu,\alpha,p),d(d^{-1}\!\circ\!\mu\circ\alpha_0\circ\ldots\circ\alpha_{n-1}\circ\alpha_{n-1}^{-1}\circ\ldots\alpha_0^{-1}(r))\big); \tag{15}$$

(note that (13) satisfies the $\cong$ clause in rep_d).

Finally, by simple rewriting and cancelling out the inverse matrices in (15) we obtain the conclusion:

$$\mathsf{rep}\big(\overline{h}(\mu,\alpha,p),d(d^{-1}\!\circ\!\mu\circ\alpha_0\circ\ldots\circ\alpha_{n-1}\circ\alpha_{n-1}^{-1}\circ\ldots\alpha_0^{-1}(r))\big)$$
$$= \mathsf{rep}(\overline{h}(\mu,\alpha,p),d\circ d^{-1}\!\circ\!\mu\circ\alpha_0\circ\ldots\circ\alpha_{n-1}\circ\alpha_{n-1}^{-1}\circ\ldots\alpha_0^{-1}(r))$$
$$= \mathsf{rep}(\overline{h}(\mu,\alpha,p),\mu(r)).$$

$$\square$$

Analysing the above process we observe that there are two kind of proofs that constitute two different aspect of the algorithm: a metric (or topological) layer which is dependent on the representation, and a type theoretic layer which is based on the coinductive structure of rep and $\cong$. Currently, a complete *Coq* formalisation of all the above definitions and lemmas for the chosen digit set is available in [23]. In future work we plan to make this formalisation more modular by formalising a theory of admissible digit sets. However, even after developing such a theory the type theoretic layer of the proofs (including the above coinductive proof) need not be changed as it is independent of the chosen digit set.

5.2 Quadratic Algorithm

The procedure for the correctness of the quadratic algorithm is quite similar to the case of homographic algorithm, only the proof itself is more meticulous. First we define a function $\overline{\delta}_q\colon \Pi(\xi\colon \mathbb{T})(\alpha,\beta\colon \mathbf{DIG}^\omega).E_q(\xi,\alpha,\beta) \longrightarrow \mathbb{N}$ that outputs the number of steps to the next emission step. We can prove the properties similar to those of $\overline{\delta}_h$.

The main auxiliary lemma in this case is the following.

Lemma $\overline{q}_{em}$: $\forall (\xi\colon \mathbb{T})\,(\alpha\ \beta\colon \mathtt{DIG}^{\omega})\,(p\colon P_q\ \xi\ \alpha\ \beta)$, $\exists n\colon \mathbb{N}\ \exists d\colon \mathtt{DIG}$,
$P_1\ (d^{-1}\circ\xi\langle\alpha_0\circ\ldots\circ\alpha_{n-1}, \beta_0\circ\ldots\circ\beta_{n-1}\rangle)\ \alpha|_n\ \beta|_n\ \wedge$
$\mathbf{Incl}(\xi\langle\alpha_0\circ\ldots\circ\alpha_{n-1}, \beta_0\circ\ldots\circ\beta_{n-1}\rangle, d)\ \wedge$
$\forall\ p',\ \overline{q}\ \xi\ \alpha\ \beta\ p\ \cong$
$\qquad\qquad \mathbf{Cons}\ d\ (\overline{q}\ (d^{-1}\circ\xi\langle\alpha_0\circ\ldots\circ\alpha_{n-1}, \beta_0\circ\ldots\circ\beta_{n-1}\rangle)\ \alpha|_n\ \beta|_n\ p')$.

Note that $\xi\langle\alpha_0\circ\ldots\circ\alpha_{n-1}, \beta_0\circ\ldots\circ\beta_{n-1}\rangle$ denotes the new tensor after n absorption steps, i.e., after n applications of $\bullet_1$ and $\bullet_2$.

We also need a result on refining tensors which is immediately provable from the definition of refining and **Incl**.

Lemma 2. *If ξ is refining and $\mathbf{Incl}(\xi, d)$ then $d^{-1}\circ\xi$ is a refining tensor.*

From these we can prove the correctness of quadratic algorithm. In particular we do *not* need any additional property of rep apart from those that were used for the homographic algorithm. The proof is quite similar to the proof of Theorem 1 and is formalised in *Coq* [23], and so we do not mention the proof here.

Theorem 2. *Let $\xi \in \mathbb{T}$, $\alpha, \beta \in \mathbf{DIG}^{\omega}$, $r_1, r_2 \in \mathbb{R}$ and $p\colon P_a\ \mu\ \alpha\ \beta$. If rep$(\alpha, r_1)$ and rep(β, r_2) then rep$(\overline{q}(\xi, \alpha, \beta, p), \xi(r_1, r_2))$.*

Note that the proofs assume the existence of a productivity predicate P_h (resp. P_q). Deriving this property depends on the specific metric properties of each tensor and Möbius map. This is natural because homographic and quadratic algorithms are partial functions. It is known that the algorithms are productive for refining Möbius maps and tensors [27]. Our future work is to derive this result for our inductive productivity predicates P_h and P_q.

6   Conclusions and Further Work

We have shown the correctness of homographic and quadratic algorithms on a stream representation of real numbers in $[-1, 1]$. Following the general setup of [21, § 5] the method is easily extensible to any admissible digit set for any compact proper subinterval of the extended real numbers $[-\infty, +\infty]$. Our correctness proofs use an inductive productivity predicate and a coinductive predicate rep that relates $\mathbf{DIG}^{\omega}$ and $[-1, 1]$. We have built on top of the earlier work [22] on using the coinductive machinery of *Coq* proof assistant to formalise functions on infinite objects and coinductive proofs. In particular we base our treatment of coinductive functions on their cofixed point equations. These exploit the inherent infinite nature of streams by adhering to $\cong$ which is a bisimulation relation and is more suitable than the the inductive (Leibniz) equality. The coinductive arguments themselves are independent of *Coq* and can be formalised in any proof assistant that accommodates coinductive types.

There are two direction perceivable for future work. The more immediate future work would be to continue the *Coq* formalisation of the algorithms, by developing a fully modular framework that axiomatises the properties of representations and refining maps that are needed for the formalisation. Each specific representation would then be portable into our formalisation if a suitable interface is satisfied. This will pave the way for applying our formalisation to more efficient representations such as the one used by Bertot [5] or Edalat–Potts [12]. The big picture would be to continue working on the formalisation of the Edalat–Potts normalisation algorithm. This would require a further development of our general corecursion technique to deal with nested corecursive functions and would require a notion similar to induction–recursion. Recent work by Setzer on combining induction–recursion and general recursion seems to open new possibilities for our work in this directions [28].

Acknowledgements. The author wishes to thank the anonymous referees for their useful comments.

References

[1] Barwise, J., Moss, L.: Vicious Circles: On the Mathematics of Non-Wellfounded Phenomena. CSLI Publications, Stanford (1996)

[2] Beckmann, A., Berger, U., Löwe, B., Tucker, J.V. (eds.): CiE 2006. LNCS, vol. 3988. Springer, Heidelberg (2006)

[3] Bertot, Y.: CoInduction in Coq. In: Lecture Notes of TYPES Summer School 2005, August 1526 2005, Göteborg, Sweden. vol II (2005) http://www.cs.chalmers.se/Cs/Research/Logic/TypesSS05/Extra/lectnotes_vol2.pdf [cited 31 January 2007]

[4] Bertot, Y.: Filters on coinductive streams, an application to Eratosthenes sieve. In: Urzyczyn, P. (ed.) TLCA 2005. LNCS, vol. 3461, pp. 102–115. Springer, Heidelberg (2005)

[5] Bertot, Y.: Affine functions and series with co-inductive real numbers. Math. Structures Comput. Sci. (to appear)

[6] Bove, A., Capretta, V.: Nested general recursion and partiality in type theory. In: Boulton, R.J., Jackson, P.B. (eds.) TPHOLs 2001. LNCS, vol. 2152, pp. 121–135. Springer, Heidelberg (2001)

[7] Ciaffaglione, A., Di Gianantonio, P.: A certified, corecursive implementation of exact real numbers. Theoret. Comput. Sci. 351(1), 39–51 (2006)

[8] The Coq Development Team. The Coq Proof Assistant Reference Manual, Version 8.0. LogiCal Project (April 2004) http://coq.inria.fr/doc/main.html [cited 31 January 2007)

[9] Coquand, T.: Infinite objects in type theory. In: Barendregt, H., Nipkow, T. (eds.) TYPES 1993. LNCS, vol. 806, pp. 62–78. Springer, Heidelberg (1994)

[10] Dubois, C., Donzeau-Gouge, V.V.: A step towards the mechanization of partial functions: domains as inductive predicates. In: Kerber, M. (ed.) Proc. Workshop on Mechanization of Partial Functions, July 5, 1998, Lindau, Germany, pp. 53–62 (1998), available at ftp://ftp.cs.bham.ac.uk/pub/authors/M.Kerber/98-CADE-WS/dubois-donzeau.ps.gz [cited 31 January 2007]

[11] Edalat, A., Heckmann, R.: Computing with Real Numbers. In: Barthe, G., Dybjer, P., Pinto, L., Saraiva, J. (eds.) APPSEM 2000. LNCS, vol. 2395, pp. 193–267. Springer, Heidelberg (2002)

[12] Edalat, A., Potts, P.J.: A new representation for exact real numbers. In: Brookes, S., Mislove, M. (eds.) Mathematical Foundations of Programming Semantics, 13th Annual Conference (MFPS XIII), Carnegie Mellon University, Pittsburgh, PA, USA, March 23–26, 1997. Electron. Notes Theor. Comput. Sci, vol. 6, pp. 23–26. Elsevier, Amsterdam (1997)

[13] Escardó, M.H., Simpson, A.K.: A universal characterization of the closed Euclidean interval. In: Proceedings of the 16th Annual IEEE Symposium on Logic in Computer Science, pp. 115–128. IEEE Computer Society Press, Los Alamitos (2001)

[14] Ghani, N., Hancock, P., Pattinson, D.: Continuous functions on final coalgebras. In: Ghani and Power [15]

[15] Ghani, N., Power, J. (eds.): Proceedings of 8th International Workshop on Coalgebraic Methods in Computer Science, CMCS 2006. Electron. Notes Theor. Comput. Sci, vol. 164(1). Elsevier, Amsterdam (2006)

[16] Gibbons, J.: Metamorphisms: Streaming representation-changers. Sci. Comput. Program. (to appear)

[17] Giménez, E.: Un Calcul de Constructions Infinies et son Application a la Verification des Systemes Communicants. PhD thesis PhD 96-11, Laboratoire de l'Informatique du Parallélisme, Ecole Normale Supérieure de Lyon (December 1996)

[18] Gosper, R.W.: HAKMEM, Item 101 B. February 29, 1972. MIT AI Laboratory Memo No. 239, http://www.inwap.com/pdp10/hbaker/hakmem/cf.html#item101b [cited 31 January 2007]

[19] Hou, T.: Coinductive proofs for basic real computation. In: Beckmann, et al. pp. 221–230 [2]

[20] Jacobs, B., Rutten, J.: A tutorial on (co)algebras and (co)induction. Bull. Eur. Assoc. Theor. Comput. Sci. EATCS 62, 222–259 (1997)

[21] Niqui, M.: Formalising Exact Arithmetic: Representations, Algorithms and Proofs. PhD thesis, Radboud Universiteit Nijmegen (September 2004)

[22] Niqui, M.: Coinductive field of exact real numbers and general corecursion. In: Ghani and Power, pp. 121–139 [15]

[23] Niqui, M.: Files under Coq 8.1gamma (January 2007) http://www.cs.ru.nl/~milad/ETrees/coinductive-field/ [cited 31 January 2007]

[24] Niqui, M.: Productivity of Edalat–Potts exact arithmetic in constructive type theory. Theory Comput. Syst. 30 pages (to appear)

[25] Pavlović, D., Escardó, M.H.: Calculus in coinductive form. In: Proceedings of the 13th Annual IEEE Symposium on Logic in Computer Science, pp. 408–417 (1998)

[26] Pavlović, D., Pratt, V.: On coalgebra of real numbers. In: Jacobs, B., Rutten, J. (eds.) Coalgebraic Methods in Computer Science, CMCS'99. Electron. Notes Theor. Comput. Sci, vol. 19, pp. 103–117. Elsevier, Amsterdam (2000)

[27] Potts, P.J.: Exact Real Arithmetic using Möbius Transformations. PhD thesis, University of London, Imperial College (July 1998)

[28] Setzer, A.: Partial recursive functions in Martin-Löf type theory. In: Beckmann, et al. pp. 505–515 [2]

[29] Vuillemin, J.E.: Exact real computer arithmetic with continued fractions. IEEE Trans. Comput. 39(8), 1087–1105 (1990)

A   *Coq* Terms Needed for Formalisation of the General Corecursive Version of Homographic Algorithm

Inductive E_h: M $\to$ DIG$^\omega$ $\to$ Prop :=
$|E_{hL}$: $\forall(\mu\colon\text{M})(\alpha\colon\text{DIG}^\omega)$, $\textbf{Incl}(\mu,\text{L})$ $\to$ E_h μ α
$|E_{hR}$: $\forall(\mu\colon\text{M})(\alpha\colon\text{DIG}^\omega)$, $\textbf{Incl}(\mu,\text{R})$ $\to$ E_h μ α
$|E_{hM}$: $\forall(\mu\colon\text{M})(\alpha\colon\text{DIG}^\omega)$, $\textbf{Incl}(\mu,\text{M})$ $\to$ E_h μ α
$|E_{hab}$: $\forall(\mu\colon\text{M})(\alpha\colon\text{DIG}^\omega)$, $\neg\textbf{Incl}(\mu,\text{L})$ $\to$ $\neg\textbf{Incl}(\mu,\text{R})$ $\to$ $\neg\textbf{Incl}(\mu,\text{M})$ $\to$
$\qquad\qquad\qquad\qquad E_h$ $(\mu\circ(\text{hd }\alpha))$ $(\text{tl }\alpha)$ $\to$ E_h μ α.

Lemma $\textbf{Incl}_{\text{dec}}$: $\forall$ $(\mu\colon\text{M})$ $(d\colon\text{DIG})$, $\textbf{Incl}(\mu,d)$ $\oplus$ $\neg\textbf{Incl}(\mu,d)$.
Lemma E_{hab}_inv: $\forall(\mu\colon\text{M})(\alpha\colon\text{DIG}^\omega)$,
$\qquad\quad\neg\textbf{Incl}(\mu,\text{L})$ $\to$ $\neg\textbf{Incl}(\mu,\text{R})$ $\to$ $\neg\textbf{Incl}(\mu,\text{M})$ $\to$ E_h μ α $\to$
$\qquad\qquad\qquad\qquad\qquad E_h$ $(\mu\circ(\text{hd }\alpha))$ $(\text{tl }\alpha)$.

Fixpoint $\overline{m}_h(\mu\colon\text{M})(\alpha\colon\text{DIG}^\omega)(t\colon E_h$ μ $\alpha)\{\text{struct } t\}\colon\text{DIG}*(\text{M}*\text{DIG}^\omega)\mathtt{:=}$
match $\textbf{Incl}_{\text{dec}}(\mu,\text{L})$ with
| left $\_\Rightarrow$ $\langle\text{L},\langle\text{L}^{-1}\circ\mu,\alpha\rangle\rangle$
| right $t_l\Rightarrow$
  match $\textbf{Incl}_{\text{dec}}(\mu,\text{R})$ with
  | left $\_\Rightarrow$ $\langle\text{R},\langle\text{R}_{-1}\circ\mu,\alpha\rangle\rangle$
  | right $t_r\Rightarrow$
    match $\textbf{Incl}_{\text{dec}}(\mu,\text{M})$ with
    | left $\_\Rightarrow$ $\langle\text{M},\langle\text{M}_{-1}\circ\mu,\alpha\rangle\rangle$
    | right $t_m\Rightarrow$ $\overline{m}_h$ $(\mu\circ(\text{hd }\alpha))$ $(\text{tl }\alpha)$ $(E_{hab}$_inv μ α t_l t_r t_m $t)$
    end
  end
end.

Inductive Ψ_h: N$\to$ M $\to$ DIG$^\omega$ $\to$ Prop :=
$|\Psi_{h0}$: $\forall(\mu\colon\text{M})(\alpha\colon\text{DIG}^\omega)$, E_h μ α $\to$ Ψ_h 0 μ α
$|\Psi_{hS}$: $\forall(n\colon\text{N})(\mu\colon\text{M})$ $(\alpha\colon\text{DIG}^\omega)$ $(t\colon E_h$ μ $\alpha)$,
$\qquad\quad\Psi_h$ n $(\pi_{23}(\overline{m}_h$ μ α $t))$ $(\pi_{33}(\overline{m}_h$ μ α $t))$ $\to$ Ψ_h $(S$ $n)$ μ α.

Inductive P_h: M $\to$ DIG$^\omega$ $\to$ Prop :=
$|P_{hab}$: $\forall(\mu\colon\text{M})(\alpha\colon\text{DIG}^\omega)$, $(\forall(n\colon\text{N})$, Ψ_h $(S$ $n)$ μ $\alpha)$ $\to$ P_h μ α.

(* Proofs of the following lemmas are based on the inverses of the
   constructors of Ψ_h and the proof irrelevance of $\overline{m}_h$ *)
Lemma $P_h\_E_h$: $\forall(\mu\colon\text{M})(\alpha\colon\text{DIG}^\omega)$, P_h μ α $\to$ E_h μ α.
Lemma $\overline{m}_h\_P_h$: $\forall(\mu\colon\text{M})(\alpha\colon\text{DIG}^\omega)(t\colon E_h$ μ $\alpha)$,
  let $\mu':=\pi_{23}(\overline{m}_h$ μ α $t)$ in
    let $\alpha':=\pi_{33}(\overline{m}_h$ μ α $t)$ in P_h μ α $\to$ P_h μ' α'.

Using Intersection Types for Cost-Analysis of Higher-Order Polymorphic Functional Programs

Hugo R. Simões[1,2], Kevin Hammond[1], Mário Florido[2], and Pedro Vasconcelos[2]

[1] School of Computer Science,
University of St Andrews,
St Andrews, KY16 9SS, UK
{hs1,kh}@cs.st-andrews.ac.uk
[2] University of Porto, DCC & LIACC,
Rua do Campo Alegre, 1021/1055,
4169-007 Porto, Portugal
{hrsimoes,amf,pbv}@ncc.up.pt

Abstract. This paper presents a system of cost derivation for higher-order and polymorphic functional programs based on a notion of sized types and exploiting a type-and-effect system approach. The paper gives an operational semantics of cost for a simple strict functional language in terms of λ-calculus β-reduction steps and introduces type rules describing cost effects. The type system is based on intersection types. The use of discrete polymorphism (intersection types) instead of the usual parametric polymorphism approach improves the analysis and solves, in many cases, the "size aliasing problem" that has been identified as a limitation on previous type-and-effect approaches. Finally we provide a proof of the soundness of our effect system with respect to the cost semantics.

1 Introduction

Obtaining good-quality information concerning runtime costs (whether space or time) is important to many systems engineering activities, including compiler or database optimization, parallel computing, and real-time systems. Many of these activities require predictive information, acquired automatically at compile-time.

This paper defines a *type-and-effect system* [23] based on intersection types to derive costs of programs for a simple strict, polymorphic and higher-order functional language. We use a modern *type-and-effect system* [23], in which a rank-2 intersection type system [10] is extended by "effects" describing the cost of evaluating individual language constructs in terms of λ-calculus β-reduction steps. The meaning of these effects is given by a formal operational semantics.

We have previously described a type-and-effect system allowing the derivation of *upper bounds* on program costs [25]. One problem that can arise with this and other similar analyses based on the standard Hindley-Milner [22], is that of *size aliasing* [25], whereby a single polymorphic type variable may capture different cost properties but the same type. In extreme cases, this leads to complete loss of cost information.

T. Altenkirch and C. McBride (Eds.): TYPES 2006, LNCS 4502, pp. 221–236, 2007.

Intersection types originate in the works of Barendregt, Coppo and Dezani [7,4]. These systems are capable of typing terms where sharing of variables with non-unifiable types may cause non-typification in other type systems.

Type inference for Intersection Type Systems is undecidable in general [7,24]. However there are decidable restrictions which are useful for type assignment in programming languages [28,11,8,29,19,20,5,10,12]. Since our primary motivation is the automatic derivation of cost information, it is crucial to maintain decidability, and we have therefore chosen to use rank-2 intersection types. We show in this paper that such a system solves the *size aliasing* problem in many cases. Moreover, the system is capable of deriving cost information in other situations where the classical approach will fail. In order to show the correctness of our analysis, we provide a soundness proof for our type system with respect to our cost semantics. We also provide a number of worked examples covering important features, including polymorphism and higher-orderness, and demonstrating our solution to size aliasing.

In this paper some theorems are presented without detailed proofs. Complete proofs of every theorem can be found in an extended version [27].

This paper is structured as follows: in the following section we discuss the related work. After that, in Section 3, we introduce the language and the type-and-effect system used in our approach. In Section 4 we show examples of cost derivations, illustrating some advantages of using intersection type systems. Correctness of the system with respect to the cost semantics is found in Section 5 and we conclude in Section 6.

2   Related Work

Type-and-effect systems [23] are a well-known technique for automatic program analysis. They have a number of advantages over earlier techniques, including avoiding the need for the construction of specialised inference engines that may be required by abstract interpretation approaches, for example, providing compositional analyses, simplifying fixpoint determination, and simplifying the construction of soundness proofs through analogy with similar and well-understood proofs for the underlying type system. The work we describe here extends our own earlier work on cost analysis [25] which used a type-and-effect system based on Hindley-Milner types to expose constraints on sized types [18] for higher-order, recursive functional programs, to provide improved quality of analysis in some important cases.

While intersection types have previously been used in abstract interpretation and type-based analysis [13,2], as far as we are aware our work is the first to use intersection types and effects for cost analysis.

Other formally-based cost analysis systems related to the one we describe here include that of Reistad and Gifford [26] which analyses costs for Lisp expressions; Grobauer's work on extracting cost recurrences for Dependent ML [16]; and Chin and Khoo's work on calculating sized types [6]. Reistad and Gifford's system handles higher-order functions through "latent costs" as we have done here, and is partially based on the "time system" by Dornic et al. [14]. Grobauer's system [16] also has some similarities to ours: size-annotated types are used to capture size information. The primary difference from our work is that Dependent ML is first-order rather than higher-order. Finally, Chin and Khoo [6] describe a type-inference based algorithm for computing size information expressed in terms of Presburger formulae, for a higher-order, strict, functional language with lists, tuples and general non-recursive data constructors.

In all these systems, size aliasing is present when dealing with higher-order functions. In the work presented we show that a system with a higher degree of polymorphism, based on intersection types, avoids size aliasing in many cases.

3   Language and Type System

In this section we introduce $\mathcal{L}$, a very simple functional language, which is intended solely as a vehicle to explore static analysis for cost determination. $\mathcal{L}$ is strict, polymorphic, and higher-order, with lists as its only compound data type. We first define the syntax and operational semantics of $\mathcal{L}$, before discussing the construction of our rank-2 intersecton type-and-effect cost model.

3.1   Language Syntax

Given a countable set of variables $x \in \mathbf{Var}$, natural numbers $n \in \mathbb{N}$, and primitive operations $\mathsf{op}_1 \in \mathbf{Prim}_1$, $\mathsf{op}_2 \in \mathbf{Prim}_2$, the terms of $\mathcal{L}$ are inductively defined by the following grammar:

$$e ::= \quad x \mid n \mid \mathsf{true} \mid \mathsf{false} \mid [] \mid e_1 :: e_2$$
$$\mid \quad \mathsf{op}_1(e) \mid \mathsf{op}_2(e_1, e_2) \mid \lambda x.e \mid e_1 \, e_2$$
$$\mid \quad \mathsf{if} \ e_1 \ \mathsf{then} \ e_2 \ \mathsf{else} \ e_3$$

Note that function definitions are *Curried* (i.e. abstract a single argument at a time) and function application is binary. Constructors and primitive operations are both *saturated*, i.e. they are restricted to the correct number of arguments; partial applications can be obtained for these forms, if required, using λ-abstractions. The two sets $\mathbf{Prim}_1$ and $\mathbf{Prim}_2$ of unary and binary primitive operations include arithmetic on naturals and lists projections. Discussion of these operations is deferred to Section 3.5.

We require the usual notational conventions on variables: the expression '$\lambda x.e$' is a *binder* for x; the scope of this binder is the sub-expression e. An occurrence of a variable that is in the scope of a binder is said to be *bound*, otherwise it is *free*. An expression where all variable occurrences are bound is *closed*. We will follow the notation, terminology and conventions presented by Barendregt in [3]. Finally, the notation $e[e'/x]$ represents the expression that results from replacing all free occurrences of x in e by e'.

3.2 Language Semantics

We use a call-by-value operational semantics for $\mathcal{L}^1$. We first define a subset of $\mathcal{L}$ expressions considered to be *result values* for computations:

$$e_w ::= n \mid \mathsf{true} \mid \mathsf{false} \mid [] \mid e_{w_1}{::}e_{w_2} \mid \lambda x.e$$

The semantics of $\mathcal{L}$ is given by a *single-step reduction relation* $e \to e'$. Figure 1 depicts the top level reduction rules; these define a pre-relation '$\rightharpoonup$' that is then extended to the full single-step reduction '$\to$' by a suitable use of *evaluation contexts* defined below (as in [15]) so as to propagate evaluation of sub-expressions in call-by-value order. Note that since reductions are not performed under λs, result values (e_w) are *weak normal forms* in the underlying λ-calculus.

Definition 1. Evaluation contexts E take the form

$$
\begin{aligned}
E ::=\ & [] \mid E\,e \mid e_w\,E \\
& \mid\ \mathsf{if}\ E\ \mathsf{then}\ e_1\ \mathsf{else}\ e_2 \\
& \mid\ \mathsf{op}_1(E) \mid \mathsf{op}_2(E,e) \mid \mathsf{op}_2(e_w,E)
\end{aligned}
$$

Note that E is a context with exactly one hole in it, and that this hole is not inside the scope of any identifier. We write $E[e]$ for the expression that has the hole in E replaced by e, and similarly $E[E']$ for the evaluation context that comes from replacing the hole in E by E'.

Definition 2. The reduction relation $\to$ is defined by $E[e] \to E[e']$ provided $e \rightharpoonup e'$ holds according to Figure 1.

We write '$e \not\to$' to mean $\neg\exists e' : e \to e'$. We require two preliminary results concerning the reduction semantics:

Lemma 1. If e_w is a value, then $e_w \not\to$.

Proof: By case-analysis of the rules in Figure 1 we conclude that $e_w \not\rightharpoonup$; the result for $\to$ then follows by simple induction on the structure of the evaluation context. □

Lemma 2. $\to$ is deterministic, i.e. if $e \to e'$ and $e \to e''$, then $e' = e''$.

Proof: $\rightharpoonup$ is deterministic because at most one rule of Figure 1 can apply at each step and it is determined by the structure of e; the result for $\to$ follows from an induction on the structure of the evaluation context. □

---

[1] While it would be interesting to also explore call-by-need, and we are, in fact, investigating a new cost model and analyses for call-by-need functional programs, the complexity of the semantics is considerably greater than that presented here. In this paper, we have therefore opted for the simpler and better understood cost model associated with call-by-value.

$[\text{beta}_\rightarrow]\ (\lambda x.e)\,e_w \rightarrow e[e_w/x]$

$\quad [\text{if1}_\rightarrow]\ \text{if true then } e_1 \text{ else } e_2 \rightarrow e_1$

$\quad [\text{if2}_\rightarrow]\ \text{if false then } e_1 \text{ else } e_2 \rightarrow e_2$

$\quad [\text{add}_\rightarrow]\ \text{add}(n, m) \rightarrow n',\ \text{where } n' = n + m$

$\quad [\text{pred}_\rightarrow]\ \text{pred}(n) \rightarrow n',\ \text{where } n' = \max(0, n - 1)$

$[\text{eq1}_\rightarrow]\ \text{eq}(n, m) \rightarrow \text{true, if } n = m$

$[\text{eq2}_\rightarrow]\ \text{eq}(n, m) \rightarrow \text{false, if } n \neq m$

$[\text{head}_\rightarrow]\ \text{head}(e_{w_1}{::}e_{w_2}) \rightarrow e_{w_1}$

$[\text{tail}_\rightarrow]\ \text{tail}(e_{w_1}{::}e_{w_2}) \rightarrow e_{w_2}$

$[\text{null1}_\rightarrow]\ \text{null}([\,]) \rightarrow \text{true}$

$[\text{null2}_\rightarrow]\ \text{null}(e_{w_1}{::}e_{w_2}) \rightarrow \text{false}$

Fig. 1. Top-level reduction relation for $\mathcal{L}$

3.3   Cost Syntax

The size and cost *values* for $\mathcal{L}$-terms are elements of the set $\overline{\mathbb{N}} = \{0, 1, 2 \ldots, \omega\}$ of natural numbers together with a top element ω. The usual ordering $\leq$ on naturals extends to $\overline{\mathbb{N}}$ by taking $x \leq \omega$ for all $x \in \overline{\mathbb{N}}$. Assuming a countable set **ZVar** of *cost variables*, the set **ZExp** of *cost expressions* is generated by the following grammar:

$$l \ \in \ \mathbf{ZVar}$$
$$n \ \in \ \overline{\mathbb{N}}$$
$$z \ ::= \ l \ | \ n \ | \ z_1 + z_2.$$

We designate the set of cost variables occurring in an expression z by $\text{ZV}(z)$. We can now define our cost model for $\mathcal{L}$.

Definition 3 (*Cost model*). Let e be a closed $\mathcal{L}$-expression; If $e \rightarrow^* e'$ and $e' \nrightarrow$ (where $\rightarrow^*$ is the reflexive transitive closure of $\rightarrow$) then the *cost of reduction* of e is the number of $[\text{beta}_\rightarrow]$ reductions[2] in $e \rightarrow^* e'$ (this number is well defined because of Lemma 2).

If the reduction from e diverges, i.e. there is an infinite sequence $(e_n)_{n\geq 0}$ such that $e_0 = e$ and for all $n \geq 0$, $e_n \rightarrow e_{n+1}$, then the cost of reduction is ω (standing for infinite cost).

Note that $e' \nrightarrow$ does *not* imply e' must be a value; it can also be an *erroneous reduction sequence*, for example, $\text{head}([\,]) \nrightarrow$. This means we cost both confluent and erroneous reduction sequences with finite cost and divergence with infinite cost[3].

---

[2] Although other metrics could have been used (see [9] for example), we have chosen to count β-reductions for simplicity.

[3] Our intuition is that in an implementation, erroneous reduction sequences will induce termination, perhaps through an exception.

3.4 Sized Types

$\mathcal{L}$ uses a notion of *sized types* [18] applied to *rank 2 intersection types* [10] where types for naturals and lists have a superscript specifying an upper bound for its *size* and function types have a *latent cost* [26] attached to the function arrow. This latent cost is an upper bound on the *cost* of evaluating the function body.

Definition 4. Given a countable set **TVar** of *type variables*, the set of *simple sized types* ($\mathbf{T_0}$), ranged over by u, and the set of *rank 2 sized intersection types* ($\mathbf{T_2}$), ranged over by v, are defined inductively by the following grammar:

$$\alpha \;\in\; \mathbf{TVar}$$
$$z \;\in\; \mathbf{ZExp}$$
$$u ::= \alpha \;\mid\; \mathsf{Bool} \;\mid\; \mathsf{Nat}^z \;\mid\; \mathsf{List}^z u \;\mid\; u_1 \xrightarrow{z} u_2 \,.$$
$$v ::= u \;\mid\; u_1 \wedge \ldots \wedge u_n \xrightarrow{z} v \,.$$

The constructor $\wedge$ binds more tightly than $\rightarrow$ and we consider $\wedge$ to be associative, commutative, and idempotent.

Since sizes are attached to types and these may be embedded within other types, it is possible to describe the sizes of the elements of a structure as well as the structure itself, e.g.: $\mathsf{List}^5\left(\mathsf{Nat}^{10}\right)$ denotes a list whose length is at most 5 with natural numbers not greater than 10 as elements.

3.5 Type System

Figure 2 and Figure 3 show the sized type system that derives judgements of the form,

$$A \vdash e : v \,\&\, z$$

which can be informally read as "under type assumptions A, the expression e admits type v and z is an upper bound for the cost of e". A *type environment* A is a set of *type assumptions* of the form $x : u_1 \wedge \ldots \wedge u_n$, where $n \geq 1$, such that every identifier x can occur at most once in A. The expression $\mathrm{dom}(A)$ denotes the *domain* of A, which is the set of identifiers found in A. We write A_1, A_2 for the environment $A_1 \cup A_2$ where it is assumed that $\mathrm{dom}(A_1) \cap \mathrm{dom}(A_2) = \emptyset$, and $A, x : t$ as short for $A, \{x : t\}$. Given two type environments A_1 and A_2 we write $A_1 \wedge A_2$ to denote the type environment

$$\{x : (u_1 \wedge \ldots \wedge u_n) \wedge (u'_1 \wedge \ldots \wedge u'_m) \mid x : u_1 \wedge \ldots \wedge u_n \in A_1 \text{ and}$$
$$x : u'_1 \wedge \ldots \wedge u'_m \in A_2\} \cup$$
$$\{x : u_1 \wedge \ldots \wedge u_n \in A_1 \mid x \notin \mathrm{dom}(A_2)\} \cup$$
$$\{x : u'_1 \wedge \ldots \wedge u'_m \in A_2 \mid x \notin \mathrm{dom}(A_1)\}$$

The system represents a straightforward extension of Damiani's rules for the λ-core of his language [10] and for consistency uses the same notation for types. Note that:

$$\frac{}{\{x:u\} \vdash x : u \ \& \ 0} \ [Var_{\wedge 2st}] \qquad \frac{n \in \mathbb{N}}{\emptyset \vdash n : \mathsf{Nat}^n \ \& \ 0} \ [Nat_{\wedge 2st}] \qquad \frac{b \in \{true, false\}}{\emptyset \vdash b : \mathsf{Bool} \ \& \ 0} \ [Bool_{\wedge 2st}]$$

$$\frac{x \in FV(e) \quad A, x : u_1 \wedge ... \wedge u_n \vdash e : v \ \& \ z}{A \vdash \lambda x.e : u_1 \wedge ... \wedge u_n \xrightarrow{z} v \ \& \ 0} \ [Abs_{\wedge 2st}]$$

$$\frac{x \notin FV(e) \quad u \in \mathbf{T}_0 \quad A \vdash e : v \ \& \ z}{A \vdash \lambda x.e : u \xrightarrow{z} v \ \& \ 0} \ [AbsVac_{\wedge 2st}]$$

$$\frac{A_0 \vdash e_1 : u_1 \wedge ... \wedge u_n \xrightarrow{z_3} v \ \& \ z_1 \quad (\forall i \in \{1, ..., n\}) \ A_i \vdash e_2 : u_i \ \& \ z_2}{A_0 \wedge A_1 \wedge ... \wedge A_n \vdash e_1 \ e_2 : v \ \& \ 1 + z_1 + z_2 + z_3} \ [App_{\wedge 2st}]$$

$$\frac{A_0 \vdash e_0 : \mathsf{Bool} \ \& \ z_0 \quad A_1 \vdash e_1 : u \ \& \ z \quad A_2 \vdash e_2 : u \ \& \ z}{A_0 \wedge A_1 \wedge A_2 \vdash \text{if } e_0 \text{ then } e_1 \text{ else } e_2 : u \ \& \ z_0 + z} \ [If_{\wedge 2st}]$$

$$\frac{A_1 \vdash e : v \ \& \ z \quad A_2 \leq_1 A_1}{A_2 \vdash e : v \ \& \ z} \ [Weak_{\wedge 2st}]$$

$$\frac{A \vdash e : v_1 \ \& \ z \quad v_1 \leq_2 v_2}{A \vdash e : v_2 \ \& \ z} \ [SubT_{\wedge 2st}] \qquad \frac{A \vdash e : v \ \& \ z_1 \quad z_1 \leq z_2}{A \vdash e : v \ \& \ z_2} \ [SubE_{\wedge 2st}]$$

Fig. 2. Typing rules for the core $\mathcal{L}$ expressions

- The three non-structural rules $[Weak_{\wedge 2st}]$, $[SubT_{\wedge 2st}]$ and $[SubE_{\wedge 2st}]$ allow *weakening, subtyping* and *subeffecting*, respectively and make use of the sub-typing relations, $\trianglelefteq$, $\leq_1$ and $\leq_2$, defined in Figure 4. We write $A_1 \leq_1 A_2$ to mean that $\text{dom}(A_1) = \text{dom}(A_2)$ and for every assumption $x : u'_1 \wedge . . \wedge u'_m \in A_2$ there is an assumption $x : u_1 \wedge . . \wedge u_n \in A_1$ such that $u_1 \wedge . . \wedge u_n \leq_1 u'_1 \wedge . . \wedge u'_m$. The subtype system is not structural, for example, $u_1 \trianglelefteq u_2$ does not imply that $\mathsf{List}^{z_1} u_1 \trianglelefteq \mathsf{List}^{z_2} u_2$ (since there is no relationship between the size of a structure and the elements of that structure).
- We have two rules for typing an abstraction $\lambda x.e$, $[Abs_{\wedge 2st}]$ and $[AbsVac_{\wedge 2st}]$, corresponding respectively to the two cases $x \in FV(e)$ and $x \notin FV(e)$. In both rules, the cost of evaluating the body of a lambda abstraction is the latent cost of the function type; the cost for the actual abstraction is zero.
- In the $[App_{\wedge 2st}]$ rule we add the latent cost of the function to the costs of obtaining the function and argument, plus a constant 1 to count the cost of the β-reduction (this is the only rule where a positive cost is added). Note that this rule allows using different typing for each expected type of the argument, but each typing may require subeffecting to ensure the same cost z_2.

$$\frac{u \in \mathbf{T}_0}{\emptyset \vdash [] : \mathsf{List}^0 u \; \& \; 0} \; [Nil_{\wedge 2st}]$$

$$\frac{A_1 \vdash e_1 : u \; \& \; z_1 \quad A_2 \vdash e_2 : \mathsf{List}^z u \; \& \; z_2}{A_1 \wedge A_2 \vdash e_1 {::} e_2 : \mathsf{List}^{1+z} u \; \& \; z_1 + z_2} \; [Cons_{\wedge 2st}]$$

$$\frac{A_1 \vdash e_1 : \mathsf{Nat}^{z_1} \; \& \; z_1' \quad A_2 \vdash e_2 : \mathsf{Nat}^{z_2} \; \& \; z_2'}{A_1 \wedge A_2 \vdash \mathsf{add}(e_1, e_2) : \mathsf{Nat}^{z_1 + z_2} \; \& \; z_1' + z_2'} \; [Add_{\wedge 2st}]$$

$$\frac{A \vdash e : \mathsf{Nat}^z \; \& \; z'}{A \vdash \mathsf{pred}(e) : \mathsf{Nat}^{z-1} \; \& \; z'} \; [Pred_{\wedge 2st}]$$

$$\frac{A_1 \vdash e_1 : \mathsf{Nat}^{z_1} \; \& \; z_1' \quad A_2 \vdash e_2 : \mathsf{Nat}^{z_2} \; \& \; z_2'}{A_1 \wedge A_2 \vdash \mathsf{eq}(e_1, e_2) : \mathsf{Bool} \; \& \; z_1' + z_2'} \; [Eq_{\wedge 2st}]$$

$$\frac{A \vdash e : \mathsf{List}^z u \; \& \; z'}{A \vdash \mathsf{head}(e) : u \; \& \; z'} \; [Head_{\wedge 2st}]$$

$$\frac{A \vdash e : \mathsf{List}^z u \; \& \; z'}{A \vdash \mathsf{tail}(e) : \mathsf{List}^{z-1} u \; \& \; z'} \; [Tail_{\wedge 2st}]$$

$$\frac{A \vdash e : \mathsf{List}^z u \; \& \; z'}{A \vdash \mathsf{null}(e) : \mathsf{Bool} \; \& \; z'} \; [Null_{\wedge 2st}]$$

Fig. 3. Typing rules for natural and list primitives

$$\frac{u \trianglelefteq u'}{u \leq_2 u'} \; [\mathsf{simple}_{\leq_2}] \qquad \frac{u_1' \wedge ... \wedge u_m' \leq_1 u_1 \wedge ... \wedge u_n \quad v \leq_2 v' \quad z \leq z'}{u_1 \wedge ... \wedge u_n \xrightarrow{z} v \leq_2 u_1' \wedge ... \wedge u_m' \xrightarrow{z'} v'} \; [\mathsf{rank2}_{\leq_2}]$$

$$\frac{n \geq m \quad \exists i_1, ..., i_m \in \{1, ..., n\} : u_{i_1} \trianglelefteq u_1', ..., u_{i_m} \trianglelefteq u_m'}{u_1 \wedge ... \wedge u_n \leq_1 u_1' \wedge ... \wedge u_m'} \; [\mathsf{rank1}_{\leq_1}] \qquad \frac{u = u'}{u \trianglelefteq u'} \; [\mathsf{reflex}_\trianglelefteq]$$

$$\frac{z \leq z'}{\mathsf{Nat}^z \trianglelefteq \mathsf{Nat}^{z'}} \; [\mathsf{nat}_\trianglelefteq] \qquad \frac{z \leq z' \quad u \trianglelefteq u'}{\mathsf{List}^z u \trianglelefteq \mathsf{List}^{z'} u'} \; [\mathsf{list}_\trianglelefteq] \qquad \frac{u_1' \trianglelefteq u_1 \quad u_2 \trianglelefteq u_2' \quad z \leq z'}{u_1 \xrightarrow{z} u_2 \trianglelefteq u_1' \xrightarrow{z'} u_2'} \; [\mathsf{abs}_\trianglelefteq]$$

Fig. 4. Subtyping relations $\trianglelefteq$, $\leq_1$ and $\leq_2$

- In the rule for $[If_{\wedge 2st}]$ we require that both branches admit the same type and cost, which may require using the $[Weak_{\wedge 2st}]$, $[SubT_{\wedge 2st}]$ and $[SubE_{\wedge 2st}]$ rules for one or both branches.

4 Examples

The next examples show a possible derivation for a selected expression, using the rules of Figure 2 and Figure 3.

The examples take advantage of the underlying rank-2 intersection type system and either were not possible to type or had a less precise type if our type system was based on a standard Hindley-Milner one.

Example 1: Conditionals

The example depicted in Figure 5 illustrates the use of conditionals combined with polymorphism.

Note that function f must be able to handle arguments of type Nat *and* Bool and since f was not explicitly declared polymorphic in "$\lambda f\,p.\text{if }p\text{ then }f\,1\text{ else }f\,true$" the whole expression cannot be typed by a standard Hindley-Milner type system.

Example 2: Size Aliasing

Consider expression "$\lambda f\,x.f\,(f\,x)$", representing function *twice* (a function that applies its first argument twice to its second argument), and expression "$\lambda y.\text{add}(y,1)$", representing function *succ* (a function that given a natural number returns its successor). One way to represent the type of function *succ* using *sized types* is $\text{Nat}^n \to \text{Nat}^{n+1}$; meaning *succ* is a function that takes an element of $\{m \in \mathbb{N} \mid 0 \leq m \leq n\}$ and returns an element of $\{m \in \mathbb{N} \mid 0 \leq m \leq n+1\}$.

We should expect "$(\lambda f\,x.f\,(f\,x))\,(\lambda y.\text{add}(y,1))$" (i.e, *twice succ*) to represent a function that given a natural number, y, returns the successor of the successor of y, i.e, $y+2$. Therefore we should also expect the sized type of *twice* applied to *succ* to be $\text{Nat}^n \to \text{Nat}^{n+2}$ and in fact we can derive such type using our system (as shown in Figure 6), but if we had based our type system on a standard Hindley-Milner type system, then the best type for *twice succ* would have been $\text{Nat}^\omega \to \text{Nat}^\omega$ (where ω represents some unbounded natural) which although correct, is not precise.

The reason for this loss of precision is that the Hindley-Milner type for the first argument of twice is $\alpha \to \alpha$ and at some point we would need to unify $\alpha \to \alpha$ with $\text{Nat}^n \to \text{Nat}^{n+1}$, the sized type of *succ*. From this unification we would obtain the constraint $n = n+1$, which can only be solved in $\overline{\mathbb{N}}$ if $n = \omega$.

5 Correctness

5.1 Semantic Correctness

In order to prove semantic correctness for our analysis we first state a few preliminary results.

$$(3.2) \begin{cases} \dfrac{true \in \{true, false\}}{\emptyset \vdash true : \mathsf{Bool} \ \& \ 0} \ [Bool_{\wedge 2st}] \end{cases}$$

$$(3.1) \begin{cases} \dfrac{}{\{f : \mathsf{Bool} \xrightarrow{z_1} \alpha_1\} \vdash f : \mathsf{Bool} \xrightarrow{z_1} \alpha_1 \ \& \ 0} \ [Var_{\wedge 2st}] \end{cases}$$

$$\left. \dfrac{(3.1) \qquad (3.2)}{\{f : \mathsf{Bool} \xrightarrow{z_1} \alpha_1\} \vdash f \, true : \alpha_1 \ \& \ 1+z_1} \ [App_{\wedge 2st}] \right\} (3)$$

$$(2.2) \begin{cases} \dfrac{1 \in \mathbb{N}}{\emptyset \vdash 1 : \mathsf{Nat}^1 \ \& \ 0} \ [Nat_{\wedge 2st}] \end{cases}$$

$$(2.1) \begin{cases} \dfrac{}{\{f : \mathsf{Nat}^1 \xrightarrow{z_1} \alpha_1\} \vdash f : \mathsf{Nat}^1 \xrightarrow{z_1} \alpha_1 \ \& \ 0} \ [Var_{\wedge 2st}] \end{cases}$$

$$\left. \dfrac{(2.1) \qquad (2.2)}{\{f : \mathsf{Nat}^1 \xrightarrow{z_1} \alpha_1\} \vdash f \, 1 : \alpha_1 \ \& \ 1+z_1} \ [App_{\wedge 2st}] \right\} (2)$$

$$\left. \dfrac{}{\{p : \mathsf{Bool}\} \vdash p : \mathsf{Bool} \ \& \ 0} \ [Var_{\wedge 2st}] \right\} (1)$$

$$\dfrac{(1) \qquad (2) \qquad (3)}{\begin{array}{c} \{f : (\mathsf{Nat}^1 \xrightarrow{z_1} \alpha_1) \wedge (\mathsf{Bool} \xrightarrow{z_1} \alpha_1), p : \mathsf{Bool}\} \vdash \\ \text{if } p \text{ then } f \, 1 \text{ else } f \, true : \alpha_1 \ \& \ 1+z_1 \end{array}} \ [If_{\wedge 2st}]$$

$$\dfrac{}{\begin{array}{c} \{f : (\mathsf{Nat}^1 \xrightarrow{z_1} \alpha_1) \wedge (\mathsf{Bool} \xrightarrow{z_1} \alpha_1)\} \vdash \\ \lambda p.\text{if } p \text{ then } f \, 1 \text{ else } f \, true : \mathsf{Bool} \xrightarrow{1+z_1} \alpha_1 \ \& \ 0 \end{array}} \ [Abs_{\wedge 2st}]$$

$$\dfrac{}{\emptyset \vdash \lambda f \, p.\text{if } p \text{ then } f \, 1 \text{ else } f \, true : (\mathsf{Nat}^1 \xrightarrow{z_1} \alpha_1) \wedge (\mathsf{Bool} \xrightarrow{z_1} \alpha_1) \xrightarrow{0} \mathsf{Bool} \xrightarrow{1+z_1} \alpha_1 \ \& \ 0} \ [Abs_{\wedge 2st}]$$

Fig. 5. A type derivation for '$\lambda f \, p.\text{if } p \text{ then } f \, 1 \text{ else } f \, true$'

Proposition 1. In any valid judgement $A \vdash e : v \ \& \ z$ we have $\mathrm{dom}(A) = \mathrm{FV}(e)$

Proposition 2. Suppose A_1, A_1', A_2, A_2' are type environments such that $A_1 \leq_1 A_1'$ and $A_2 \leq_1 A_2'$. Then $A_1 \wedge A_2 \leq_1 A_1' \wedge A_2'$.

Proposition 3. Suppose $A_1, ..., A_n$, where $n \geq 1$, are type environments such that $\mathrm{dom}(A_1) = ... = \mathrm{dom}(A_n)$. Then $A_1 \wedge ... \wedge A_n \leq_1 A_{i_1} \wedge ... \wedge A_{i_m}$, for all subset $\{i_1, ..., i_m\}$ of $\{1, ..., n\}$, $m \geq 1$.

Proposition 4. Suppose A_1, A_1', A_2, A_2' are type environments such that $A_2' \subseteq A_2$ and $A_1' = \{x : t \in A_2 \mid x \in \mathrm{dom}(A_1)\}$. Then $A_1 \wedge A_1' \wedge A_2' \subseteq A_1 \wedge A_2$ and $A_1 \wedge A_1' \wedge A_2' \leq_1 A_1 \wedge A_2'$.

To prove the subject reduction property for the top-level reduction relation, $\rightharpoonup$, we also need the next two lemmas.

(3.2) $\left\{ \dfrac{}{\emptyset \vdash 1 : \mathsf{Nat}^1 \ \& \ 0} \ [Nat_{\wedge 2st}] \right.$

(3.1) $\left\{ \dfrac{}{\{y : \mathsf{Nat}^{n+1}\} \vdash y : \mathsf{Nat}^{n+1} \ \& \ 0} \ [Var_{\wedge 2st}] \right.$

$$\left. \dfrac{\dfrac{(3.1) \qquad (3.2)}{\{y : \mathsf{Nat}^{n+1}\} \vdash \mathsf{add}(y, 1) : \mathsf{Nat}^{n+2} \ \& \ 0} \ [Add_{\wedge 2st}]}{\emptyset \vdash \lambda y.\mathsf{add}(y, 1) : \mathsf{Nat}^{n+1} \xrightarrow{0} \mathsf{Nat}^{n+2} \ \& \ 0} \ [Abs_{\wedge 2st}] \right\} (3)$$

$$\left. \dfrac{\dfrac{\dfrac{}{\{y : \mathsf{Nat}^n\} \vdash y : \mathsf{Nat}^n \ \& \ 0} \ [Var_{\wedge 2st}] \quad \dfrac{}{\emptyset \vdash 1 : \mathsf{Nat}^1 \ \& \ 0} \ [Nat_{\wedge 2st}]}{\{y : \mathsf{Nat}^n\} \vdash \mathsf{add}(y, 1) : \mathsf{Nat}^{n+1} \ \& \ 0} \ [Add_{\wedge 2st}]}{\emptyset \vdash \lambda y.\mathsf{add}(y, 1) : \mathsf{Nat}^n \xrightarrow{0} \mathsf{Nat}^{n+1} \ \& \ 0} \ [Abs_{\wedge 2st}] \right\} (2)$$

(1.2.2) $\left\{ \dfrac{}{\{x : \mathsf{Nat}^n\} \vdash x : \mathsf{Nat}^n \ \& \ 0} \ [Var_{\wedge 2st}] \right.$

(1.2.1) $\left\{ \dfrac{}{\{f : \mathsf{Nat}^n \xrightarrow{0} \mathsf{Nat}^{n+1}\} \vdash f : \mathsf{Nat}^n \xrightarrow{0} \mathsf{Nat}^{n+1} \ \& \ 0} \ [Var_{\wedge 2st}] \right.$

(1.2) $\left\{ \dfrac{(1.2.1) \qquad (1.2.2)}{\{f : \mathsf{Nat}^n \xrightarrow{0} \mathsf{Nat}^{n+1}, x : \mathsf{Nat}^n\} \vdash f \, x : \mathsf{Nat}^{n+1} \ \& \ 1} \ [App_{\wedge 2st}] \right.$

(1.1) $\left\{ \dfrac{}{\{f : \mathsf{Nat}^{n+1} \xrightarrow{0} \mathsf{Nat}^{n+2}\} \vdash f : \mathsf{Nat}^{n+1} \xrightarrow{0} \mathsf{Nat}^{n+2} \ \& \ 0} \ [Var_{\wedge 2st}] \right.$

$$\left. \dfrac{\dfrac{\dfrac{(1.1) \qquad (1.2)}{\begin{array}{c}\{f : (\mathsf{Nat}^n \xrightarrow{0} \mathsf{Nat}^{n+1}) \wedge (\mathsf{Nat}^{n+1} \xrightarrow{0} \mathsf{Nat}^{n+2}), x : \mathsf{Nat}^n\} \\ \vdash f\,(f\,x) : \mathsf{Nat}^{n+2} \ \& \ 2\end{array}} \ [App_{\wedge 2st}]}{\begin{array}{c}\{f : (\mathsf{Nat}^n \xrightarrow{0} \mathsf{Nat}^{n+1}) \wedge (\mathsf{Nat}^{n+1} \xrightarrow{0} \mathsf{Nat}^{n+2})\} \vdash \\ \lambda x.f\,(f\,x) : \mathsf{Nat}^n \xrightarrow{2} \mathsf{Nat}^{n+2} \ \& \ 0\end{array}} \ [Abs_{\wedge 2st}]}{\emptyset \vdash \lambda f\,x.f\,(f\,x) : (\mathsf{Nat}^n \xrightarrow{0} \mathsf{Nat}^{n+1}) \wedge (\mathsf{Nat}^{n+1} \xrightarrow{0} \mathsf{Nat}^{n+2}) \xrightarrow{0} \mathsf{Nat}^n \xrightarrow{2} \mathsf{Nat}^{n+2} \& 0} \ [Abs_{\wedge 2st}] \right\} (1)$$

$$\dfrac{(1) \qquad (2) \qquad (3)}{\emptyset \vdash (\lambda f\,x.f\,(f\,x))\,(\lambda y.\mathsf{add}(y, 1)) : \mathsf{Nat}^n \xrightarrow{2} \mathsf{Nat}^{n+2} \ \& \ 1} \ [App_{\wedge 2st}]$$

Fig. 6. A type derivation for '$(\lambda f\,x.f\,(f\,x))\,(\lambda y.\mathsf{add}(y, 1))$'

Our first lemma states that if a value admits a type, then it admits zero cost:

Lemma 3. Let e_w be a value; if $A \vdash e_w : v \ \& \ z$ then $A \vdash e_w : v \ \& \ 0$.

Proof: By induction on the structure of e_w combined with case-analysis on the the rules of Figure 2 and Figure 3 (by analogy with [1]). $\qquad\qquad \square$

Our second lemma states that we can replace a variable in a typing context by an expression of the correct type provided the expression has zero cost:

Lemma 4. If $A, y : u_1' \wedge ... \wedge u_m' \vdash e : v \,\&\, z$ and $(\forall j \in \{1, ..., m\})\, A_j' \vdash e' : u_j' \,\&\, 0$ then $A \wedge A_1' \wedge ... \wedge A_m' \vdash e[e'/y] : v \,\&\, z$.

Proof: By induction on the length of the type derivation tree (see [27]).   □

Our first semantic correctness result is a *subject reduction property* stating that types are maintained by single-step reductions. The first theorem deals with top-level reductions only.

Theorem 1 (Subject reduction for $\rightharpoonup$). If $e \rightharpoonup e'$ and $A \vdash e : v \,\&\, z$, then also $A' \vdash e' : v \,\&\, z$ where $A' \subseteq A$.

Proof: By induction in the inference tree for $A \vdash e : v \,\&\, z$ and by case analysis on the last rule applied (see [27]).   □

We now lift the subject reduction result to reductions occurring in an arbitrary evaluation context. In order to do so, we first present some auxiliary results on inference trees.

The following definition allow us to be precise about what it means for a judgement to occur "at the address indicated by the hole in E" and at the same time we shall record the "depth" of the judgement in order to facilitate proofs by induction.

Definition 5. Consider judgements $jdg' = A' \vdash e' : v' \,\&\, z'$, $jdg = A \vdash e : v \,\&\, z$ and some evaluation context E.

We say that jdg' occurs at E with depth 0 in the inference tree for jdg if $E = [\,]$ and $jdg' = jdg$ (implying that $e = E[e']$).

We say that jdg' occurs at E with depth 1 in the inference tree for jdg if $e = E[e']$ and the last rule applied in the inference tree for jdg is *either*

- $[Weak_{\wedge 2st}]$, $[SubT_{\wedge 2st}]$ or $[SubE_{\wedge 2st}]$, with jdg' as the leftmost premise and $E = [\,]$; or
- $[App_{\wedge 2st}]$ with jdg' as the leftmost premise and E of the form $[\,]\,e_2$; or
-

$$
\dfrac{
\begin{array}{c}
A_0 \vdash e_{w_1} : u_1 \wedge ... \wedge u_{j-1} \wedge v' \wedge u_{j+1} \wedge ... \wedge u_n \xrightarrow{z_3} v \,\&\, z_1 \\[4pt]
A' \vdash e' : v' \,\&\, z' \\[4pt]
(\forall i \in \{1, ..., j-1, j+1, ..., n\})\, A_i \vdash e' : u_i \,\&\, z_2
\end{array}
}{
A \vdash e : v \,\&\, z
}\ [App_{\wedge 2st}] \qquad , j \in \{1, ..., n\}
$$

where $v' = u_j \in \mathbf{T}_0$, $A' = A_j$, $z' = z_2$, $A = A_0 \wedge A_1 \wedge ... \wedge A_n$, $z = 1 + z_1 + z_2 + z_3$ and E has the form $e_{w_1}[\,]$ implying $e = e_{w_1} e'$; or

- $[If_{\wedge 2st}]$, with jdg' as the leftmost premise and E of the form if $[\,]$ then e_1 else e_2; or

- $[Pred_{\wedge 2st}]$, $[Head_{\wedge 2st}]$, $[Tail_{\wedge 2st}]$ or $[Null_{\wedge 2st}]$, with jdg' as premise and E of the form $\mathsf{op}_1([\,])$, where op_1 is the corresponding unary operation; or
- $[Add_{\wedge 2st}]$ or $[Eq_{\wedge 2st}]$, with jdg' as the leftmost premise and E of the form $\mathsf{op}_2([\,], e_2)$, where op_2 is the corresponding binary operation; or
- $[Add_{\wedge 2st}]$ or $[Eq_{\wedge 2st}]$, with jdg' as the rightmost premise and E of the form $\mathsf{op}_2(e_{w_1}, [\,])$, where op_2 is the corresponding binary operation.

We say that jdg' occurs at E with depth $d > 1$ in the inference tree for jdg if there exists $d_1, d_2 < d$ with $d_1 + d_2 = d$, evaluation contexts E_1, E_2 with $E_2[E_1] = E$, and judgement jdg'', such that jdg' occurs at E_1 with depth d_1 in the inference tree for jdg'', and jdg'' occurs at E_2 with depth d_2 in the inference tree for jdg.

Lemma 5. Given a judgement $jdg = (A \vdash E[e] : v \,\&\, z)$, there exists (at least one) judgement jdg' of the form $A' \vdash e : v' \,\&\, z'$ such that jdg' occurs at evaluation context E in the inference tree for jdg.

Proof: By analogy with the proof in [1], p.197–198. ☐

Lemma 6. Suppose that the judgement $jdg' = (A' \vdash e : v' \,\&\, z')$ occurs at E in the inference tree of judgement $jdg = (A \vdash E[e] : v \,\&\, z)$. If e_r is such that $A'_1 \vdash e_r : v' \,\&\, z'$, where $A'_1 \subseteq A'$, then also $A_1 \vdash E[e_r] : v \,\&\, z$, where $A_1 \subseteq A$.

Proof: By induction in the depth d at which jdg' occurs at E in the inference tree for jdg (see [27]). ☐

We can now state and prove the subject reduction for the single step reduction relation $\rightarrow$:

Theorem 2 (Subject reduction for $\rightarrow$). If $e_1 \rightarrow e_2$ and $A_1 \vdash e_1 : v \,\&\, z$, then $A_2 \vdash e_2 : v \,\&\, z$ where $A_2 \subseteq A_1$.

Proof: From Definition 2 we assume the existence of some E, e'_1, e'_2 such that $e_1 = E[e'_1]$, $e_2 = E[e'_2]$ and $e'_1 \rightarrow e'_2$. By Lemma 5 there exists A'_1, v' and z' such that $A'_1 \vdash e'_1 : v' \,\&\, z'$ occurs at E in the inference tree of $A_1 \vdash E[e'_1] : v \,\&\, z$.

Theorem 1 then gives $A'_2 \vdash e'_2 : v' \,\&\, z'$ where $A'_2 \subseteq A'_1$ and by Lemma 6 we obtain $A_2 \vdash E[e'_2] : v \,\&\, z$ where $A_2 \subseteq A_1$ which is as desired since $e_2 = E[e'_2]$. ☐

Note that subject redution normally does not hold in a rank-2 intersection type system with two rules for abstraction. However, our subject reduction result can be seen as a *weak subject reduction*, since, according to our semantics, reductions are not performed under λs.

5.2   Correctness of the Cost Analysis

We now prove the correctness of the cost analysis by relating the effect derived in the type judgement to the number of β-reductions in the operational semantics. We revisit Theorems 1 and 2 and show that the cost decreases by one for each β-redex.

Lemma 7. If $A \vdash (\lambda x.e)\, e_w : v \,\&\, z$, then there exists z' such that $1+z' \leq z$ and $A' \vdash e[e_w/x] : v \,\&\, z'$ where $A' \subseteq A$.

Proof: Examining the case [beta$_\rightarrow$] in the proof of Theorem 1 we had $z = 1+z_1+z_2+z_3$, $A = A_2 \wedge A'_1 \wedge ... \wedge A'_n$ with $A_1 \wedge A'_{p_1} \wedge ... \wedge A'_{p_m} \vdash e[e_w/x] : v_1 \,\&\, z'_3$ and $z'_3 \leq z_3$. Taking $z' = z'_3$ yields $1+z' \leq 1+z_3 \leq z$ as required.   □

Lemma 8. If $e \rightarrow e'$ is a β-reduction and $A \vdash e : v \,\&\, z$ then there exists z' such that $1+z' \leq z$ and $A' \vdash e' : v \,\&\, z'$ where $A' \subseteq A$.

Proof: By lifting Lemma 7 to $\rightarrow$ (analogous to the proof of Theorem 2).   □

Theorem 3 (Correctness of cost analysis). Consider a $\mathcal{L}$-expression e for which we derive a type judgement $A \vdash e : v \,\&\, z$ with a finite cost ($z < \omega$). Then z is an upper bound on the number of β-reductions in the evaluation of e.

Proof: Using Lemma 8 (see [27]).   □

6   Conclusions and Further Work

In this paper we have presented a *type-and-effect system* [23] using intersection types to derive the costs of programs written in a simple higher-order and polymorphic functional language.

The main novelty of our approach is the use of intersection types to construct a cost analysis using a cost model built on a formal notion of sized type. We have shown both how our approach can be used to improve the quality of analysis, and to overcome the size aliasing problem in certain prototypical cases. Size aliasing is an instance of a general problem in analyses built using type-and-effect systems, where these are based on a simple Hindley-Milner polymorphism, and we anticipate that the approach described here is also applicable to domains other than cost analysis.

Since our main objective in this work was to study size aliasing and determine whether quality improvements could be made in cost analysis, we have focused on the higher-order polymorphic aspects of $\mathcal{L}$. One important omission is treatment of recursion. We are currently in the process of extending $\mathcal{L}$ with a fixpoint operator analogously to the approach used in Vasconcelos' forthcoming PhD thesis, extending the type-and-effect system and completing the associated proofs. We do not, however, anticipate any major technical problems arising from this work. While it may seem that size and time analysis are orthogonal in our system, note that the system is ready to include built-in functions relating size with time (e.g. a function with type $\mathsf{Nat}^n \xrightarrow{n} \mathsf{Nat}^n$).

Our work is undertaken in the context of a strict, purely functional language, producing cost information in the form of data structure sizes. We are extending this work in a number of ways. Firstly, we are studying extensions of this work to a call-by-need semantics, based on extending Launchbury's semantics for graph reduction [21] with appropriate cost information; and secondly, we are investigating more realistic metrics in the form of stack and heap memory allocations, and in terms of concrete time costs. Finally, although this work is restricted to a purely functional programming notation, recent work by Hofmann and Jost [17] has shown that it is possible to incorporate both assignment and some object-oriented features into a similar cost framework. We anticipate also exploring these directions in due course.

Acknowledgements

This work has been generously supported by EU Framework VI grant IST-2004-510255 (EmBounded), by EPSRC Grant EPC/0001346, by "Fundação para a Ciência e Tecnologia" grant SFRH/BD/17096/2004 and by a Support Fellowship from the Royal Society of Edinburgh.

References

1. Amtoft, T., Nielson, F., Nielson, H.R.: Type and Effect Systems: Behaviours for Concurrency. Imperial College Press (1999)
2. Banerjee, A., Jensen, T.: Modular control-flow analysis with rank 2 intersection types. Mathematical. Structures in Comp. Sci. 13(1), 87–124 (2003)
3. Barendregt, H.: The Lambda Calculus. Its Syntax and Semantics. Studies in Logic and the Foundations of Mathematics, vol. 103, North-Holland (1984)
4. Barendregt, H., Coppo, M., Dezani-Ciancaglini, M.: A filter lambda model and the completeness of type assignment. The Journal of Symbolic Logic 48(4), 931–940 (1983)
5. Carlier, S., Polakow, J., Wells, J.B., Kfoury, A.J.: System e: Expansion variables for flexible typing with linear and non-linear types and intersection types. In: Schmidt, D. (ed.) ESOP 2004. LNCS, vol. 2986, pp. 294–309. Springer, Heidelberg (2004)
6. Chin, W.-N., Khoo, S.-C.: Calculating Sized Types. Higher-Order and Symbolic Computing 14(2,3) (2001)
7. Coppo, M., Dezani-Ciancaglini, M.: An extension of the basic functionality theory for the λ-calculus. Notre-Dame Journal of Formal Logic 21(4), 685–693 (1980)
8. Coppo, M., Giannini, P.: Principal types and unification for simple intersection type systems. Information and Computation 122(1) (1995)
9. Lago, U.D., Martini, S.: An invariant cost model for the lambda calculus. In: Beckmann, A., Berger, U., Löwe, B., Tucker, J.V. (eds.) CiE 2006. LNCS, vol. 3988, pp. 105–114. Springer, Heidelberg (2006)
10. Damiani, F.: Rank 2 intersection types for local definitions and conditional expressions. ACM Transactions On Programming Languages and Systems 25(4), 401–451 (2003)
11. Damiani, F., Giannini, P.: A decidable intersection type system based on relevance. In: Theoretical Aspects of Computer Science. LNCS, Springer, Heidelberg (1994)
12. Damiani, F.: Rank-2 Intersection and Polymorphic Recursion. In: Urzyczyn, P. (ed.) TLCA 2005. LNCS, vol. 3461, pp. 146–161. Springer, Heidelberg (2005)
13. Davies, R., Pfenning, F.: Intersection types and computational effects. In: ICFP, pp. 198–208 (2000)
14. Dornic, V., Jouvelot, P., Gifford, D.K.: Polymorphic Time Systems for Estimating Program Complexity. ACM Letters on Prog. Lang. and Systems 1(1), 33–45 (1992)

15. Felleisen, M., Friedman, D.P.: Control Operators, the SECD-Machine and the λ-Calculus. In: Wirsing, M. (ed.) Formal Description of Programming Concepts III, pp. 193–217. Elsevier, Amsterdam (1986)
16. Grobauer, B.: Cost Recurrences for DML Programs. In: Proc. 2001 ACM Intl. Conf. on Functional Programming – ICFP 2001, Florence, Italy (September 2001)
17. Hofmann, M., Jost, S.: Type-based amortised heap-space analysis (for an object-oriented language). In: Sestoft, P. (ed.) ESOP 2006 and ETAPS 2006. LNCS, vol. 3924, pp. 22–37. Springer, Heidelberg (2006)
18. Hughes, R.J.M., Pareto, L., Sabry, A.: Proving the Correctness of Reactive Systems using Sized Types. In: Proc. 1996 ACM Symposium on Principles of Programming Languages – POPL '96, St Petersburg, FL (January 1996)
19. Jim, T.: Rank 2 type systems and recursive definitions. Technical Report MIT/LCS/TM-531, Massachusetts Institute of Technology, Laboratory for Computer Science (November 1995)
20. Kfoury, A.J., Wells, J.B.: Principality and decidable type inference for finite-rank intersection types. In: Assaf, J. (ed.) popl99, pp. 161–174. ACM Press, New York (1999)
21. Launchbury, J.: A Natural Semantics for Lazy Evaluation. In: Proc. 1993 ACM Symp. on Principles of Prog. Langs. – POPL '93, pp. 144–154 (1993)
22. Milner, A.J.R.G.: A Theory of Type Polymorphism in Programming. J. Computer System Sciences 17(3), 348–375 (1976)
23. Nielson, F., Nielson, H., Hankin, C.: Principles of Program Analysis. Springer, Heidelberg (1999)
24. Pottinger, G.: A type assignement for the strongly normalizable terms. In: Hindley, J.R., Seldin, J.P. (eds.) To H.B. Curry, Essays in Combinatory Logic, Lambda-calculus and Formalism, pp. 561–577. Academic Press, San Diego (1980)
25. Portillo, A.J.R., Hammond, K., Loidl, H.-W., Vasconcelos, P.: Cost analysis using automatic size and time inference. In: Peña, R., Arts, T. (eds.) IFL 2002. LNCS, vol. 2670, pp. 232–247. Springer, Heidelberg (2003)
26. Reistad, B., Gifford, D.K.: Static Dependent Costs for Estimating Execution Time. In: Proc. 1994 ACM Conference on Lisp and Functional Programming – LFP '94, pp. 65–78, Orlando, FL (June 1994)
27. Simoes, H.R., Hammond, K., Florido, M., Vasconcelos, P.: Using intersection types for cost-analysis of higher-order polymorphic functional programs. Technical report (2006) http://www.dcs.st-and.ac.uk/~kh/papers/itypes_cost.pdf
28. van Bakel, S.: Intersection Type Disciplines in Lambda Calculus and Applicative Term Rewriting Systems. PhD thesis, Department of Computer Science, University of Nijmegen (1993)
29. van Bakel, S.: Rank 2 intersection type assignment in term rewriting systems. Fundam. Inform. 26(2), 141–166 (1996)

Subset Coercions in Coq

Matthieu Sozeau

Université Paris Sud, CNRS, Laboratoire LRI, UMR 8623, Orsay, F-91405
INRIA Futurs, ProVal, Parc Orsay Université, F-91893
`sozeau@lri.fr`

Abstract. We propose a new language for writing programs with dependent types on top of the Coq proof assistant. This language permits to establish a phase distinction between writing and proving algorithms in the Coq environment. Concretely, this means allowing to write algorithms as easily as in a practical functional programming language whilst giving them as rich a specification as desired and proving that the code meets the specification using the whole Coq proof apparatus. This is achieved by extending conversion to an equivalence which relates types and subsets based on them, a technique originating from the *"Predicate subtyping"* feature of PVS and following mathematical convention. The typing judgements can be translated to the Calculus of (Co-)Inductive Constructions (Cic) by means of an interpretation which inserts coercions at the appropriate places. These coercions can contain existential variables representing the propositional parts of the final term, corresponding to proof obligations (or PVS type-checking conditions). A prototype implementation of this process is integrated with the Coq environment.

1 Introduction

There are many means to program in the Coq environment [1]. One can write programs as in ML and prove properties about them separately, losing the possibility of using dependent types in specifications, or give a rich type expressing them as a goal and use the proof tactics to solve it, producing a corresponding program by the Curry-Howard isomorphism but having much less control on its algorithmic essence. It is however difficult to mix the two methods (writing code and proving interactively) using a rich specification. Indeed, when using simple terms and types (ML) or even complex terms and types (Coq), we can have decidable type-checking. However, when using simple terms to represent inhabitants of complex types, we immediately get undecidability of type-checking, as the terms do not give enough information. Consider for example the function `tail` which returns the tail of a non-empty list. In ML:

```
let tail = function hd :: tl -> tl | [] -> assert(false)
```

This function is partial, its domain is reduced to non-empty lists. In Coq, we would rather write the following:

T. Altenkirch and C. McBride (Eds.): TYPES 2006, LNCS 4502, pp. 237–252, 2007.
© Springer-Verlag Berlin Heidelberg 2007

```
Definition tail (l : list A) : option (list A) :=
  match l with hd :: tl => Some tl | [] => None end.
```

The option type is the usual way to encode partiality in COQ. However we would more naturally constrain the argument l to non-empty lists to be more faithful to the original ML code, and also attach a property to the result to express more precisely what `tail` does:

```
Definition tail (l : list A) : (l <> []) ->
  { l' : list A | exists a, a :: l' = l }.
```

Now the definition's body requires some heavy plumbing of the code which is not affordable when programming. Moreover, the user is forced to give a proof term justifying that l <> [] when calling the function. We propose a solution to overcome these difficulties, based on the *Predicate subtyping* [2] technique of PVS [3]. It is separated into two phases. First, we have a weak, *decidable* type-checking procedure which does not require proofs to be present in the code when constructing objects of a subset type. In our new language RUSSELL, the following is a well-typed term:

```
Program Definition tail ( l : list A | l <> [] ) :
  { l' : list A | exists a, a :: l' = l } :=
  match l with hd :: tl -> tl | [] -> [] end.
```

The specification shows that we are defining a partial function and enforces a relation between input and output, using a dependent type, yet the code remains as simple as the ML definition. This is only possible because we do not require the user to write proofs in the code. After type-checking, there is an automatic elaboration into partial COQ terms, which collects obligations the user has to prove. In our example, the proof assistant will ask the user to prove that:

1. The list tl has the property exists a, a :: tl = hd :: tl, and
2. In the context where l is a non-empty list and l = [], the list [] has the property exists a, a :: [] = [] (which should be obvious as the context is contradictory).

This solution also provides facilities to express properties with a more mathematical flavour using subsets, bridging a gap between mathematical convention and type theory.

The PROGRAM tactic by C. Parent [4] had the same goal as ours but a different method for achieving it. It was strongly linked to the extraction mechanism included in COQ, both theoretically and practically. Sketching the mechanism, she defined a weakened extraction operation on CIC terms which could be inverted because it left enough information in the extracted term to rebuild a partial proof. The mechanism, while general and theoretically well thought out, required some heuristics and did not integrate smoothly with the COQ environment. In particular it lacked the pervasiveness our method has, being applicable in a wide variety of situations in the proof assistant environment.

Instead of trying to find a general method for synthesizing proofs from programs in the Calculus of (Co-)Inductive Constructions, we have integrated a method which permits to link strong specifications and purely algorithmical code. This method, known as *Predicate subtyping* in the PVS system, has been used with great success and fits naturally with how we write specifications using the subset type [5]. The main contribution of this paper is to show how the *Predicate subtyping* method can be adapted in a proof assistant with proof terms, which formally justifies the extension in the first place.

The remaining of the article is organised as follows: in section 2 we present a type system based on Cic, which integrates subset "subtyping", and prove decidability of type-checking. Then, we show how it relates to Cic by means of translations between judgements of the two systems in section 3. Next we present a prototype implementation in section 4, and finally we give our conclusions on this work and discuss future directions in section 5.

2 Russell

The name of our language is an homage to the mathematician Bertrand Russell who discovered the famous paradox of set theory with the unrestricted comprehension axiom. In this theory, it is possible to construct the set $X = \{x \mid x \notin x\}$ whose definition is circular. Clearly, if $X \in X$ then $X \notin X$ and if $X \notin X$ then $X \in X$, hence we have an inconsistency. Russell was one of the pioneers of type theory when he devised a set theory with a restricted comprehension axiom that permitted to create subsets only from already defined subsets, forbidding the definition of X.

In Coq, the distinction between informative and propositional parts of a term is formalised by the Set/Prop sorts. In Russell, we have special support for propositions appearing in subset types. By delimiting the use of propositions, we can separate code from proof.

This idea is already present in mathematics. When you have an element of subset $\{x \in S \mid P\}$, you can freely forget about the property P and use any operation which is defined on S. Conversely, when you want to use an operation defined on a subset, say $f : \{x : \mathbb{N} \mid P\} \rightarrow X$, you usually prove *first* that you apply it only to elements having the desired property. For example $(f\ 2)$ is a correct application only if $(P[2/x])$ is provable. In the context of formal development of programs, such a workflow is not entirely satisfactory because it forces one to create objects and prove properties about them *at the same time*. In Coq, we are forced to apply f to an object (elt nat $\lambda x : \mathtt{nat}.P\ 2\ p$) where p is a proof of $(P[2/x])$ (Figure 1 presents the definition of the subset type in Coq). We would like to be able to prove that our usage of partial functions is correct only after the program is written. Subsets are particularly well-suited in this respect because they separate the objects we want to manipulate and their associated properties. Similar treatments of subsets as a construct to separate informative and propositional parts of an object include the work by Nordström et al. [5] and subset types in NuPRL [6].

$$\text{SUBSET} \ \frac{\Gamma \vdash A : \mathtt{Set} \quad \Gamma, x : A \vdash P : \mathtt{Prop}}{\Gamma \vdash \{\, x : A \mid P \,\} : \mathtt{Set}}$$

$$\text{ELEMENT} \ \frac{\Gamma \vdash a : A \quad \Gamma \vdash p : P[a/x] \quad \Gamma \vdash \{\, x : A \mid P \,\} : \mathtt{Set}}{\Gamma \vdash \mathrm{elt}\ A\ (\lambda x : A.P)\ a\ p : \{\, x : A \mid P \,\}}$$

$$\text{SUBSET-}\sigma_1 \ \frac{\Gamma \vdash t : \{\, x : A \mid P \,\}}{\Gamma \vdash \sigma_1\ t : A} \qquad \text{SUBSET-}\sigma_2 \ \frac{\Gamma \vdash t : \{\, x : A \mid P \,\}}{\Gamma \vdash \sigma_2\ t : P[\sigma_1\ t/x]}$$

Fig. 1. Subset type in CIC

2.1 From *Predicate Subtyping* to Subset Equivalence

The *Predicate subtyping* mechanism [2] is an extension of the PVS type system which internalises this idea. Concretely, this means that the following rules are derivable in PVS:

$$\frac{\Gamma \vdash t : \{x : T \mid P\}}{\Gamma \vdash t : T} \qquad \frac{\Gamma \vdash t : T \quad \Gamma \vdash P[t/x]}{\Gamma \vdash t : \{x : T \mid P\}}$$

The first one formalises the fact that an object of a subset based on T is an object of type T. The second one permits using an object of type T as an object of type $\{x : T \mid P\}$, but it generates a *type-checking condition* $\Gamma \vdash P[t/x]$ which will need to be discharged later. Effectively, the typing algorithm of PVS collects the conditions that must be satisfied for the term to be accepted as a valid definition. However, the acceptance criteria can be rather large in PVS. By design, when proving in PVS, the trusted code base (TCB hereafter) is the entire system, not only the typing system but also the various decision procedures and tactics used to build proofs automatically or interactively. It is nonetheless a widely used proof assistant and the predicate subtyping feature has apparently helped to build a consequent library of certified code. On the other hand, CoQ has a small TCB and greater expressiveness but less automation and methodology to build certified programs. We capitalise on the PVS success to make CoQ more usable for this kind of tasks, and as we will see, it will have other benefits.

2.2 A Weaker Type System

To formalise this idea in CoQ, we simply weaken the type system so that it doesn't require the terms to contain the proof components for objects of subset types. This permits to have a simple language for code while retaining the richness of CoQ's specification language. Once we have a derivation in this new type system, we can translate it to a partial CoQ derivation, where the missing parts are represented by metavariables. It can then be completed by instantiating these holes with actual proofs.

$$\text{WF-EMPTY} \frac{}{\vdash [] \ \textbf{wf}} \qquad \text{WF-VAR} \frac{\Gamma \vdash A : s}{\vdash \Gamma, x : A \ \textbf{wf}} \ s \in \mathcal{S} \land x \notin \Gamma$$

$$\text{VAR} \frac{\vdash \Gamma \ \textbf{wf} \quad x : A \in \Gamma}{\Gamma \vdash x : A} \qquad \text{AXIOM} \frac{\vdash \Gamma \ \textbf{wf}}{\Gamma \vdash s_1 : s_2} \ (s_1, s_2) \in \mathcal{A}$$

$$\text{PROD} \frac{\Gamma \vdash T : s_1 \quad \Gamma, x : T \vdash U : s_2}{\Gamma \vdash \Pi x : T.U : s_2}$$

$$\text{ABS} \frac{\Gamma \vdash \Pi x : T.U : s \quad \Gamma, x : T \vdash M : U}{\Gamma \vdash \lambda x : T.M : \Pi x : T.U} \qquad \text{APP} \frac{\Gamma \vdash f : \Pi x : V.W \quad \Gamma \vdash u : V}{\Gamma \vdash (fu) : W[u/x]}$$

$$\text{SUM} \frac{\Gamma \vdash T : s \quad \Gamma, x : T \vdash U : s}{\Gamma \vdash \Sigma x : T.U : s} \ s \in \{\text{Prop}, \text{Set}\}$$

$$\text{PAIR} \frac{\Gamma \vdash \Sigma x : T.U : s \quad \Gamma \vdash t : T \quad \Gamma \vdash u : U[t/x]}{\Gamma \vdash (t, u)_{\Sigma x : T.U} : \Sigma x : T.U}$$

$$\text{PI-1} \frac{\Gamma \vdash t : \Sigma x : T.U}{\Gamma \vdash \pi_1 t : T} \qquad \text{PI-2} \frac{\Gamma \vdash t : \Sigma x : T.U}{\Gamma \vdash \pi_2 t : U[\pi_1 t/x]}$$

$$\text{CONV} \frac{\Gamma \vdash t : U \quad \Gamma \vdash U \equiv_{\beta\pi} T : s}{\Gamma \vdash t : T}$$

Fig. 2. CIC typing judgement

$$\text{COERCE} \frac{\Gamma \vdash t : U \quad \Gamma \vdash U \rhd T : s}{\Gamma \vdash t : T}$$

$$\text{SUBSET} \frac{\Gamma \vdash U : \text{Set} \quad \Gamma, x : U \vdash P : \text{Prop}}{\Gamma \vdash \{ \ x : U \mid P \ \} : \text{Set}}$$

Fig. 3. RUSSELL new rules

Type system. RUSSELL's type system is based on the Calculus of Inductive Constructions (figure 2) [7], with sigma types but without universes. This restriction may be removed in future work, but causes no problem for our main purpose which is programming. We omit inductive constructs here, as they can be considered as constants and leave their treatment as future work. The judgement $\Gamma \vdash t : T$ means t is a well-typed term of type T in environment Γ.

Following the presentation as a Pure Type System, the set of sorts $\mathcal{S}$ is defined as $\{\text{Set}, \text{Prop}, \text{Type}\}$. As usual, we let s, s_i for $i \in \mathbb{N}$ range over sorts. The axioms are $\mathcal{A} = \{(\text{Set}, \text{Type}), (\text{Prop}, \text{Type})\}$ and the set of rules $\mathcal{R}$ is defined by the functional relation $\forall s_1 s_2, (s_1, s_2, s_2) \in \mathcal{R}$. We allow products $\Pi x : A.B$ where $A : \text{Prop}$ and $B : \text{Set}$ but the user is encouraged to encode them as $\Pi x : \{\_ : \text{unit} \mid A\}.B$ (with unit the type containing a single element tt) to benefit from the subtyping mechanism which we will define later.

The reduction rules of this system are the standard β and π for projections. We denote by $x^\downarrow$ the weak head normal form of x. We use a standard judgemental equality $\Gamma \vdash T \equiv_{\beta\pi} U : s$ meaning T, U are well-typed, convertible terms of sort s in context Γ.

We allow to form dependent sums $\Sigma x : U.V$ (SUM, PAIR, PI-1, PI-2 rules) only when $U, V : \mathtt{Set}$ or $U, V : \mathtt{Prop}$. The first sum represents dependent pairs, useful when defining functions returning a tuple, the latter represents dependent pairs of propositions, which is most often used as conjunction of propositions. The dependent pair with $U : \mathtt{Set}$ and $V : \mathtt{Prop}$ is the subset type which we distinguish (rule SUBSET). The last possible pair $U : \mathtt{Prop}, V : \mathtt{Set}$ is forbidden because it corresponds to a pair where the last component, which is informative, may depend on a particular proof of the proposition U. This is quite contrary to the mantra "Computations do not depend on proofs" which governs our programming language. If there is no dependence then the components can be swapped.

In RUSSELL, the conversion rule CONV is replaced by a new subsumption rule COERCE (figure 3 on the previous page) which will implement the subset equivalence. The judgement $\Gamma \vdash T \triangleright U : s$ means T is equivalent to U in environment Γ, both sorted with s. The essence of our equivalence is to identify subset types if they have equivalent supports, hence we have the following property:

Proposition 1 (Subset erasure and consistency). *If we erase subset types from* RUSSELL *terms and rules, leaving only the supports, we get a valid* CIC *term and derivation, hence* $\bot$ *is not provable in* RUSSELL.

Proof. By erasing subsets in RUSSELL rules, the SUBSET rule becomes admissible and the COERCE rule becomes CONV. By eliminating the subset type in a term we get a valid CIC term. Hence to each RUSSELL derivation corresponds a CIC derivation.

Equivalence. We have renamed the technique from *Predicate subtyping* to *subset equivalence* because we have a symmetric relation, contrary to usual subtyping relations. It also conveys the idea that it can include the usual $\beta\pi$-conversion directly in the judgement. The judgement $\Gamma \vdash T \triangleright U : s$ (figure 4 on the facing page) reads T is equivalent to U in environment Γ, both being sorted by s.

The rule $\triangleright$-CONV integrates $\beta\pi$-conversion in the judgement. We use a judgemental equality here, which will be refined by the usual conversion relation later. The $\triangleright$-TRANS rule ensures that our judgement builds an *equivalence* and has proper modularity. It is trivial to check that we have symmetry given the symmetry of the definitional equality. The next two rules ($\triangleright$-PROD and $\triangleright$-SUM) do context closure for dependent products and sums. It is remarkable that we use contravariance for domains in the $\triangleright$-PROD rule, not restricting to invariance as in PVS. It is accessory here, as we could have used *co*variance and still get the same judgements because we have symmetry. However, it will become important when we create coercions (see figure 8 on page 247).

The really interesting rules are $\triangleright$-SUBSET and $\triangleright$-PROOF. The first one allows to use an object of a subset type as an object of its support type. The later allows (maybe abusively) to consider an object of any type as an object of any

$$\triangleright\text{-}\textsc{Conv}\ \frac{\Gamma \vdash T \equiv_{\beta\pi} U : s}{\Gamma \vdash T \triangleright U : s} \qquad \triangleright\text{-}\textsc{Trans}\ \frac{\Gamma \vdash S \triangleright T : s \quad \Gamma \vdash T \triangleright U : s}{\Gamma \vdash S \triangleright U : s}$$

$$\triangleright\text{-}\textsc{Prod}\ \frac{\Gamma \vdash U \triangleright T : s_1 \quad \Gamma, x : U \vdash V \triangleright W : s_2}{\Gamma \vdash \Pi x : T.V \triangleright \Pi x : U.W : s_2}$$

$$\triangleright\text{-}\textsc{Sum}\ \frac{\Gamma \vdash T \triangleright U : s \quad \Gamma, x : T \vdash V \triangleright W : s}{\Gamma \vdash \Sigma x : T.V \triangleright \Sigma y : U.W : s}\ s \in \{\mathtt{Set}, \mathtt{Prop}\}$$

$$\triangleright\text{-}\textsc{Subset}\ \frac{\Gamma \vdash U \triangleright V : \mathtt{Set} \quad \Gamma, x : U \vdash P : \mathtt{Prop}}{\Gamma \vdash \{\, x : U \mid P \,\} \triangleright V : \mathtt{Set}}$$

$$\triangleright\text{-}\textsc{Proof}\ \frac{\Gamma \vdash U \triangleright V : \mathtt{Set} \quad \Gamma, x : V \vdash P : \mathtt{Prop}}{\Gamma \vdash U \triangleright \{\, x : V \mid P \,\} : \mathtt{Set}}$$

Fig. 4. RUSSELL conversion

subset based on this type. We must check that the property P is well-formed, but we do not care about its provability.

Properties. We have proved some of the metatheory for this system in CoQ [8]. We have assumed that it is strongly normalising (SN) but we have proved subject reduction (SR) for it: if $\Gamma \vdash t : T$ and $t \to_{\beta\pi} t'$ then $\Gamma \vdash t' : T$ (our proof of SR does not depend on SN). Gang Chen [9] has studied various type systems from the λ-cube extended by subtyping or coercive subtyping [10], including the Calculus of Constructions, and proved such results as SN and SR for them. It requires a very careful analysis of the system to avoid cycles in the proof due to the presence of dependent types and conversion. However we preferred to adapt the method of Robin Adams [11] to prove SR, because Gang Chen's method seemed very tied to the peculiarities of the system he studied.

So, apart from strong normalisation, we have shown all the usual structural and metatheoretic properties of a dependent programming language, like weakening, thinning and substitution, stability by context coercion, etc. We stop here on the subject of theoretical properties of this first system, as it gives no new insights for our purpose and we will focus more on the algorithmic system's properties. Besides, the details of the formalisation and the proof fall outside the scope of this paper.

Algorithmic Typing

We use a standard transformation to get an algorithm from our typing rules, removing the conversion (or subsumption) rule (here COERCE) and integrating it to the premises of the other rules. Our algorithmic judgement $\Gamma \vdash_\bullet t : T$ (figure 5 on the following page) reads: t has type T in environment Γ. We need to introduce the notion of support for subset types in order to define our

$$\text{App} \ \frac{\Gamma \vdash_\bullet f : T \quad \mu_\bullet(T) = \Pi x : V.W \quad \Gamma \vdash_\bullet u : U \quad \Gamma \vdash_\bullet U \triangleright_\bullet V : s}{\Gamma \vdash_\bullet (fu) : W[u/x]}$$

$$\text{Pair} \ \frac{\Gamma \vdash_\bullet t : T' \quad \Gamma \vdash_\bullet T' \triangleright_\bullet T : s}{\dfrac{\Gamma \vdash_\bullet \Sigma x : T.U : s \quad \Gamma \vdash_\bullet u : U' \quad \Gamma \vdash_\bullet U' \triangleright_\bullet U[t/x] : s}{\Gamma \vdash_\bullet (t,u)_{\Sigma x:T.U} : \Sigma x : T.U}}$$

$$\text{Pi-1} \ \frac{\Gamma \vdash_\bullet t : S \quad \mu_\bullet(S) = \Sigma x : T.U}{\Gamma \vdash_\bullet \pi_1 \, t : T} \qquad \text{Pi-2} \ \frac{\Gamma \vdash_\bullet t : S \quad \mu_\bullet(S) = \Sigma x : T.U}{\Gamma \vdash_\bullet \pi_2 \, t : U[\pi_1 \, t/x]}$$

Fig. 5. RUSSELL algorithmic typing, new rules

algorithmic system. Indeed, when typing an application, we need to ensure that the object we apply can be seen as an object of a product type modulo the equivalence.

$$\mu_\bullet(x) \Rightarrow \mu_\bullet(U) \text{ if } x^\downarrow = \{ \, x : U \mid P \, \}$$
$$\mu_\bullet(x) \Rightarrow x \qquad \text{otherwise}$$

Fig. 6. $\mu_\bullet()$ definition

The next thing to do is to construct a decidable judgement for equivalence given two types. We denote it by $\Gamma \vdash_\bullet T \triangleright_\bullet U : s$ (figure 7). In our case, we have to check that definitional equality is decidable, which is the case as we can replace the judgemental equality with the usual untyped convertibility $\equiv_{\beta\pi}$ (we proved $\Gamma \vdash T \equiv_{\beta\pi} U : s \Leftrightarrow T \equiv_{\beta\pi} U \wedge \Gamma \vdash T, U : s$), which is decidable on well typed terms. We also need to eliminate the transitivity rule which is not syntax-directed. That is the purpose of the following theorem:

Theorem 1 (Admissibility of transitivity). *If* $\Gamma \vdash_\bullet T \triangleright_\bullet U : s$ *and* $\Gamma \vdash_\bullet U \triangleright_\bullet V : s$, *then* $\Gamma \vdash_\bullet T \triangleright_\bullet V : s$.

Proof. By induction on the sum of the depths of the two derivations.

Finally, we need to restrict the application of the $\triangleright$-CONV rule when no other rules apply and the two types are in head normal form. We can then prove:

Theorem 2 (Decidability of algorithmic typing). $\exists T, \Gamma \vdash_\bullet t : T$, $\Gamma \vdash_\bullet t : T$ *and* $\Gamma \vdash_\bullet T \triangleright_\bullet U : s$ *are decidable problems.*

Proof. Rules are syntax-directed.

We denote by $\mathbf{type}_\Gamma(t)$ the function which returns the type of term t in context Γ such that $\Gamma \vdash_\bullet t : \mathbf{type}_\Gamma(t)$, when it exists.

As usual, we have the following correspondence between the two systems:

Theorem 3 (Soundness). *If* $\Gamma \vdash_\bullet t : T$ *then* $\Gamma \vdash t : T$. *If* $\Gamma \vdash_\bullet T \triangleright_\bullet U : s$ *then* $\Gamma \vdash T \triangleright U : s$.

$$\rhd\text{-Conv} \frac{T \equiv_{\beta\pi} U \quad \Gamma \vdash_\bullet T, U : s}{\Gamma \vdash_\bullet T \rhd_\bullet U : s} \quad T = T^\downarrow \wedge T \neq \Pi, \Sigma, \{|\} \wedge U = U^\downarrow$$

$$\rhd\text{-}\downarrow \frac{\Gamma \vdash_\bullet T^\downarrow \rhd_\bullet U^\downarrow : s \quad \Gamma \vdash_\bullet T, U : s}{\Gamma \vdash_\bullet T \rhd_\bullet U : s} \quad T \neq T^\downarrow \vee U \neq U^\downarrow$$

$$\rhd\text{-Prod} \frac{\Gamma \vdash_\bullet U \rhd_\bullet T : s_1 \quad \Gamma, x : U \vdash_\bullet V \rhd_\bullet W : s_2}{\Gamma \vdash_\bullet \Pi x : T.V \rhd_\bullet \Pi x : U.W : s_2}$$

$$\rhd\text{-Sum} \frac{\Gamma \vdash_\bullet T \rhd_\bullet U : s \quad \Gamma, x : T \vdash_\bullet V \rhd_\bullet W : s}{\Gamma \vdash_\bullet \Sigma x : T.V \rhd_\bullet \Sigma x : U.W : s} \quad s \in \{\mathsf{Set}, \mathsf{Prop}\}$$

$$\rhd\text{-Proof} \frac{\Gamma \vdash_\bullet T \rhd_\bullet U : \mathsf{Set}}{\Gamma \vdash_\bullet T \rhd_\bullet \{\, x : U \mid P \,\} : \mathsf{Set}} \quad T = T^\downarrow$$

$$\rhd\text{-Subset} \frac{\Gamma \vdash_\bullet U \rhd_\bullet T : \mathsf{Set}}{\Gamma \vdash_\bullet \{\, x : U \mid P \,\} \rhd_\bullet T : \mathsf{Set}}$$

Fig. 7. RUSSELL algorithmic equivalence

Theorem 4 (Completeness). *If $\Gamma \vdash t : T$ then there exists T', s so that $\Gamma \vdash_\bullet t : T'$ and $\Gamma \vdash_\bullet T' \rhd_\bullet T : s$. If $\Gamma \vdash T \rhd U : s$ then $\Gamma \vdash_\bullet T \rhd_\bullet U : s$.*

Finally, we can state the desired property of our typing system:

Corollary 1 (Decidability of declarative typing). $\Gamma \vdash t : T$ *is decidable.*

2.3 From CIC to RUSSELL

We have presented a calculus based on CIC with a stronger equivalence but with restricted typing rules. We will now see how RUSSELL and CIC relate formally. We can build a forgetful map from terms of CIC to RUSSELL (interpreting inductives as constants). It bears relation to the ϵ extraction function defined by Werner in [12]. Essentially, RUSSELL terms do not contain or manipulate logical information attached to objects of subset types, hence we must forget about it when translating.

We define the forgetful map $()^\circ$ from CIC terms to RUSSELL terms plus a distinguished object $\perp$ as an homomorphism on terms except for the following cases:

$$
\begin{aligned}
(\sigma_1\ t)^\circ &= t^\circ \\
(\mathsf{elt}\ T\ P\ t\ p)^\circ &= t^\circ \\
(\sigma_2\ t)^\circ &= \perp
\end{aligned}
$$

Definition 1 (Definedness). $()^\circ$ *is defined on t if t° does not contain $\perp$.*

We can now prove the following:

Theorem 5 (Forgetful map correctness). *If $\Gamma \vdash_{CCI} t : T$ then $\Gamma^\circ \vdash t^\circ : T^\circ$ if $()^\circ$ is defined on Γ, t and T.*

Practically, this means we can use almost all existing definitions in the CoQ environment. This map is involutive and the identity on RUSSELL terms as they do not contain the constructor elt or subset projections. Moreover, this erasure function will be defined on all CoQ terms elaborated from RUSSELL. Indeed, the second projection of a subset element will only appear in the second component of a subset element. Otherwise, we would have used the projection directly in the RUSSELL term, where it does not exist. In fact, as we can derive the judgement $0 : \{ x : \mathbb{N} \mid x \neq 0 \}$ in RUSSELL, allowing the second projection would permit to derive $0 \neq 0$, an inconsistency. Hence we left only the subset type forming rule (figure 3 on page 241) in RUSSELL, while the introduction and first projection of subset elements are internalized by the coercion judgement.

3   From RUSSELL to CIC?

We now build an interpretation $[\![t]\!]_\Gamma$ of RUSSELL terms t in environment Γ into CIC terms. We will check at the end that it respects the $()^\circ$ operator in the sense that if t is well-typed in Γ then $[\![t]\!]_\Gamma^\circ = t$.

Our interpretation will build a full-fledged CIC term from its algorithmic skeleton and a rich type written in RUSSELL. Obviously, we cannot infer proof terms where they are needed in CIC, but we can build a *partial* term, leaving typed holes where proofs are needed. Hence we add a rule to build existential variables (or metavariables) in the target calculus ($\vdash_?$ denotes the new system's typing judgement):

$$\frac{\Gamma \vdash_? P : \mathtt{Prop}}{\Gamma \vdash_? ?_P : P}$$

We restrict it to objects of type Prop because we consider that the informative part of the algorithm has been entirely given in the original term.

We are ready to build the interpretation, which should have the following property:

Theorem 6 (Interpretation correctness). *If $\Gamma \vdash_\bullet t : T$ then $[\![\Gamma]\!] \vdash_? [\![t]\!]_\Gamma : [\![T]\!]_\Gamma$.*

The proof of this proposition is the main technical contribution of this work, it also represented the main difficulty. The remainder of this section is organised as follows: first we define the explicit coercion derivation algorithm which will permit to put proof obligations in the terms, then we define the interpretation of terms which is mutually recursive with the previous algorithm. Finally we present a proof sketch of the aforementioned theorem.

3.1   Explicit Coercions

The derivation of explicit coercions (figure 8 on the facing page) is based on the algorithmic equivalence derivations. The side-conditions of the rules have not changed, so we omit them for better readability. The judgement $\Gamma \vdash_{CCI}$

$c : T \rhd_\bullet U : s$ builds a coercion c from $[\![T]\!]_\Gamma$ to $[\![U]\!]_\Gamma$ in context $[\![\Gamma]\!]$. We use the RUSSELL types to drive the derivation but the resulting object c will be a well-typed CoQ term (lemma 1). We denote $\mathbf{coerce}_\Gamma \ T \ U$ the function which builds c given Γ, T and U.

$$\rhd\text{-Conv} \ \frac{T \equiv_{\beta\pi} U}{\Gamma \vdash_{CCI} \bullet : T \rhd_\bullet U : s} \qquad \rhd\text{-}\!\downarrow \ \frac{\Gamma \vdash_{CCI} c : T^\downarrow \rhd_\bullet U^\downarrow : s}{\Gamma \vdash_{CCI} c : T \rhd_\bullet U : s}$$

$$\rhd\text{-Prod} \ \frac{\Gamma \vdash_{CCI} c_1 : U \rhd_\bullet T : s_1 \quad \Gamma, x : U \vdash_{CCI} c_2 : V \rhd_\bullet W : s_2}{\Gamma \vdash_{CCI} \lambda x : [\![U]\!]_\Gamma. c_2[\bullet \ (c_1[x])] : \Pi x : T.V \rhd_\bullet \Pi x : U.W : s_2}$$

$$\rhd\text{-Sum} \ \frac{\Gamma \vdash_{CCI} c_1 : T \rhd_\bullet U : s \quad \Gamma, x : T \vdash_{CCI} c_2 : V \rhd_\bullet W : s}{\Gamma \vdash_{CCI} (c_1[\pi_1 \ \bullet], c_2[\pi_2 \ \bullet][\pi_1 \ \bullet/x])_{[\![\Sigma x:U.W]\!]_\Gamma} : \Sigma x : T.V \rhd_\bullet \Sigma x : U.W : s}$$

$$\rhd\text{-Subset} \ \frac{\Gamma \vdash_{CCI} c : U \rhd_\bullet T : \mathsf{Set} \quad \Gamma \vdash_\bullet \{ \, x : U \mid P \, \} : \mathsf{Set}}{\Gamma \vdash_{CCI} c[\sigma_1 \ \bullet] : \{ \, x : U \mid P \, \} \rhd_\bullet T : \mathsf{Set}}$$

$$\rhd\text{-Proof} \ \frac{\Gamma \vdash_{CCI} c : T \rhd_\bullet U : \mathsf{Set} \quad \Gamma \vdash_\bullet \{ \, x : U \mid P \, \} : \mathsf{Set}}{\Gamma \vdash_{CCI} \mathsf{elt} \quad c \ ?_{[\![P]\!]_{\Gamma,x:U}[c/x]} : T \rhd_\bullet \{ \, x : U \mid P \, \} : \mathsf{Set}}$$

Fig. 8. Coercion derivation

Coercions are formalised as multi-holes evaluation contexts, with $\bullet$ the denotation of a hole. We define the instantiation $C[x]$ of a context C by a term x as simultaneous substitution of x for every $\bullet$ in C. Hence if C is an evaluation context and x a term, $C[x]$ is a term (i.e., it cannot contain $\bullet$). The holes will denote the object to which the coercion is applied.

As conversion will be preserved by the interpretation $[\![]\!]$, the rule $\rhd$-Conv derives an empty coercion, as the target system will be able to derive itself that $[\![T]\!]_\Gamma \equiv [\![U]\!]_\Gamma$. Similarly for $\rhd$-$\!\downarrow$, we rely on the target system's conversion rule so that the coercion c of codomain $[\![U^\downarrow]\!]_\Gamma$ can be seen as an object of type $[\![U]\!]_\Gamma$.

Next, we have the rules for products and sums, which compose coercions. In the $\rhd$-Prod rule, we first coerce the argument to an object of type $[\![T]\!]_\Gamma$. Then we apply it to the coerced function, resulting in an object $\bullet \ (c_1[x])$ of type $[\![V]\!]_{\Gamma,x:T}[c_1[x]/x]$. We will see later that this type is equivalent to $[\![V]\!]_{\Gamma,x:U}$, so we can apply the second coercion and get an object of type $[\![W]\!]_\Gamma$. There is a little twist when coercing the second component of a dependent sum. Indeed, we must simultaneously substitute the first projection of the coerced sum in the second coercion which was typed in environment $\Gamma, x : T$ and also instantiate the coercion with the second component. Clearly, the $\bullet$ of the former shouldn't be instantiated with the later, so we must first instantiate c_2 and then substitute $\pi_1 \ \bullet$ for x (x cannot appear in $\pi_2 \ t$ where $t : \Sigma x : T.V$).

Now for the original part, the $\rhd$-Subset rule simply inserts a projection before applying the inductively defined coercion on the support of the subset. On the other hand, $\rhd$-Proof creates an object of a subset using its sole constructor

elt (with type parameters omitted for brevity). The object part is given by the coercion to the support of the target subset while the proof part is promised using a metavariable of type $[\![P]\!]_{\Gamma,x:U}[c/x]$ which we will turn later into a proof obligation.

3.2   Interpretation of Terms

The interpretation of terms from RUSSELL to CIC? (figure 9) is a straightforward recursive traversal of algorithmic typing derivations. In other, more practical words, we can do interpretation of terms simultaneously with typing in our implementation, just like for the existing coercion system of COQ. We describe

$$
\begin{aligned}
[\![x]\!]_\Gamma &= x \\
[\![s]\!]_\Gamma &= s \qquad\qquad\qquad s \in \{\mathsf{Set}, \mathsf{Prop}, \mathsf{Type}\} \\
[\![\Pi x : T.U]\!]_\Gamma &= \Pi x : [\![T]\!]_\Gamma.[\![U]\!]_{\Gamma,x:T} \\
[\![\{\, x : U \mid P \,\}]\!]_\Gamma &= \{\, x : [\![U]\!]_\Gamma \mid [\![P]\!]_{\Gamma,x:U} \,\} \\
[\![\Sigma x : T.U]\!]_\Gamma &= \Sigma x : [\![T]\!]_\Gamma.[\![U]\!]_{\Gamma,x:T} \\
[\![\lambda x : \tau.v]\!]_\Gamma &= (\lambda x : [\![\tau]\!]_\Gamma.[\![v]\!]_{\Gamma,x:\tau})
\end{aligned}
$$

$$
\begin{aligned}
[\![f\ u]\!]_\Gamma \;=\; &\mathbf{let}\ F = \mathbf{type}_\Gamma(f)\ \mathbf{and}\ U = \mathbf{type}_\Gamma(u)\ \mathbf{in} \\
&\mathbf{let}\ (\Pi x : V.W) = \mu_\bullet(F)\ \mathbf{in} \\
&\mathbf{let}\ \pi = \mathbf{coerce}_\Gamma\ F\ (\Pi x : V.W)\ \mathbf{in} \\
&\mathbf{let}\ c = \mathbf{coerce}_\Gamma\ U\ V\ \mathbf{in} \\
&(\pi[[\![f]\!]_\Gamma])\ (c[[\![u]\!]_\Gamma])
\end{aligned}
$$

$$
\begin{aligned}
[\![(t, u)_{\Sigma x:T.U}]\!]_\Gamma \;=\; &\mathbf{let}\ T' = \mathbf{type}_\Gamma(t)\ \mathbf{in} \\
&\mathbf{let}\ ct = \mathbf{coerce}_\Gamma\ T'\ T\ \mathbf{in} \\
&\mathbf{let}\ U' = \mathbf{type}_\Gamma(u)\ \mathbf{in} \\
&\mathbf{let}\ cu = \mathbf{coerce}_\Gamma\ U'\ U[t/x]\ \mathbf{in} \\
&(ct[[\![t]\!]_\Gamma], cu[[\![u]\!]_\Gamma])_{[\![\Sigma x:T.U]\!]_\Gamma}
\end{aligned}
$$

$$
\begin{aligned}
[\![\pi_i\ t]\!]_\Gamma \;=\; &\mathbf{let}\ T = \mathbf{type}_\Gamma(t)\ \mathbf{in} \qquad\qquad i \in \{1, 2\} \\
&\mathbf{let}\ \Sigma x : V.W = \mu_\bullet(T)\ \mathbf{in} \\
&\mathbf{let}\ c = \mathbf{coerce}_\Gamma\ T\ (\Sigma x : V.W)\ \mathbf{in} \\
&\pi_i\ c[[\![t]\!]_\Gamma]
\end{aligned}
$$

Fig. 9. Interpretation of terms

the case of application here; others are similar. First we get the RUSSELL types F and U of the function f and argument u by calling the typing function on both terms. Then we must ensure that F can indeed be seen as a product using the $\mu_\bullet()$ operator. We can build the coercion between F and this product and between U and its domain. We finally instantiate these coercions by their corresponding interpreted objects and return their application. Well-typedness of a term in RUSSELL is a sufficient condition for $[\![t]\!]$ to be defined.

3.3   A Little More Expressiveness

One may wonder how can we interpret RUSSELL terms into a system with only $\beta\pi$-equivalence and still be a simple extension of CIC. Indeed, consider the following example. Suppose we have two types A, B with coercions: $c : A \rhd_\bullet B$ and $d : B \rhd_\bullet A$. Additionally, let $P : A \to Prop$ with introduction terms $p : \Pi x : A, P\,x$ and $q : \Pi x : B, P\,x$. Consider the following well-typed RUSSELL term $p\,x =_P x\,q\,x$ in context $\Gamma = x : A$. By interpretation into CIC, we have $[\![P]\!]_\Gamma = P$, $[\![p]\!]_\Gamma = p$ and $[\![q]\!]_\Gamma = q : \Pi x : B, P\,d[x]$. Now the interpretation of the equality gives: $p\,x =_P x\,[\![q\,x]\!]_\Gamma$ where $[\![q\,x]\!]_\Gamma \equiv q\,c[x]$. The right hand side has type $P\,d[c[x]]$ which must be convertible with $P\,x$ for the whole term to be well-typed. This indicates that coercions must be unique (so that $d \circ c \equiv id_A \equiv \bullet$), which implies that we have to include η conversion in the definitional equality. For example if $A = \Pi x : X.Y$ and $B = \Pi x : X'.Y'$ with $X = X'$ and $Y = Y'$, we have $d \circ c = \lambda x : X.\bullet\,x \circ \lambda x : X'.\bullet\,x = \lambda x : X.\bullet\,x$ which must be definitionally equal to $\bullet$. Similarly, we must add η rules for all introduction terms of our language to ensure uniqueness of coercions. It is an important property of our system, as it corresponds in practice to the determinism of proof-obligations generation. The complete equational theory of CIC? is given in figure 10. It includes η, surjective pairing for dependent sums and subsets (ρ) and proof-irrelevance for the second component of subset objects (τ) and is sufficient to ensure uniqueness of coercions. Benjamin Werner has studied the addition of proof-irrelevance (rule σ) in the Calculus of Constructions [12]. It is not a trivial extension and it has far-reaching consequences on the model of the calculus [13]. We direct the reader to these papers for further information on the subject.

$$
\begin{array}{lll}
(\beta) & (\lambda x : X.e)\,v & \equiv e[v/x] \\
(\pi_i) & \pi_i\,(e_1, e_2)_T & \equiv e_i \\
(\sigma_i) & \sigma_i\,(\text{elt}\,E\,P\,e_1\,e_2) & \equiv e_i \\
(\eta) & (\lambda x : X.e\,x) & \equiv e & \text{if } x \notin FV(e) \\
(\rho) & (\pi_1\,e, \pi_2\,e)_{\Sigma x:X.Y} & \equiv e & \text{if } e : \Sigma x : X.Y \\
(\tau) & \text{elt}\,E\,P\,(\sigma_1\,e)\,(\sigma_2\,e) & \equiv e & \text{if } e : \{\,x : E \mid P\,\} \\
(\sigma) & \text{elt}\,E\,P\,t\,p & \equiv \text{elt}\,E\,P\,t'\,p' \text{ if } t \equiv t'
\end{array}
$$

Fig. 10. Equational theory of CIC?

3.4   Properties

The correctness proof of the translation is very involved, so we will only sketch it here. A report [14] is available (in French) and a mechanically checked proof is under development.

We first prove reflexivity, symmetry and transitivity of the coercion derivation algorithm, which extends previous proofs for the algorithmic system with properties on the generated coercions. Then we show substitutivity of the interpretation: if $\Gamma, x : U, \Delta \vdash t : T$ then $[\![t[u/x]]\!]_{\Gamma,\Delta[u/x]} = [\![t]\!]_{\Gamma,x:U,\Delta}[[\![u]\!]_\Gamma/x]$. We extend the notion of coercion to contexts next and show stability of judgements under context coercion. It is then possible to prove a lemma about commutativity of substitutions with the interpretation and coercions. Then we get an important corollary: if $\Gamma \vdash u : U$, $\Gamma \vdash_{CCI} c : U \rhd_\bullet V : s$ and $\Gamma, x : V \vdash T : s$ then $\llbracket T[u/x] \rrbracket_\Gamma \equiv \llbracket T \rrbracket_{\Gamma, x:V}[c[\llbracket u \rrbracket_\Gamma]/x]$. Essentially, it shows that coercions inserted by the interpretation depend only on the context and hence could be added later using substitution instead. This allows us to show that equivalence is conserved by interpretation and the following lemma:

Lemma 1 (Coercion derivation correctness). *If $\Gamma \vdash_{CCI} c : T \rhd_\bullet U : s$ then* $\llbracket \Gamma \rrbracket \vdash_{CCI} \lambda x : \llbracket T \rrbracket_\Gamma.c[x] : \llbracket T \rrbracket_\Gamma \to \llbracket U \rrbracket_\Gamma$.

Finally we can show that our interpretation is correct (theorem 6).

4   Implementation

This mechanism has been implemented in COQ and tested on simple examples (some are available on the author's website, including safe list operations and a head normal form definition). We actually generate coercions simultaneously with typing in the implementation, which is sound thanks to the previous proofs. Our typing is a clone of COQ's original typing algorithm, so we benefit from all the features of COQ including implicit variables, notations and the existing coercion system. The prototype also contains support for strucural and well-founded `Fixpoint` definitions. The type-checker transforms pattern-matching so that an equality between the matched term and the pattern is present in the typing context of each branch and it allows the user to put explicit holes in the term, for example:

```
Program Definition tail ( l : list A | l <> [] ) : list A :=
    match l with hd :: tl -> tl | [] -> False_rec (list A) _ end.
```

In the second branch we have $l = []$ and $l \neq []$, hence we can prove $\bot$ and derive an object of type `list` A.

It must be stressed that we do not lose any information when generating obligations as we have the whole term context at hand, avoiding the complicated proof obligation generation of PVS [3]. Here we would have $l : \{ l : \text{list } A \mid l \neq [] \}, Heql : \sigma_1 \, l = []$. We have tactics that clean the goal and context (deconstructing subsets and simplifying to present $l : \text{list } A, H : l \neq [], Heql : l = []$ to the user) but it does not match the usability of PVS yet.

5   Conclusion

We have developed a new language for writing programs in the COQ proof assistant which allows the user to specify complex programs while keeping the corresponding code simple. It is a first step towards interpretation of ML code into COQ to build certified programs. While the system and its accompanying proofs may seem complex, the implementation is short, simple and has been tested with success on simple examples. This work is also a second proof of the meaningfulness of the Predicate Subtyping feature of PVS and may help build a common interface for the two provers.

5.1   Related Work

Other efforts to build certified programs using type theory include CAYENNE [15] and EPIGRAM [16]. CAYENNE offers dependent types and general recursion at the expense of a non-terminating type checker; it can be thought of as a testbed for developing languages with dependent types but is not so much aimed at building certified programs. EPIGRAM ought to become a complete programming language with dependent types at its heart. Instead of using a phase distinction to separate coding and proving, EPIGRAM is based on an interactive, 2-dimensional editing process where code and proofs are written incrementally to get a complete program. Using type annotations and with the help of the editor for structuring, one is able to write programs with precise specifications. However it is not obvious if it could be made usable by programmers because of the paradigm change and scaling issues, time will tell which approach is more beneficial for writing and reasonning about dependently typed programs. The DML language [17] is more akin to comparison with our solution. It provides a way to use a restricted set of dependent types in ML programs, carefully chosen so as to keep type-checking decidable with the help of an automatic prover. Our method should subsume this one as it is perfectly possible to integrate automatic proof tools with CoQ to discharge the generated obligations. There exists many more systems like Xi's Applied Type Systems and Tim Sheard's Omega language which explore the design space of dependently typed languages.

5.2   Further Work

Much has to be done to smoothly integrate RUSSELL into CoQ. The treatment of existential variables of CoQ has to be improved, the type inference algorithm needs some tuning to become more similar to ML and an integration of proof-irrelevance in the kernel, while rarely needed in practice, is necessary to have a robust system. We also hope to extend this mechanism of proof-obligations generation to other constructs, notably (co-)inductive types. Finally, we intend to use this system as a basis for interpretation of more general ML programs in CoQ, using a monadic translation or some kind of effects to reflect imperative constructs like exceptions and references.

Acknowledgements. We would like to thank Christine Paulin-Mohring for directing this work and Jean-Christophe Filliâtre for discussions on previous versions of this paper. We thank the anonymous referees for their insightful comments.

References

1. Bertot, Y., Castéran, P.: Interactive Theorem Proving and Program Development. Springer, Heidelberg (2004)
2. Shankar, N., Owre, S.: Principles and pragmatics of subtyping in PVS. In: Bert, D., Choppy, C., Mosses, P.D. (eds.) WADT 1999. LNCS, vol. 1827, pp. 37–52. Springer, Heidelberg (2000)
3. Owre, S., Shankar, N.: The formal semantics of PVS. Technical Report SRI-CSL-97-2, Computer Science Laboratory, SRI International, Menlo Park, CA (1997)
4. Parent, C.: Synthesizing proofs from programs in the Calculus of Inductive Constructions. In: Möller, B. (ed.) MPC 1995. LNCS, vol. 947, pp. 351–379. Springer, Heidelberg (1995)
5. Nordström, B., Petersson, K., Smith, J.M.: Programming in Martin-Löf's Type Theory. Oxford University Press, Oxford (1990)
6. Constable, R.L., Allen, S.F., Bromley, H., Cleaveland, W., Cremer, J., Harper, R., Howe, D.J., Knoblock, T., Mendler, N., Panangaden, P., Sasaki, J.T., Smith, S.F.: Implementing Mathematics with the Nuprl Development System. Prentice-Hall, Englewood Cliffs (1986)
7. Coquand, T., Huet, G.: The Calculus of Constructions. Inf. Comp. 76, 95–120 (1988)
8. Sozeau, M.: Russell Metatheoretic Study in Coq, experimental development (2006) http://www.lri.fr/~sozeau/research/russell/proof.en.html
9. Chen, G.: Sous-typage, Conversion de Types et Élimination de la Transitivité. PhD thesis, Université Paris VII, Laboratoire d'Informatique de l'École Normale Supérieure, Paris (1998)
10. Luo, Z.: Coercive subtyping in type theory. In: van Dalen, D., Bezem, M. (eds.) CSL 1996. LNCS, vol. 1258, pp. 276–296. Springer, Heidelberg (1997)
11. Adams, R.: Pure Type Systems with Judgemental Equality. Journal of Functional Programming 16, 219–246 (2006)
12. Werner, B.: On the strength of proof-irrelevant type theories. In: 3rd International Joint Conference on Automated Reasoning (2006)
13. Miquel, A., Werner, B.: The not so simple proof-irrelevant model of CC. In: Geuvers, H., Wiedijk, F. (eds.) TYPES 2002. LNCS, vol. 2646, pp. 240–258. Springer, Heidelberg (2003)
14. Sozeau, M.: Coercion par prédicats en Coq. Master's thesis, Université Paris VII, LRI, Orsay, extended version - (2005) http://www.lri.fr/~sozeau/research/russell/report.pdf
15. Augustsson, L.: Cayenne—A language with dependent types. In: ACM SIGPLAN International Conference on Functional Programming (ICFP), Baltimore, Maryland, pp. 239–250 (1998)
16. McBride, C., McKinna, J.: The view from the left. J. Funct. Program. 14(1), 69–111 (2004)
17. Xi, H.: Dependent Types in Practical Programming. PhD thesis, Carnegie Mellon University, Pittsburgh, Pennsylvania (1998)

A Certified Distributed Security Logic for Authorizing Code

Nathan Whitehead

Department of Computer Science
University of California, Santa Cruz
nwhitehe@cs.ucsc.edu

Abstract. In previous work we have proposed a distributed security logic for authorizing code. To gain assurance about the correctness of the implementation of our system, we now present a series of security logics of increasing expressive power leading up to our logic. We encode each logic in Coq, develop an algorithm for deciding queries, and prove properties about the algorithm in Coq. By using Coq's automatic extraction mechanism, we are able to gain a high assurance about the resulting reference monitor implementations. Following this strategy yields reference monitors fully certified at the source code level for Datalog, Binder, Binder with a general extension mechanism, and a logic that combines Binder and the calculus of co-inductive constructions.

1 Introduction

The reference monitor in a computer security system with access control decides which requests to allow. Every access the system handles will go through the reference monitor, so ensuring its correctness is of paramount importance. Since the original requirements for access control systems were developed in the 1970s [1] it has been recognized that verifying the reference monitor is an important part of designing a secure access control system.

The most pressing access control problem today is knowing when code is safe to execute. Code may be provided through mechanisms such as applets, plug-ins, and on-line patches. Standard protection mechanisms such as digital signatures help determine where code comes from, but cannot answer the question of whether the code is malicious or not. A signature from someone I do not know and trust is meaningless. Proof-carrying code [18] answers this problem by requiring that code producers provide proofs of correctness for their code. Recipients of the code check the proof and know the code is safe without needing to trust the code provider. However, proofs are not always possible or appropriate.

Previously we have proposed a distributed security logic, BCiC, for authorizing proof-carrying code [26, 25]. Our logic extends Binder [8] with the ability to reason about proofs and program properties. The logic allows fine-grained combinations of proofs and digital signatures. We implemented a prototype of the logic for testing purposes, but after discovering subtle, hard to exercise errors in the implementation we decided that a more high-assurance approach was

T. Altenkirch and C. McBride (Eds.): TYPES 2006, LNCS 4502, pp. 253–268, 2007.

needed. If one cannot trust the implementation of the logic, it is worthless as the basis of a secure system.

In this paper we present a series of security logics of increasing expressive power leading up to BCiC. We encode each logic in Coq, develop an algorithm for deciding queries, then prove properties about the algorithm. Using Coq's automatic program extraction mechanism we extract working implementations of the decision algorithms which satisfy the properties. This strategy yields a series of reference monitors fully certified at the source level for Datalog, Binder, Binder with a general purpose extension mechanism, and BCiC, which combines Binder and the calculus of co-inductive constructions.

1.1 Related Work

BCiC is a continuation of research into proof-carrying code [18]. Techniques such as typed assembly language [17] and foundational proof-carrying code [2] are the base for generating the proofs about code that BCiC uses, but a system for generating and checking proofs is not by itself an access control system for untrusted code. BCiC is an access control logic that incorporates proofs as a way of trusting code, together with a mechanism for reasoning about statements made by other principals as an alternate way code might be trusted. The goal of BCiC is to extend the idea of proof-carrying code and incorporate it into an access control framework.

There is a long line of work that has been put into designing and implementing secure reference monitors, and consequently many different techniques have been applied to the problem. The earliest approaches involved writing detailed specifications of the desired behavior of the reference monitor, then proving properties by hand about the specification. The reference monitor would then be implemented to closely follow the specification. Stanford's provably secure operating system follows this approach [19]. This method has the following disadvantages:

- The implementation may not correctly follow the specification.
- Proofs about the specification may contain errors.
- The properties that are proved might not ensure correctness.

Gutmann argues that formal reasoning is the wrong tool for increasing trust in security applications [9]. He cites numerous examples where code that was "formally proved correct" failed. He develops a methodology based on dynamic program assertions and informal reasoning that is designed to be easily understood by parties not acquainted with the code. We agree that formal methods are not a panacea, but we do believe that they have an important role to play in the development of correct software. With the advent of effective and reliable tools such as Coq [23] machine checkable formal proofs about code are becoming not only possible but practical.

One technique that has proven especially effective at producing correct code is program extraction. By the Curry-Howard isomorphism, proofs and programs are isomorphic. Program extraction uses the isomorphism to automatically convert correct proofs into correct programs. Program extraction in Coq has many advantages, such as:

- The implementation comes directly from the proof.
- The proofs are all machine-checked.
- Coq is expressive enough to encode full correctness properties.

Extraction for Coq was developed by Paulin-Mohring [21] then further refined by Letouzey [14]. Coq and its extraction mechanism have been used to generate certified algorithms for many problems such as a compiler back-end [13], binary decision diagrams, unification, and binary search trees. These and other examples are available at the Coq user contributions page[1].

Our work is an example of using Coq extraction to produce certified code from mathematical proofs. It is also an example of a system that *uses* proofs in its operation. Even for access control applications that do not need proofs, the reference monitors we develop along the way are useful as general purpose access control models and are a reasonable starting point for developing other customized extensions for applications that require them.

Another example of a certified reference monitor is DHARMA [6]. DHARMA is a formally verified reference monitor for a distributed delegation logic. It was verified in PVS [20] then automatically extracted into Lisp code. Our work differs in several ways. First, the logic of DHARMA is based on access control lists and bounded delegation. Our base logic, Datalog, is more general purpose and less specific to one type of access control (e.g., one cannot define new predicates in a DHARMA policy). Our final logic also allows access control policies to include references to proofs and properties of code, which DHARMA does not attempt to do.

There are many authorization policy languages that have been designed for different authorization tasks. Binder is part of a family of authorization logics based on Datalog with varying features and decidability properties. Other examples include SD3 [12], Delegation Logic [15], and RT [16]. BCiC extends Binder with the ability to reason about proofs in the calculus of co-inductive constructions in such a way that there is still a reasonably simple decision procedure for the logic. There are many ways to extend Datalog and remain decidable; Jaffar and Maher have a survey of methods based on constraint solving [11]. SecPAL [4] is a recent example of an authorization logic based on Datalog with constraints.

Proof-carrying authentication [3] is a method of doing authentication inspired by proof-carrying code. In proof-carrying authentication the authentication policy is described using predicates defined in Twelf [22]. Requests for access to resources come paired with an explicit proof that the access should be granted according to the policy. These proofs are necessary because the policy logic is undecidable. We choose instead to work with decidable logics. Even when we incorporate the calculus of co-inductive constructions in our logic, we add restrictions to the terms that may be constructed in order to keep the logic decidable.

---

[1] http://coq.inria.fr/contribs/extraction-eng.html

Decidable logics have the advantage that the reference monitor is both sound and complete. For proof-carrying authentication, the reference monitor is only sound. Because the requester must construct a proof that their request should be granted, they must know the entire policy of the granter. This requirement limits the applicability of the logic. Decidable logics have no such restriction.

1.2   Organization

We start by describing Datalog, Binder, and BCiC in Section 2. We show how access control works in these logics and present small example policies for motivation. Next in Section 3 we discuss how the logics were formalized in Coq, what we proved about them, and some of the properties of the automatically extracted code. We conclude in Section 4 by considering the work that remains to be done.

2   The Logics

2.1   Datalog

Datalog is Horn logic with all function symbols of arity 0, i.e. constants only [24]. A Datalog program contains facts and rules of inference. For the case of access control, we call Datalog programs *policies*. Queries run against the policy and either succeed or fail. Datalog is similar to Prolog, but unlike Prolog it does not have negation, function symbols, or any meta-logical constructs such as cut.

The syntax of Datalog terms, atomic formulas, and formulas is as follows:

$$t ::= \qquad \qquad \qquad \textbf{term:}$$
$$c \qquad \qquad \qquad \textit{constant}$$
$$x \qquad \qquad \qquad \textit{variable}$$
$$a ::= \qquad \qquad \qquad \textbf{atomic formula:}$$
$$p(t_1, \ldots, t_n) \qquad \qquad \textit{predicate}$$
$$f ::= \qquad \qquad \qquad \textbf{formula:}$$
$$a :\text{-} a_1, \ldots, a_n \qquad \qquad \textit{clause}$$

In addition we place a restriction on clauses that no variable appears in a unless it appears somewhere in $a_1, \ldots, a_n$. This is a standard restriction that simplifies the analysis of the logic without restricting expressivity [24].

A policy is a list of formulas, and a query against a policy is a ground atomic formula. When a query q succeeds against a policy r we will write $r \vdash q$, or q is derivable from r. Let s be a substitution over atomic formulas that maps variables to constants. Then $r \vdash q$ is defined by the single rule:

$$\frac{(a :\text{-} a_1, \ldots, a_n) \in r \qquad r \vdash s(a_i) \quad \text{for } i = 1, \ldots, n}{r \vdash s(a)}$$

Datalog is a suitable match for many access control policies because it is relatively simple but expressive enough to capture concepts such as categories, roles, and subgroup inclusion.

Consider a University computer center with administrators who wish to control which programs may be executed by different users in order to protect the integrity of the center. The administrators categorize users into a hierarchy of groups, then give each group permission to execute various programs. Datalog facts in the policy specify which group each user belongs to, and Datalog clauses denote the subgroup relationships.

```
student(nathan).
student(avik).
faculty(martin).
admin(jordan).
user(X)  :- student(X).
user(X)  :- faculty(X).
user(X)  :- admin(X).
mayrun(X, matlab)  :- user(X).
mayrun(X, netconfig)  :- admin(X).
```

Fig. 1. Example Datalog policy for a university computer center

To decide if person p should be allowed to run program g, the reference monitor must decide if mayrun(p, g) is derivable from the policy in Figure 1. This is not an entirely trivial decision. There are no restrictions on policies, so the subgroup relationships could describe a complicated directed graph with cycles. Deciding which groups someone belongs to is deciding graph reachability.

2.2   Binder

Binder differs syntactically from Datalog only in atomic formulas:

$$
\begin{array}{lr}
a ::= & \textbf{atomic formula:} \\
\quad p(t_1, \ldots, t_n) & \textit{bare predicate} \\
\quad t \textbf{ says } p(t_1, \ldots, t_n) & \textit{quoted predicate}
\end{array}
$$

Derivations in Binder are defined by the same rule as for Datalog.

Binder adds the notion of importing and exporting facts from one context to another. This allows policies to define mechanisms for distributed authentication and authorization. The general idea is that if Alice has a policy with the fact f, then this fact will be exported from Alice's policy using public-key encryption and then imported into another context as $Alice^{pk}$ **says** f, where $Alice^{pk}$ is a constant that represents Alice's public key. For more details and examples see the original paper on Binder [8].

Notice that **says** cannot be nested arbitrarily deep. This means that if Alice signs a statement f and Bob receives that statement and passes it along to Carol, Carol can only know that Alice signed f and/or that Bob signed f. Carol can never know that Bob knows that Alice signed f. The restriction on nesting is natural for access control applications and also convenient because it makes demonstrating decidability easier.

Continuing our computer center example, we can now write the distributed policy shown in Figure 2 for deciding who may run which applications.

```
trusted(rootkey).
trusted(X) :- Y says trusted(X), trusted(Y).
registrar(X) :- Y says registrar(X), trusted(Y).
admin(X) :- Y says admin(X), trusted(Y).
student(X) :- Y says student(X), registrar(Y).
user(X) :- student(X).
user(X) :- admin(X).
mayrun(X, matlab) :- user(X).
mayrun(X, labtools) :- user(X), Y says mayrun(X, labtools), admin(Y).
```

Fig. 2. Example of a distributed authorization policy in Binder

The first rule of Figure 2 defines a trusted root key from which all other trust derives. The trusted root key signs other keys which are then also trusted. The rule `trusted(X) :- Y says trusted(X), trusted(Y)` allows unbounded delegation of trust. Trusted keys determine who is the registrar, and the registrar determines who is a student. Students are users, and users may run `matlab`. For `labtools`, the person must be a user and an administrator must explicitly authorize them.

2.3   BCiC

BCiC is an extension of Binder [8] with terms from the Calculus of (co)Inductive Constructions [7,5]. The motivation is that instead of merely relying on signed statements and rules in a distributed access control policy, policies may depend on explicit proofs about programs. In BCiC a policy can allow the execution of a program if it is asserted to be safe by a trusted authority *or* the code producer provides a proof of its safety.

In order to prove decidability of BCiC we present a slightly simplified version of BCiC here as compared to our original description [25]. We discuss the differences in more detail in Section 3.4.

The syntax of BCiC adds CiC terms paired with their signature as an option for terms.

$$
\begin{array}{lr}
t ::= & \text{\textbf{term:}} \\
\quad c & \textit{constant} \\
\quad x & \textit{variable} \\
\quad (s, o) & \textit{CiC signature and term}
\end{array}
$$

CiC terms can represent objects, types, kinds, theorems and proofs in the standard way.

In addition to bare predicates and predicates quoted with **says**, in BCiC we have the **sat** predicate. The atom $\textbf{sat}(s, p, (s', t))$ means that in CiC signature s property p holds for the CiC term t and that $s' \subseteq s$. The notation $s' \subseteq s$ means that signature s' is a sub-signature of s; every definition contained in s' is also contained in s. BCiC atomic formulas have the syntax:

$$
\begin{array}{lr}
a ::= & \text{\textbf{atomic formula:}} \\
\quad p(t_1, \ldots, t_n) & \textit{predicate} \\
\quad t \text{ \textbf{says} } p(t_1, \ldots, t_n) & \textit{quoted predicate} \\
\quad \textbf{sat}(s, p, t) & \textit{proof} \\
\quad t \text{ \textbf{says} } \textbf{sat}(s, p, t) & \textit{quoted proof}
\end{array}
$$

In addition to the Datalog derivation rule, BCiC has a rule for deriving **sat** statements. The notation $s \vdash_{CiC} o : T$ means that in CiC signature s, the object o has type or kind T.

$$
\frac{
\begin{array}{ccc}
s \vdash_{CiC} p : T \rightarrow Prop & & s' \vdash_{CiC} o : T \\
s \vdash_{CiC} T : Set & & s' \subseteq s \\
\multicolumn{3}{c}{\exists \pi \in r \,.\, s \vdash_{CiC} \pi : (p \ o)}
\end{array}
}{
r \vdash \textbf{sat}(s, p, (s', o))
}
$$

The restrictions $s \vdash_{CiC} T : Set$ and $s' \vdash_{CiC} o : T$ also ensure that s and s' are well-formed signatures. Note that the proof object π is restricted to being in r, the policy. This restriction means that BCiC only considers proofs that have been imported into the policy. If a proof does not exist in the policy, it has no effect on the logic. Without this restriction BCiC would be hopelessly undecidable. Trying to decide when proof terms exist in CiC is equivalent to trying to automatically prove arbitrarily difficult mathematical theorems.

Proofs are represented in our framework as explicit terms mentioned somewhere in the policy. If Alice is the code producer and she proves that her code o has property p, she will have a term π of type $p \ o$. She introduces this term by exporting the statement `proof((s, π))`.

We now extend the previous example to BCiC. The computer center administration just got access to a new supercomputer. For performance reasons the operating system cannot provide effective memory protection between programs. If one program crashes, it can interfere with all the results being calculated by other users. In addition, programs can dynamically declare how much processing power they need and the department will be billed accordingly. Because of these

```
trusted(rootkey).
trusted(X) :- Y says trusted(X), trusted(Y).
mayrunonsupercomputer(X, Y) :- user(X), safe(Y), economical(Y).
safe(X) :- Y says safe(X), trusted(Y).
economical(X) :- Y says economical(X), trusted(Y).
safe(X) :- sat(S_safety, memsafe, X).
economical(X) :- sat(S_economy, cheap, X).
```

Fig. 3. An example BCiC policy for a supercomputer

factors, the administrators require all programs that run on the supercomputer be memory safe and not request an inordinate amount of processing power without prior authorization. To enforce these requirements they construct the policy in Figure 3.

The signature S_safety defines the assembly language syntax for the supercomputer along with the predicate memsafe which guarantees memory safety. The signature S_economy defines the assembly language syntax and a very simple proposition cheap that states programs must begin by allocating a reasonable amount of processing power and then may never reallocate more. To run their programs on the supercomputer, principals must be authorized users and either a trusted principal must vouch that the program is safe and economical, or someone must produce a proof that the program satisfies memsafe and cheap.

3   Formalization in Coq

The goals of formalizing BCiC in Coq were twofold. First, by formalizing the logic in Coq we were able to prove decidability of the logic with a high degree of confidence that the proof was correct. Second, we could then use Coq's automatic program extraction to extract a certified decision procedure from the proof. Our strategy was to start with pure Datalog and prove that it is decidable. From there we modified the proof to apply to Binder. To formalize full BCiC we first created an extension mechanism for Binder and proved decidability, then instantiated the extension appropriately to make the extended Binder be equivalent to BCiC. An advantage of this approach is that we can instantiate the same extension mechanism with a variety of other logics without rewriting all the proofs. For example, we can instantiate the extension with LF [10] instead of CiC and the proofs about extended Binder still hold.

3.1   Datalog and Binder

Formalizing Datalog and Binder in Coq was relatively straightforward. Here is the definition of Binder, using natural numbers to name constants, variables, predicates, and principals.

```
Inductive term : Set :=
| const : nat -> term
| var : nat -> term.

Inductive atomic : Set :=
| bare : nat -> list term -> atomic
| says : nat -> nat -> list term -> atomic.

Inductive form : Set :=
| clause : atomic -> list atomic -> form.
```

After defining substitutions, the semantics of derivations for both Datalog and Binder is defined by:

```
Inductive der : list form -> atomic -> Prop :=
| modus_ponens :
  forall (LF : list form)(F : form)(S : substitution),
    In F LF ->
    forallelts
      (fun x => der LF (multisubs_atomic S x)) (body F) ->
    der LF (multisubs_atomic S (head F)).
```

The main theorem we prove about Datalog and Binder is that derivability is decidable, which is stated:

```
Theorem der_dec:
  forall (LF : list form)(A : atomic),
    no_head_vars_form_list LF ->
    {der LF A} + {~ der LF A}.
```

The condition no_head_vars_form_list LF expresses the restriction that variables that appear in the heads of clauses in the policy must also appear in the body of the clause.

The notation {A}+{B} indicate the sumbool operator, which has kind $Prop \rightarrow Prop \rightarrow Set$. An object of type {A}+{B} is either a proof of A or a proof of B. We use {A}+{B} instead of A∨B because of the way program extraction behaves. During program extraction, elements of a proof of kind $Prop$ are erased, while elements of kind Set are not. The expression A∨B has kind $Prop \rightarrow Prop \rightarrow Prop$, so if our theorem were stated using it, program extraction would yield an empty program. The expression {A}+{B} has kind Set rather than $Prop$, which means that it will not be ignored during program extraction. The contents of the left and right arguments, however, are in $Prop$, so they will be ignored. What this means is that extracting from the proof of this theorem will yield an algorithm that when given a policy and a goal, will return left or right as its decision.

The extracted algorithm will not remember the proof about why its answer is correct as it computes, avoiding unnecessary runtime overhead.

It is well-known that Datalog is decidable [24]. Our proof starts by considering all the possible atomic formulas $p(t_1, \ldots, t_n)$ that could be derived using the derivation rule from a given policy r.

Lemma 1. *If $r \vdash p(t_1, \ldots, t_n)$, then the predicate p and arity n appear in r.*

Proof. Every derivable atomic formula must be of the form $s(p(t_1, \ldots, t_n))$ where $p(t_1, \ldots, t_n) :- a_1, \ldots, a_n \in r$. Substitutions do not change predicates or arities, so the lemma follows. $\square$

Lemma 2. *If $p(t_1, \ldots, t_n)$ has no variables and $r \vdash p(t_1, \ldots, t_n)$, then the terms $t_1, \ldots, t_n$ all appear in r.*

Proof. The proof is by induction over derivations and uses the fact that variables may not appear in the heads of clauses of the policy without appearing in the body of the clause. $\square$

These lemmas together imply that there is a finite universe of ground atomic formulas A such that $\forall a . (r \vdash a \implies a \in A)$. In our formalization, we define a function to calculate A and prove that every derivation must be included in it.

The next step is to decide which elements of A are in fact derivable. In Coq we define an extension function `extend : list form → list atomic → list atomic` that when given a policy and a set of derivable ground atomic formulas, returns the new ground atomic formulas that are derivable in one step from the inputs. We prove that `extend` is sound and complete with respect to Datalog derivations. The soundness and completeness lemmas are encoded as:

```
Lemma sound :
  forall (F : form)(LF : list form)(A : atomic)(LA : list atomic),
    In F LF -> ground_atomic_list LA = true ->
    In A (extend F LA) -> forallelts (der LF) LA ->
    der LF A.
```

```
Lemma complete :
  forall (F : form)(LF : list form)(LA : list atomic)
         (s : substitution),
    no_head_vars_form_list LF -> ground_atomic_list LA = true ->
    In F LF ->
    forallelts (fun x => In (multisubs_atomic s x) LA) (body F) ->
    In (multisubs_atomic s (head F)) (extend F LA).
```

The function **extend** is monotonically increasing and bounded by A, so it must have a fixed point. The fixed point θ of **extend** is exactly the set of ground atomic formulas that are derivable from r. Soundness shows that elements in θ are derivable, and completeness shows that every derivable element is in θ.

Binder is decidable in almost the same way as Datalog is. The proof follows the Datalog proof almost exactly except for extra cases in atomic formulas which are trivial.

3.2 Encoding BCiC

Instead of directly encoding BCiC with a deep embedding of CiC into Coq and then trying to prove theorems, we instead formalized a general extension mechanism for Binder. Given an extension set, the extended version of Binder allows terms to refer to extended objects.

$t ::=$	**term:**
c	*constant*
x	*variable*
e	*extended term*

The extension mechanism adds an generic extension predicate to atomic formulas. The **ext** predicate is indexed by natural numbers to allow multiple predicates over the same extension set.

$a ::=$	**atomic formula:**
$p(t_1, \ldots, t_n)$	*bare predicate*
t **says** $p(t_1, \ldots, t_n)$	*quoted predicate*
$\mathbf{ext}(n, t)$	*extension*
t **says** $\mathbf{ext}(n, t)$	*quoted extension*

The extension mechanism takes an extension set **extset**, a proof that equality is decidable over **extset**, a proposition **extder** over the extension set, a function **extset** that extracts all the relevant extended terms from a policy, and a proof that **extder** is decidable for all possible results of **extset**. Given these arguments, we prove that the extended version of Binder is decidable. By proving properties about Binder with the extension mechanism, we can instantiate the extension mechanism in different ways without re-proving anything about Binder. Each instantiation of the extension mechanism only requires creating a decision procedure for the extension type. We will encode BCiC by instantiating the extension mechanism with CiC terms.

For the extension mechanism, we take the extension set and other auxiliary proofs to be parameters of the extension. We do this in Coq by using structured environments. For example, the following code fragment shows how the extension set **extset** and a decision procedure for equality between elements of **extset** is defined.

```
Section ExtendedBinder.
  Variable extset : Set.
  Variable eq_extset_dec :
    forall (E1 E2 : extset), {E1 = E2}+{E1 <> E2}.
  .
  . (proofs)
  .
End ExtendedBinder.
```

The rules for derivation in extended Binder are:

```
Inductive der : list form -> atomic -> Prop :=
  | modus_ponens :
    forall (LF : list form)(F : form)(S : substitution),
      In F LF ->
      forallelts
        (fun x => der LF (multisubs_atomic S x)) (body F) ->
      der LF (multisubs_atomic S (head F)).
  | extension_der :
    forall (LF : list form)(n : nat)(e : extset),
      In E (extset LF) ->
      extder n E ->
      der LF (ext n (extconst E)).
```

Besides proving the decidability of extended Binder, we prove a small result about conservativity. An extension is conservative if `extder n E` must be true when `ext(n, E)` is derivable from a policy. For policies without `ext(n, E)` in the heads of clauses, we prove all extensions are conservative. Of course if a policy is allowed to conclude `ext(n, E)` then the theorem would not hold. For the case of BCiC, conservativity means that by mixing Binder and CiC we do not somehow make the proof logic of CiC inconsistent and able to prove anything.

3.3 CiC as an Extension

To create BCiC we instantiate the extension mechanism. The extension set is defined to be tuples consisting of a CiC signature and a CiC term. Given a BCiC policy r we can extract all the signatures and CiC terms that appear in r. Combining these elements in all possible ways gives us `extset`.

To define `extder`, the proposition over the extension set, we examine all the signatures that appear in r and find all possible propositions defined by the signatures. There will be a finite number of signature-proposition pairs; label them with natural numbers and call the function γ. Using this labeling, the nth extension proposition `extder n (s', o)` is defined to mean that $s' \subseteq s$ and $\exists \pi \in r.s \vdash_{CiC} \pi : (p\,o)$ where $\gamma(n) = (s, p)$. Deciding when the proposition extder n (s', o) is true involves examining all the CiC terms in extset r and making a call to Coq to see if they have the correct type. It also involves checking for signature inclusion, which is easy.

In our formalization of BCiC, we do not formalize CiC itself. Instead we assume there is an external type of CiC terms and signatures, an external typing judgement over CiC terms, an external proof that the typing judgement is decidable, and a function that extracts all the identifiers from a signature. Our instantiation of the Binder extension uses these external types and functions to model BCiC. During program extraction, places where these external items are used are converted to function calls of undefined names. The intention is that these calls are really calls to Coq in the implementation. Our current implementation of BCiC instantiates these datatypes and functions as hand-coded OCaml routines that communicate directly with Coq running as a subprocess through Unix named pipes. It should also be possible to include the logical type-checking core of Coq directly in the implementation but we have not done so.

3.4   Limitations of the Encoding

The version of BCiC presented here is slightly simplified from our original presentation [25]. The previous version allowed arbitrary CiC terms and propositions intermixed with Binder variables at any position. To argue that a reasonably complete algorithm for that version of BCiC existed, we imposed a restriction on the CiC terms and Binder variables that could appear in them, then proved that repeated applications and instantiations of these terms yielded a finite universe.

In this paper we restrict the intermixing of CiC terms and Binder variables even more. We do not allow Binder variables to appear anywhere inside of CiC terms, we only allow Binder variables to instantiate to CiC terms. An example of a policy that is allowed in the previous version of BCiC and is disallowed in the current version is a policy that a program is safe if it is linked with a dynamic module that somehow enforces the safety condition.

The policy might say safe(Q) :- sat(link dynchecker P Q). The predicate link takes three arguments, two programs that are to be linked together and the resulting output program. This policy is not directly expressible in the version of BCiC so far presented because the P and Q Binder variables appear inside the CiC term link dynchecker P Q. To encode these types of policies requires modifying the extension mechanism to accept multiple arguments. The policy would then be encoded as safe(Q) :- ext3(link, dynchecker, P, Q). We do not see any theoretical difficulty with adding these extensions, but have not yet done so in our formal development.

3.5   Alternate Encodings

There are many possible ways to encode BCiC in Coq. An attractive option is to somehow encode the CiC part of BCiC using Coq directly. For example, we can imagine encoding BCiC terms as follows:

```
Inductive term : Type :=
| constant : nat -> term
| cicterm : forall (T : Type), T -> term.
```

We give **term** kind *Type* instead of *Set* because Coq requires definitions containing *Type* to be of kind *Type*. This definition can encode CiC objects of any type. It seems as though it can encode its own type, but if this is attempted Coq will give a reassuring universe inconsistency error.

The problem with this encoding is that program extraction is impossible. Elements of kind *Type* are removed during program extraction, so no extracted program with terms defined in this way will be able to reason about CiC terms. Another problem with this encoding is that the CiC signature to use cannot be controlled. Choosing which signature to use for proofs is critically important when transmitting proofs between untrusted parties.

3.6   Extracted Code

The specification of Datalog took 82 lines of Coq, Binder took 85 lines, and BCiC took 137 lines. Proving decidability took 6654 lines for Datalog, 7470 lines for Binder, and 9394 lines for BCiC. The extracted algorithms were small; Datalog was 387 lines of OCaml code, of which 50 were redundant definitions of standard data structures. Binder extracted to 487 lines, and BCiC extracted to 626 lines of OCaml.

The extracted code was not particularly tuned for efficiency but neither was it grossly inefficient. Because the **extend** function is repeatedly evaluated to reach a fixed point during the decision procedure, the efficiency of **extend** largely determines the efficiency of the overall algorithm. We wrote the definition of **extend** by hand and then refined it as the proof development proceeded. During extraction, this version of **extend** is translated directly from Coq syntax to OCaml.

Our version of **extend** is inefficient in some ways. For example it checks for duplicate entries by linearly traversing a list rather than using a hash table. In other ways it is efficient; it matches clauses against ground atomic formulas using a simple unification algorithm and keeps track of variable bindings. A less efficient (but still correct) algorithm could expand every clause into the set of all possible ground instances of the clause before matching, yielding an exponential slowdown. Such an algorithm would have much simpler proofs but would be unacceptably slow. Although we have not done a formal analysis of the performance of our extracted code, our prototype implementation has performed well for all the security policy examples we have tried.

4   Conclusion

By using Coq's automatic program extraction we were able to generate certified reference monitors for a series of logics useful for access control. In the end we developed a decision procedure for Datalog, Binder, and a slightly limited version of BCiC, our proposed security logic for authorizing proof-carrying code. All the proof scripts are available online at the author's homepage[2].

There still remains work to be done bridging the gap between our formalized version of BCiC and our original specification of the logic. We are in the process of proving decidability for an encoding of BCiC that allows signatures to appear alone in terms and allows Binder variables to appear inside CiC terms. More generally, we are in the process of integrating the automatically extracted ML code with the rest of our framework to make a finished application that deals with network communication and cryptography, along with logical inference.

Acknowledgments

Thanks to my advisor, Martín Abadi, for many fruitful discussions and encouragement with this research. The questions and comments from other participants at the TYPES workshop helped refine the direction of this work in a significant way. Thanks to Katia Hayati for help proofreading and for ideas about presentation, and to the anonymous reviewers for their feedback and suggestions. This work was partly supported by the National Science Foundation under Grants CCR-0208800 and CCF-0524078.

References

1. Anderson, J.P.: Computer security technology planning study. Technical Report ESD-TR-73-51, Electronic Systems Division, Hanscom Air Force Base (October 1972) Available at http://csrc.nist.gov/publications/history/ande72.pdf.
2. Appel, A.W.: Foundational proof-carrying code. In: Proceedings of the 16th Annual Symposium on Logic in Computer Science, pp. 247–258 (June 2001)
3. Appel, A.W., Felten, E.W.: Proof-carrying authentication. In: Proceedings of the 5th ACM Conference on Computer and Communications Security, pp. 52–62 (November 1999)
4. Becker, M.Y., Gordon, A.D., Fournet, C.: SecPAL: Design and semantics of a decentralized authorization language. Technical Report MSR-TR-2006-120, Microsoft Research (September 2006)
5. Bertot, Y., Castéran, P.: Interactive Theorem Proving and Program Development. Springer, Heidelberg (2004)
6. Chander, A., Dean, D., Mitchell, J.: A distributed high assurance reference monitor. In: Zhang, K., Zheng, Y. (eds.) ISC 2004. LNCS, vol. 3225, pp. 231–244. Springer, Heidelberg (2004)
7. Coquand, T., Paulin-Mohring, C.: Inductively defined types. In: Martin-Löf, P., Mints, G. (eds.) COLOG-88. LNCS, vol. 417, Springer, Heidelberg (1990)
8. DeTreville, J.: Binder, a logic-based security language. In: Proceedings of the 2002 IEEE Symposium on Security and Privacy, pp. 105–113 (May 2002)
9. Gutmann, P.: Cryptographic Security Architecture: Design and Verification. Springer, Heidelberg (2004)

---

[2] http://www.soe.ucsc.edu/~nwhitehe/bcic

10. Harper, R., Honsell, F., Plotkin, G.: A framework for defining logics. Journal of the ACM 40(1), 143–184 (1993)
11. Jaffar, J., Maher, M.J.: Constraint logic programming: A survey. Journal of Logic Programming 19(20), 503–581 (1994)
12. Jim, T.: SD3: A trust management system with certified evaluation. In: Proceedings of the 2001 IEEE Symposium on Security and Privacy, pp. 106–115 (May 2001)
13. Leroy, X.: Formal certification of a compiler back-end. In: 33rd Annual ACM SIGPLAN-SIGACT Symposium on Principles of Programming Languages (POPL'06) (2006)
14. Letouzey, P.: Programmation fonctionnelle certifiée – L'extraction de programmes dans l'assistant Coq. PhD thesis, Université Paris-Sud (July 2004) English translation available at http://www.pps.jussieu.fr/~letouzey
15. Li, N., Grosof, B.N., Feigenbaum, J.: Delegation logic: A logic-based approach to distributed authorization. ACM Transactions on Information and System Security 6(1), 128–171 (2003)
16. Li, N., Mitchell, J.C., Winsborough, W.H.: Design of a role-based trust-management framework. In: Proceedings of the 2002 IEEE Symposium on Security and Privacy, pp. 114–130 (May 2002)
17. Morrisett, G., Walker, D., Crary, K., Glew, N.: From System F to typed assembly language. ACM Transactions on Programming Languages and Systems 21(3), 528–569 (1999)
18. Necula, G.C.: Proof-carrying code. In: Proceedings of the 24th ACM SIGPLAN-SIGACT Symposium on the Principles of Programming Languages (POPL'97), pp. 106–119 (1997)
19. Neumann, P.G., Robinson, L., Levitt, K.N., Boyer, R.S., Saxena, A.R.: A provably secure operating system. Technical Report M79-225, Stanford Research Institute (June 1975)
20. Owre, S., Rushby, J., Shankar, N.: PVS: A prototype verification system. In: 11th International Conference on Automated Deduction, pp. 748–752 (1992)
21. Paulin-Mohring, C.: Extracting F_ω's programs from proofs in the calculus of constructions. In: Principles of Programming Languages (POPL'89), pp. 89–104 (1989)
22. Pfenning, F., Schürmann, C.: System description: Twelf — A meta-logical framework for deductive systems. In: Ganzinger, H. (ed.) Automated Deduction - CADE-16. LNCS (LNAI), vol. 1632, pp. 202–206. Springer, Heidelberg (1999)
23. The Coq Development Team. The Coq proof assistant. http://coq.inria.fr/
24. Ullman, J.D.: Principles of Database and Knowledge-base Systems, vol. 2. Computer Science Press (1988)
25. Whitehead, N., Abadi, M.: BCiC: A system for code authentication and verification. In: Baader, F., Voronkov, A. (eds.) LPAR 2004. LNCS (LNAI), vol. 3452, pp. 110–124. Springer, Heidelberg (2005)
26. Whitehead, N., Abadi, M., Necula, G.: By reason and authority: A system for authorization of proof-carrying code. In: Proceedings of the 17th IEEE Computer Security Foundations Workshop, pp. 236–250 (June 2004)

Author Index

Lecture Notes in Computer Science

Sublibrary 1: Theoretical Computer Science and General Issues

For information about Vols. 1– 4474
please contact your bookseller or Springer

Vol. 4638: T. Stützle, M. Birattari, H. H. Hoos (Eds.), Engineering Stochastic Local Search Algorithms. X, 223 pages. 2007.

Vol. 4628: L.N. de Castro, F.J. Von Zuben, H. Knidel (Eds.), Artificial Immune Systems. XII, 438 pages. 2007.

Vol. 4627: M. Charikar, K. Jansen, O. Reingold, J.D.P. Rolim (Eds.), Approximation, Randomization, and Combinatorial Optimization. XII, 626 pages. 2007.

Vol. 4624: T. Mossakowski, U. Montanari, M. Haveraaen (Eds.), Algebra and Coalgebra in Computer Science. XI, 463 pages. 2007.

Vol. 4621: D. Wagner, R. Wattenhofer (Eds.), Algorithms for Sensor and Ad Hoc Networks. XIII, 415 pages. 2007.

Vol. 4619: F. Dehne, J.-R. Sack, N. Zeh (Eds.), Algorithms and Data Structures. XVI, 662 pages. 2007.

Vol. 4618: S.G. Akl, C.S. Calude, M.J. Dinneen, G. Rozenberg, H.T. Wareham (Eds.), Unconventional Computation. X, 243 pages. 2007.

Vol. 4616: A. Dress, Y. Xu, B. Zhu (Eds.), Combinatorial Optimization and Applications. XI, 390 pages. 2007.

Vol. 4614: B. Chen, M.S. Paterson, G. Zhang (Eds.), Combinatorics, Algorithms, Probabilistic and Experimental Methodologies. XII, 530 pages. 2007.

Vol. 4613: F.P. Preparata, Q. Fang (Eds.), Frontiers in Algorithmics. XI, 348 pages. 2007.

Vol. 4600: H. Comon-Lundh, C. Kirchner, H. Kirchner (Eds.), Rewriting, Computation and Proof. XVI, 273 pages. 2007.

Vol. 4599: S. Vassiliadis, M. Berekovic, T.D. Hämäläinen (Eds.), Embedded Computer Systems: Architectures, Modeling, and Simulation. XVIII, 466 pages. 2007.

Vol. 4598: G. Lin (Ed.), Computing and Combinatorics. XII, 570 pages. 2007.

Vol. 4596: L. Arge, C. Cachin, T. Jurdziński, A. Tarlecki (Eds.), Automata, Languages and Programming. XVII, 953 pages. 2007.

Vol. 4595: D. Bošnački, S. Edelkamp (Eds.), Model Checking Software. X, 285 pages. 2007.

Vol. 4590: W. Damm, H. Hermanns (Eds.), Computer Aided Verification. XV, 562 pages. 2007.

Vol. 4588: T. Harju, J. Karhumäki, A. Lepistö (Eds.), Developments in Language Theory. XI, 423 pages. 2007.

Vol. 4583: S.R. Della Rocca (Ed.), Typed Lambda Calculi and Applications. X, 397 pages. 2007.

Vol. 4580: B. Ma, K. Zhang (Eds.), Combinatorial Pattern Matching. XII, 366 pages. 2007.

Vol. 4576: D. Leivant, R. de Queiroz (Eds.), Logic, Language, Information and Computation. X, 363 pages. 2007.

Vol. 4547: C. Carlet, B. Sunar (Eds.), Arithmetic of Finite Fields. XI, 355 pages. 2007.

Vol. 4546: J. Kleijn, A. Yakovlev (Eds.), Petri Nets and Other Models of Concurrency – ICATPN 2007. XI, 515 pages. 2007.

Vol. 4545: H. Anai, K. Horimoto, T. Kutsia (Eds.), Algebraic Biology. XIII, 379 pages. 2007.

Vol. 4533: F. Baader (Ed.), Term Rewriting and Applications. XII, 419 pages. 2007.

Vol. 4528: J. Mira, J.R. Álvarez (Eds.), Nature Inspired Problem-Solving Methods in Knowledge Engineering, Part II. XXII, 650 pages. 2007.

Vol. 4527: J. Mira, J.R. Álvarez (Eds.), Bio-inspired Modeling of Cognitive Tasks, Part I. XXII, 630 pages. 2007.

Vol. 4525: C. Demetrescu (Ed.), Experimental Algorithms. XIII, 448 pages. 2007.

Vol. 4514: S.N. Artemov, A. Nerode (Eds.), Logical Foundations of Computer Science. XI, 513 pages. 2007.

Vol. 4513: M. Fischetti, D.P. Williamson (Eds.), Integer Programming and Combinatorial Optimization. IX, 500 pages. 2007.

Vol. 4510: P. Van Hentenryck, L.A. Wolsey (Eds.), Integration of AI and OR Techniques in Constraint Programming for Combinatorial Optimization Problems. X, 391 pages. 2007.

Vol. 4507: F. Sandoval, A.G. Prieto, J. Cabestany, M. Graña (Eds.), Computational and Ambient Intelligence. XXVI, 1167 pages. 2007.

Vol. 4502: T. Altenkirch, C. McBride (Eds.), Types for Proofs and Programs. VIII, 269 pages. 2007.

Vol. 4501: J. Marques-Silva, K.A. Sakallah (Eds.), Theory and Applications of Satisfiability Testing – SAT 2007. XI, 384 pages. 2007.

Vol. 4497: S.B. Cooper, B. Löwe, A. Sorbi (Eds.), Computation and Logic in the Real World. XVIII, 826 pages. 2007.

Vol. 4494: H. Jin, O.F. Rana, Y. Pan, V.K. Prasanna (Eds.), Algorithms and Architectures for Parallel Processing. XIV, 508 pages. 2007.

Vol. 4493: D. Liu, S. Fei, Z. Hou, H. Zhang, C. Sun (Eds.), Advances in Neural Networks – ISNN 2007, Part III. XXVI, 1215 pages. 2007.

Vol. 4492: D. Liu, S. Fei, Z. Hou, H. Zhang, C. Sun (Eds.), Advances in Neural Networks – ISNN 2007, Part II. XXVII, 1321 pages. 2007.

Vol. 4491: D. Liu, S. Fei, Z.-G. Hou, H. Zhang, C. Sun (Eds.), Advances in Neural Networks – ISNN 2007, Part I. LIV, 1365 pages. 2007.

Vol. 4490: Y. Shi, G.D. van Albada, J.J. Dongarra, P.M.A. Sloot (Eds.), Computational Science – ICCS 2007, Part IV. XXXVII, 1211 pages. 2007.

Vol. 4489: Y. Shi, G.D. van Albada, J.J. Dongarra, P.M.A. Sloot (Eds.), Computational Science – ICCS 2007, Part III. XXXVII, 1257 pages. 2007.

Vol. 4488: Y. Shi, G.D. van Albada, J.J. Dongarra, P.M.A. Sloot (Eds.), Computational Science – ICCS 2007, Part II. XXXV, 1251 pages. 2007.

Vol. 4487: Y. Shi, G.D. van Albada, J.J. Dongarra, P.M.A. Sloot (Eds.), Computational Science – ICCS 2007, Part I. LXXXI, 1275 pages. 2007.

Vol. 4484: J.-Y. Cai, S.B. Cooper, H. Zhu (Eds.), Theory and Applications of Models of Computation. XIII, 772 pages. 2007.

Vol. 4475: P. Crescenzi, G. Prencipe, G. Pucci (Eds.), Fun with Algorithms. X, 273 pages. 2007.